Hydrocarbons in Crystalline Rocks

Special Publication reviewing procedures

The Society makes every effort to ensure that the scientific and production quality of its books matches that of its journals. Since 1997, all book proposals have been refereed by specialist reviewers as well as by the Society's Books Editorial Committee. If the referees identify weaknesses in the proposal, these must be addressed before the proposal is accepted.

Once the book is accepted, the Society has a team of Book Editors (listed above) who ensure that the volume editors follow strict guidelines on refereeing and quality control. We insist that individual papers can only be accepted after satisfactory review by two independent referees. The questions on the review forms are similar to those for *Journal of the Geological Society*. The referees' forms and comments must be available to the Society's Book Editors on request.

Although many of the books result from meetings, the editors are expected to commission papers that were not presented at the meeting to ensure that the book provides a balanced coverage of the subject. Being accepted for presentation at the meeting does not guarantee inclusion in the book.

Geological Society Special Publications are included in the ISI Index of Scientific Book Contents, but they do not have an impact factor, the latter being applicable only to journals.

More information about submitting a proposal and producing a Special Publication can be found on the Society's web site: www.geolsoc.org.uk.

It is recommended that reference to all or part of this book should be made in one of the following ways:

Petford, N. & McCaffrey, K. J. W. (eds) 2003. *Hydrocarbons in Crystalline Rocks.* Geological Society, London, Special Publications, **214**.

Sanders, C. A. E., Fullarton, L. & Clavet, S. Modelling fracture systems in extensional crystalline basement. *In*: Petford, N. & McCaffrey, K. J. W. (eds) *Hydrocarbons in Crystalline Rocks.* Geological Society, London, Special Publications, **214**, 221–236.

GEOLOGICAL SOCIETY SPECIAL PUBLICATION NO. 214

Hydrocarbons in Crystalline Rocks

EDITED BY

N. PETFORD
Kingston University, UK

and

K. J. W. McCAFFREY
University of Durham, UK

2003

Published by

The Geological Society

London

THE GEOLOGICAL SOCIETY

The Geological Society of London (GSL) was founded in 1807. It is the oldest national geological society in the world and the largest in Europe. It was incorporated under Royal Charter in 1825 and is Registered Charity 210161.

The Society is the UK national learned and professional society for geology with a worldwide Fellowship (FGS) of 9000. The Society has the power to confer Chartered status on suitably qualified Fellows, and about 2000 of the Fellowship carry the title (CGeol). Chartered Geologists may also obtain the equivalent European title, European Geologist (EurGeol). One fifth of the Society's fellowship resides outside the UK. To find out more about the Society, log on to www.geolsoc.org.uk.

The Geological Society Publishing House (Bath, UK) produces the Society's international journals and books, and acts as European distributor for selected publications of the American Association of Petroleum Geologists (AAPG), the American Geological Institute (AGI), the Indonesian Petroleum Association (IPA), the Geological Society of America (GSA), the Society for Sedimentary Geology (SEPM) and the Geologists' Association (GA). Joint marketing agreements ensure that GSL Fellows may purchase these societies' publications at a discount. The Society's online bookshop (accessible from www.geolsoc.org.uk) offers secure book purchasing with your credit or debit card.

To find out about joining the Society and benefiting from substantial discounts on publications of GSL and other societies worldwide, consult www.geolsoc.org.uk, or contact the Fellowship Department at: The Geological Society, Burlington House, Piccadilly, London W1J 0BG: Tel. +44 (0)20 7434 9944; Fax +44 (0)20 7439 8975; E-mail: enquiries@geolsoc.org.uk.

For information about the Society's meetings, consult *Events* on www.geolsoc.org.uk. To find out more about the Society's Corporate Affiliates Scheme, write to enquiries@geolsoc.org.uk.

Published by The Geological Society from:
The Geological Society Publishing House
Unit 7, Brassmill Enterprise Centre
Brassmill Lane
Bath BA1 3JN, UK

(*Orders*: Tel. +44 (0)1225 445046
Fax +44 (0)1225 442836)
Online bookshop: http://bookshop.geolsoc.org.uk

British Library Cataloguing in Publication Data
A catalogue record for this book is available from the British Library.

ISBN 1-86239-137-8

Typeset by J W Arrowsmith Ltd, Bristol, UK
Printed by the Alden Press, Oxford, UK

Distributors

USA
AAPG Bookstore
PO Box 979
Tulsa
OK 74101-0979
USA
Orders: Tel. +1 918 584-2555
Fax +1 918 560-2652
E-mail bookstore@aapg.org

India
Affiliated East-West Press PVT Ltd
G-1/16 Ansari Road, Daryaganj,
New Delhi 110 002
India
Orders: Tel. +91 11 327-9113
Fax +91 11 326-0538
E-mail affiliat@nda.vsnl.net.in

Japan
Kanda Book Trading Co.
Cityhouse Tama 204
Tsurumaki 1-3-10
Tama-shi
Tokyo 206-0034
Japan
Orders: Tel. +81 (0)423 57-7650
Fax +81 (0)423 57-7651
E-mail geokanda@ma.kcom.ne.jp

Contents

Preface

This book is the result of a two-day meeting held at Burlington House, London in February 2001 under the auspices of the Geological Society, on the theme *Hydrocarbons in Crystalline Rocks*. It attracted over 20 contributions from industry and academia and broke new ground by bringing together for the first time three specialist groups of the Society, the Volcanic and Magmatic Studies Group, the Tectonic Studies Group and the Petroleum Geology group, to address oil and gas exploration and production from basement rocks. The idea for the meeting arose from a grant awarded to Kingston University by the Japanese Vietnamese Petroleum Company (JVPC) to help with the petrological, geochemical and structural characterization of the Rang Dong oil field, offshore Vietnam. The reservoir rocks were not sediments, but granites (monzogranites, to be precise). Although it proved to be a challenging project, exploitation of these reservoirs continues to be commercially successful.

It is of course no secret that hydrocarbons can reside in crystalline rocks. But what if these play types are much more extensive than previously recognized? Despite direct evidence that hydrocarbons are present, petroleum geologists generally ignore basement rocks in their exploration plans. Similar attitudes prevail amongst our academic colleagues, many of whom teach courses on petroleum geology and exploration, yet dismiss hydrocarbons in basement rocks as insignificant curios. A good reason for this may be the lack of a coherent source of reference material for teaching purposes. If so, then this book serves to provide a remedy in part. The editors would like to express thanks to the authors, Midland Valley and Titus Murray (now an independent consultant), JVPC, in particular Takeo Aoyama, the VMSG, TSG and Petroleum Geology groups for financial support, all of the contributors at the London meeting and to the reviewers of the papers. Special thanks also go to Claire Ivison for her skill in redrafting figures.

Nick Petford
Ken McCaffrey

Hydrocarbons in crystalline rocks: an introduction

NICK PETFORD[1] & KEN McCAFFREY[2]

[1]*Centre for Earth and Environmental Science Research, Kingston University, Kingston, KT1 2EE, UK*
[2]*Department of Geological Sciences, University of Durham, Durham DH1 3LE, UK*

Commercial oil deposits in basement rocks are not geological 'accidents' but are oil accumulations which obey all the rules of oil sourcing, migration and entrapment; therefore in areas of not too deep basement, oil deposits within basement rocks should be explored with the same professional skill and zeal as accumulations in the overlying sediments. Landes *et al.* (1960), *American Association of Petroleum Geologists Bulletin*

Oil and gas fields in crystalline basement are discovered mostly by accident, usually when the well operator notices hydrocarbon shows and tests the well. However, as shown in this book, such reservoirs can be very prolific, especially if the basement rock is highly faulted or fractured (the Bach-Ho fractured granite reservoir, Vietnam, produced some 130,000 BOPD). The standard definition of crystalline basement by petroleum geologists is any metamorphic or igneous rock unconformably overlain by a sedimentary sequence. However, crystalline rocks need not be metamorphosed, nor significantly older than their sedimentary cover. Perhaps for a more appropriate definition of crystalline basement, we must again look to Landes *et al.* (1960): 'the only major difference between basement rock and the overlying sedimentary rock oil deposits is that in the former case the original oil-yielding formation (source rock) cannot underlie the reservoir'. As such, further exploration involving geological, geochemical and geophysical studies may lead to a significant revision of the definition and nature of basement rocks in a particular area, with the possibility of discovering hydrocarbon source rocks located stratigraphically within rocks previously regarded as basement. Examples of where hydrocarbons have migrated into older porous metamorphic or igneous rocks to form a basement reservoir include the volcanic reservoirs of Japan, the oil fields of Mexico and the Maracaibo Basin of Venezuela (see Schutter 2003). Although still often dismissed as exotic curios, this may be a mistake. A case in point (discussed in Koning 2003) is the Suban field, southern Sumatra. Prior to its discovery, the search for oil was confined to structural highs in Tertiary sediments. While a number of wells were drilled into basement in order to tie the top of basement into seismic data, it was presumably not thought worthwhile to investigate the basement itself for hydrocarbons. It was not until 1999 that Gulf penetrated sufficiently deeply to discover the giant Suban gas field where hydrocarbons were found in the basement rocks. Transient heat from igneous rocks can also contribute to the maturation process in sediments that have been heated rapidly by magmatic intrusion (e.g. Saxby & Stephenson 1987; Stagpoole & Funnell 2001; Schutter 2003), making excellent cap rocks (Chen *et al.* 1999).

The moral here must be that the explorationist's definition of basement rock needs to be less narrow and more responsive to new geological ideas and data (e.g. Lamb 1997). Indeed, under the right conditions, igneous rocks, either as volcanic extrusives or high-level intrusions, come as a package of heat source and reservoir rock combined. It is for these reasons that we believe crystalline basement comprising igneous rocks, and their potential for hydrocarbon reservoirs, is deserving of in-depth study. The purpose of this book is to encourage further work in this direction.

Crystalline basement and inorganic hydrocarbons

While the majority of natural hydrocarbons form through thermal decomposition of organic material and associated microbial processes, some authors have argued that their presence in crystalline rocks is proof that all hydrocarbons are non-biogenic in origin (e.g. Gold 1998). Admittedly, the idea that abiogenic hydrocarbons contribute significantly to global hydrocarbon reservoirs has proved hard to challenge, due to uncertainties in carbon isotopic signatures between both groups. However, a recent study by Sherwood Lollar *et al.* (2002) has shown conclusively that (abiogenic) hydrocarbons in crystalline rocks from the Canadian Shield differ significantly in isotopic composition from thermogenic hydrocarbons, effectively ruling

From: PETFORD, N. & MCCAFFREY, K. J. W. (eds) 2003. *Hydrocarbons in Crystalline Rocks*. Geological Society, London, Special Publications, **214**, 1–5. 0305-8719/03/$15 © The Geological Society of London.

out abiogenic hydrocarbons as a major source of oil and gas. Hydrocarbons do form inorganically via Fischer–Tropsch reactions (Anderson 1984), but only in relatively small amounts. However, the geological conditions required to promote such reactions (cooling of magma and hydrothermal systems) can result in significant alteration (e.g. serpentinization) of the host rocks, leading to the formation of a secondary porosity that may provide important migration pathways. Further detailed study of these processes may also help improve our understanding of the relationship between metals (notably U, Pb–Zn, Au, Hg and Mo) and hydrocarbons.

Thermogenic/organic hydrocarbons in igneous rocks

Hydrocarbons have been discovered in association with many different types of igneous rocks (e.g. Powers 1932). Figure 1 shows a breakdown of lithologies in which hydrocarbon deposits have been described from around the world, based on the compilation provided by Schutter (2003). While not all are of economic value, the data reveal that volcanic rocks (basalts, andesites and rhyolites) appear most closely associated with hydrocarbons, despite the fact that most large scale production is currently from granitic and associated plutonic rocks. Unfortunately, there are still insufficient data to be able to conclude whether hydrocarbons occur in some igneous rocks simply because of post-emplacement migration, or if there is something inherent in magma composition that results in preferential accumulation.

Since most hydrocarbon systems begin outside crystalline rock, this requires hydrocarbons in the adjacent sediments. Any distinction between hydrocarbons around, as well as within, igneous rocks is thus arbitrary, and exploration for hydrocarbons in igneous rocks may well create opportunities in the adjacent sediments. A case in point is the Athabasca tar sands, Canada, where the operator Uranium Power Corporation plans to re-enter a *c.* 1,770 m well on the western outskirts of Fort McMurray. Drilled originally in 1994 and considered the first North American well to target Precambrian granite as a potential hydrocarbon reservoir, the original effort stalled due to lack of funds. Oil is currently believed to

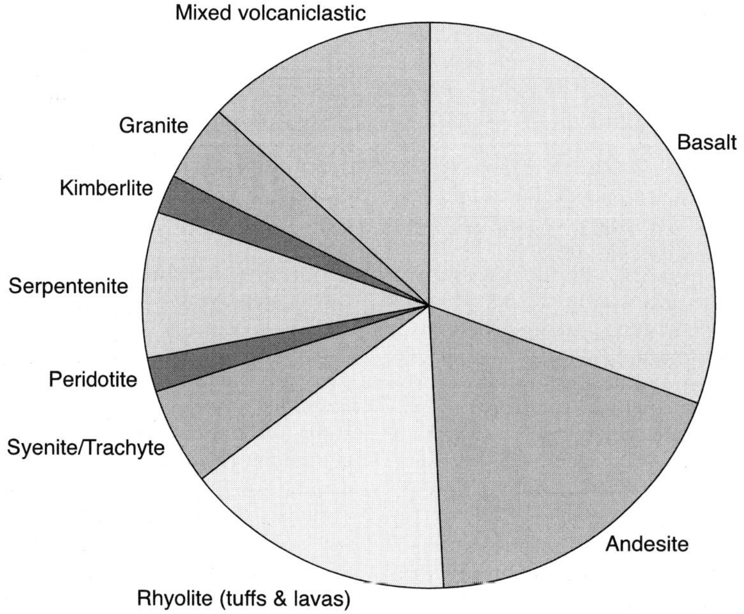

Fig. 1. The distribution of hydrocarbons in and around igneous rocks according to lithology (from Schutter 2003, Table 1). The highest reported occurrences are in basalts, followed by andesite and rhyolite tuffs and lavas. Although volcanic rocks in this survey constitute close to three-quarters of all hydrocarbon-bearing lithotypes, the majority of production and global reserves appears to be confined predominantly to fractured and weathered granitic rocks. A compilation of hydrocarbon production from fractured basement reservoirs can also be found at http://www.geoscience.co.uk/.

be trapped in fractures in the granite (*Oil & Gas Journal Online* 2002). More exotically, impact structures in basement (and sedimentary) cover may hold giant field potential. Of the 17 confirmed impact structures occurring in petroliferous areas of North America, nine are being exploited for commercial hydrocarbons. Production comes from impact-affected granites, as well as carbonate rocks and sandstones, yielding between 30 b/d to over 2 million b/d of oil and over 1.4 bcfd of gas. In some basins, the hydrocarbon systems occur beneath volcanic cover, and as well as acting as reservoirs, the igneous rocks may also provide the principal seals. For example, in the Paraná Basin of Brazil, one of the principal potential trap systems are the laccoliths and sills beneath the flood basalts. Although sub-basalt seismic imaging currently poses a technical problem, fractured sills here have produced gas, and igneous activity played an important role in the maturation process. In another example (the Phetchebun Basin, Thailand), thermal maturation of lacustrine sediments has resulted in a good sized (10 to *c.* 30 million barrel) oil field, reservoired in dolerite and sealed by lacustrine sediments, which were preferentially intruded by the rising magma. The laccolithic structure of the intrusion provides 'closure'. This is an excellent example of ways in which crystalline rocks can contribute significantly to hydrocarbon formation and accumulation.

This volume

The 12 papers in this volume cover a diverse range of topics related broadly to the theme of hydrocarbons in crystalline rocks.

The first set of papers are reviews that help to set the scene for some of the more process-oriented studies that follow. **Schutter** provides two timely and extremely thorough contributions on hydrocarbons in igneous rocks. His primary objective is to show that hydrocarbons in and around igneous rocks are not isolated anomalies, but rather are sufficiently common and orderly that exploration can be done systematically, and included in a regional exploration plan. The problem often is trying to convince those who control the finances to be less risk-adverse. A companion paper provides a broad data base identifying many of the known occurrences of hydrocarbons in and around igneous rocks. There may be more than you think! In a short contribution, **Magara** reviews the main Japanese oil producing areas that lie on the Japan Sea side of Honshu island. Although the total reserve here is small and production supplies only three-tenths of a percent of total Japanese oil consumption, the main reservoir rocks are volcanic and primary oil and gas migration seems to have taken place downward from the overlying source rocks. Marine volcanic activity since 15 Ma formed the main reservoir sections along with significant secondary porosity development. Thick and continuous deposition of organic-rich shales and mudstones followed and lower parts of these fine-grained rocks became the main source rocks. **Koning** continues in a similar vein, showing that basement rocks are important oil and gas reservoirs in various areas around the world. Such reservoirs include fractured or weathered granites, quartzites and other metamorphic rocks. In the USA, basement-derived oil production occurs in a number of areas, including California (Wilmington and Edison fields), Kansas (El Dorado and Orth fields) and Texas (Apco field). In SE Asia, basement reservoirs are the main contributor of oil production in Vietnam. Although in Indonesia, hydrocarbon production from basement rocks to date has been minimal, the recent large gas discovery in pre-Tertiary fractured granites in southern Sumatra has led to a focusing of exploration in basement reservoirs. Major oil production has also been obtained from basement reservoirs in the La Paz and Mara oil fields in Venezuela. He ends by summarizing some of the lessons learnt by companies operating in crystalline basement.

Petford reviews some of the processes contributing to the development of primary porosity in igneous rocks due to the cooling and crystallization of magma. A distinction is made between volcanic and plutonic rocks, and crystalline and granular volcanic material. The porosity in each rock type is classified according to a proposed effective length scale and geometry into diffusive (Class D) and macroscopic flow (Class F) features. Some types of primary porosity in igneous rocks are strongly time- and scale-dependent due to thermal effects associated with the emplacement of magmas. Tectonic reworking of the primary petrophysical properties of basement-forming igneous rocks may be significant in the development of regions of anisotropy and enhanced permeability. **McCaffrey *et al.*** provide a quantitative description of fracture attributes from one-dimensional samples across exposures of typical crystalline rocks. Vein thickness and fracture aperture data show predominantly power-law distributions, while vein and fracture spacing data are best described by exponential distributions with negative slopes, and appear to vary with composition in intrusive rocks. The fracture systems exhibit a range of anti-clustered to clustered patterns and densities are an order of magnitude higher for joints compared to veins. They show that thermal stress-related joint patterns are distinguishable from tectonic-related fractures in plutonic rocks, and that fracture density and clustering increases towards a major reactivated basement fault. **Ogilvie *et al.*** provide a characterization of the rough surfaces of fractures and their resulting apertures as an important step toward an improved understanding of the factors controlling fluid flow in crystalline rocks. Significantly, their tests have allowed the standard deviation of surface asperity heights, the fractal dimension and the matching parameters to be related to the resulting aperture of the fractures.

Koenders & Petford present the results of an analytical study of the mechanical effects associated with the emplacement and cooling of a magma body in the continental crust. The temperature and subsequent strain fields as a function of both position and time are calculated, with the latter providing information on the primary (cooling-related) fracture formation pattern and direction within and immediately surrounding the intrusion. Large strain jumps across the intrusion–country rock contact suggest that fracture formation will be maximized at the edges and corners of the intrusion. Low predicted strains and assumed low fracture connectivity in the centre of the intrusion imply that deformation associated with emplacement, or later tectonic motions, may be important in improving reservoir quality by providing enhanced fracture connectivity within the rock mass.

Potter & Konnerup-Madsen discuss the presence of hydrocarbons in igneous rocks, showing that while most occurrences are due to the incorporation of organic material into the magmatic system, hydrocarbons formed by inorganic processes may not be as rare as previously thought and may have implications for natural gas resources in the future. This paper reviews these occurrences and the models proposed for the generation of these hydrocarbons, concluding that the Fischer–Tropsch synthesis of hydrocarbons in igneous rocks seems to be a more applicable model for a wide variety of igneous rock types. While not dealing explicitly with hydrocarbons in crystalline rocks, the paper by **Psyrillos *et al.*** explores the important relationship between fluid flow, regional tectonics and hydrothermal alteration in granitic rocks, and complements similar studies of hydrocarbon migration in granitoid basement. They propose a new genetic model for the formation of the St Austell kaolin deposits in southwestern England, showing from fluid inclusion evidence that the kaolinization is a low-temperature hydrothermal event (50–100 °C), coincident with the oil generation window. The kaolinization appears contemporary with a major period of uplift that affected the Cornubian Massif as a consequence of offshore rifting. The most plausible fluid types for the kaolinization are either basinal brines expelled from Permo-Triassic sediments, or highly evolved meteoric waters that circulated through the sediments enclosing the pluton. The kaolinization process converted large volumes of fractured granite to a porous quartz–kaolin rock matrix. **Degnan *et al.*** provide an important crossover into the hydrogeology of low-permeability, fractured rocks. For over 20 years, intensive efforts have been underway in a number of countries to find suitable locations for underground repositories for the disposal of radioactive wastes. Such investigations have concentrated on characterizing fluid flow in low-permeability rocks, and the potential for developing and applying a breakcross-industry understanding is clear. The article summarizes the results of an eight-year study by Nirex on the detailed groundwater flow properties of a rock volume near Sellafield, northwestern England, as part of a site characterization programme to determine whether the site was suitable as a deep repository for radioactive wastes. The investigations showed that groundwater flow occurred predominantly through a limited subset of fractures, parts of which formed networks of connected channels referred to collectively as Potential Flowing Features (PFFs). These authors show how the detailed information about the geometrical and hydrogeological properties of the PFFs was used to calculate the upscaled effective parameters that are required for regional-scale flow calculations and to determine the uncertainties associated with the upscaled parameters.

Finally, **Sanders *et al.*** use observations from an extensional basin in Vietnam to simulate and analyse fracture systems typical of crystalline basement in such structural settings. Information from field observations, seismic surveys and three-dimensional structural modelling were integrated and used to build geologically realistic three-dimensional fracture networks. Their results suggest that during flexural uplift, the hanging wall is deformed significantly, containing fracture populations related to kinematic hanging wall deformation, flexural isostatic uplift and primary (cooling-related) fractures. In contrast, the footwall blocks will probably only host primary fractures. Their study brings together many important aspects set out in previous chapters (fracture density studies, surface roughness, fluid flow and knowledge of primary joint sets), and highlights the importance of a multidisciplinary approach where a proper characterization of fractured basement is needed.

We would like to thank S. Schutter, T. Koning, S. Bergman and P. Degnan (NIREX) for helpful correspondence and guidance regarding the industry perspective on hydrocarbon exploration in crystalline rocks. R. Swarbrick and J. Turner are thanked for a careful reading of the manuscript.

References

ANDERSON, R. B. 1984. *The Fischer–Tropsch Synthesis.* Academic Press, New York.

CHEN, Z., YAN, H., LI, J., ZHANG, G., ZHANG, Z. & LIU, B. 1999. Relationship between Tertiary volcanic rocks and hydrocarbons in the Liaohe basin, People's Republic of China. *American Association of Petroleum Geologists, Bulletin*, **88**, 1004–1014.

GOLD, T. 1998. *The Deep Hot Biosphere.* Copernicus, New York.

LANDES, K. K., AMORUSO, J. J., CHARLESWORTH, L. J., HEANY, F. & LESPERANCE, P. J. 1960. Petroleum resources in basement rocks. *American Association of Petroleum Geologists, Bulletin*, **44**, 1682–1691.

LAMB, C. F. 1997. Basement reservoirs—an overlooked opportunity. *Canadian Society of Petroleum Geologists and Society of Economic and Petroleum Mineralogists Joint Convention, Calgary.*

KONING, T. 2003. Oil and gas production from basement reservoirs: examples from Indonesia, USA and Vietnam. *In*: PETFORD, N. & MCCAFFREY, K. J. W. (eds) *Hydrocarbons in Crystalline Rocks.* Geological Society, London, Special Publications, **214**, 83–92.

Oil & Gas Journal Online 2002. Oil in granite concept due tests under Canada's Athabasca area. http//ogj.com.

POWERS, S. 1932. Notes on minor occurrences of oil, gas and bitumen with igneous and metamorphic rocks. *American Association of Petroleum Geologists, Bulletin*, **16**, 837–858.

SAXBY, J. D. & STEPHENSON, L. C. 1987. Effect of an igneous intrusion on oil shale at Rundle (Australia). *Chemical Geology*, **63**, 1–16.

SHUTTER, S. R. 2003. Hydrocarbon occurrence and exploration in and around igneous rocks. *In*: PETFORD, N. & MCCAFFREY, K. J. W. (eds) *Hydrocarbons in Crystalline Rocks.* Geological Society, London, Special Publications, **214**, 7–33.

SHERWOOD LOLLAR, B., WESTGATE, T. D., WARD, J. A., SLATER, G. F. & LACRAMPE-COULOUME, G. 2002. Abiogenic formation of alkanes in the Earth's crust as a minor source for global hydrocarbon reservoirs. *Nature*, **146**, 522–524.

STAGPOOLE, V. & FUNNELL, R. 2001. Arc magmatism and hydrocarbon generation in northern Taranaki Basin, New Zealand. *Petroleum Geosciences*, **7**, 255–267.

Hydrocarbon occurrence and exploration in and around igneous rocks

STEPHEN R. SCHUTTER

Subsurface Consultants & Associates, LLC, 2500 Tanglewilde, Suite 120, Houston, Texas 77063, USA (e-mail:104034.454@compuserve.com)
Current address: Murphy Exploration and Production Company, 550 Westlake Park Boulevard, Suite 1000, Houston, Texas 77079, USA (e-mail: steve_schutter@murphyoilcorp.com)

Abstract: Hydrocarbons can occur within and around igneous rocks, sometimes in commercially significant quantities. Igneous or closely associated rocks can be hydrocarbon sources in the conventional sense (biotic) as well as possibly through abiotic processes. Maturation is extremely variable, depending on the extrusive/intrusive nature of the activity and the relative importance of a deep heat source. Igneous volatiles and hydrothermal fluids may also be important in mobilizing and moving hydrocarbons. Igneous rocks can have good reservoir qualities, and they can produce their own trapping structures as well as being part of a larger feature. Many exploration methods are individually unreliable in and around igneous rocks, and an integrated approach is most effective. Seismic, magnetotelluric, gravity and magnetic surveys may all provide helpful information. Geological mapping, geochemistry and remote imagery may also be helpful. Evaluation of potentially commercial hydrocarbon accumulations requires interpretation of well logs, which may have unusual characteristics. Drill stem and production tests may also be needed for evaluation before exploration ends and development begins.

Hydrocarbons located in and around igneous rocks should be considered in any systematic exploration strategy. Igneous activity can produce distinctive source rocks, maturation and migration pathways, traps and reservoir rocks. Some of these features provide exploration opportunities where there might otherwise be none, while other prospects have been bypassed due to the presence of igneous cover. A significant number of igneous reservoirs are greater than 10 MMBOE (million barrels of oil equivalent), and while most are generally small, there are a small number of giant fields. They may occur in extensive fairways (a number of oil pools in similar trap characteristics) or as isolated occurrences. Understanding the particular conditions in and around igneous rocks may also have broader implications, particularly in terms of potential hydrocarbon sources, maturation pathways and migration mechanisms. The common association of such hydrocarbons and various metals, often in hydrothermal systems, could also improve the concepts used in metal exploration; this would be particularly true for U, Pub-Zn, Au, Hg and Mo.

There is little reason why igneous rocks, particularly those in sedimentary basins with effective source rocks, should be disregarded. There are many ways to develop porosity and permeability in igneous rocks; in some cases, they may be more porous and permeable than the adjacent sediments. They can also occur in a wide range of traps, in some cases self-produced, as with salt structures. Igneous reservoirs may not be a basis for exploration in a basin, but should be considered within a possible array of options.

Many more questions arise than answers exist concerning hydrocarbons in and around igneous rocks. This contribution attempts to establish a systematic framework for their study and the practical applications that arise. This should include consideration of the relationship to possible source rocks, the maturation history, the possible migration pathways, the possible reservoir characteristics and the type of traps likely to be present. With these aspects in mind, an exploration programme can be devised, with consideration given to eventual evaluation and engineering conditions. Here, the hydrocarbon system, as it relates to igneous rocks, is discussed first, followed by methods of commercial exploration. Exploration methodologies and statistical parameters are provided.

Igneous rocks have been overlooked in hydrocarbon exploration, largely due to their perceived lack of reservoir quality and environmental hostility to hydrocarbons. As a result, igneous rocks have never been systematically examined, reinforcing the concept that they are reservoirs only in exceptional circumstances. Unfortunately, this means that past study has been uneven and anecdotal, which is reflected in this review. Critical analysis of the entire hydrocarbon system in relation to igneous rocks is completely lacking. In addition to systematically reviewing what is known, a principal purpose

From: PETFORD, N. & MCCAFFREY, K. J. W. (eds) 2003. *Hydrocarbons in Crystalline Rocks*. Geological Society, London, Special Publications, **214**, 7–33. 0305-8719/03/$15 © The Geological Society of London.

here is to emphasize poorly known aspects of igneous-related hydrocarbon systems and provide a framework for future studies.

Scope

Clarification of the scope of this contribution is appropriate here. In terms of hydrocarbon systems and exploration, there may be little difference between porous and permeable igneous rocks and the surrounding sediments, particularly when the trap is an igneous feature. Exploration beneath igneous rocks is closely related to exploration within the igneous rocks. However, hydrocarbons in weathered basement are not considered here, as the exploration concepts are generally not related to the igneous nature of the rocks. However, hydrocarbons associated with hydrothermal systems related to igneous activity are included; they may be present in hydrothermally created fracture systems. Sedimentary facies related to igneous activity (such as atoll facies or volcaniclastic sands) are better discussed in the context of their depositional systems, which are only indirectly related to igneous rocks.

Commercial significance of hydrocarbons in igneous rocks

Igneous rocks host commercial hydrocarbon reservoirs. Many of the known reservoirs are small (as are those in sedimentary rocks), while a substantial number are in the 1 million to 10 million barrel range; a few are giants. Jatibarang, in andesitic volcanics in northwestern Java, has produced 1.2 billion barrels of oil and 2.7 TCF of gas (Kartanegara *et al.* 1996). Kudu, a 3 TCF gas field off Namibia, is in aeolian sandstones interfingering with the edge of the flood basalts of the South Atlantic volcanic passive margin (Bray *et al.* 1998). Igneous reservoirs may also occur in regional trends, similar to pinnacle reefs, so that while the individual reservoirs are small, the overall trend contains significant reserves.

Another major consideration of exploration in and around igneous rocks is the enormous underexplored region that is in this category, particularly those basins that are beneath volcanics. For example, the Siberian flood basalts cover 1.5×10^6 km^2 (Zolotukhin & Al'mukhamedov 1988) and the Paraná flood basalts of Brazil cover about 1×10^6 km^2 of sedimentary basins. A Precambrian dyke complex (which probably fed flood basalts) is 2000 km in diameter; intrusive sheets associated with another dyke swarm extend over 1.2×10^5 km^2 (Thompson 1998). Oceanic volcanic passive margins, which are similar to continental plateau basalts, cover similarly huge areas (Skogseid 2001). Felsic ash-flow tuffs may also cover large areas of sedimentary basins. The mid-Tertiary ash-flow tuffs and rhyolites of northwestern Mexico cover 2.5×10^5 km^2 (McDowell & Clabaugh 1979); similar ash-flow sheets cover large portions of the western United States.

Hydrocarbons in igneous rocks may be a valuable exploration criterion for a basin in general (Kharkiv *et al.* 1988). Several important producing regions have been initially drilled because of hydrocarbons leaking up along igneous rocks, including Mexico (Salas 1968) and the Maracaibo Basin of Venezuela (Mencher *et al.* 1953). Many areas that produce commercial hydrocarbons, such as Siberia, California, Texas and even Illinois have igneous rocks with associated hydrocarbons. This may be a positive indicator for such areas as the Columbia Basin of Washington and Oregon and the Triassic rift basins of eastern North America.

One notable related feature is the association of hydrocarbons with metal minerlization related to igneous activity, particularly mercury (Powers 1932; Sylvester-Bradley & King 1963; Peabody & Einaudi 1992; Stoffers *et al.* 1999), but also including such large, low-grade deposits as the Carlin-type gold ore bodies (Gize 1986; Ilchik *et al.* 1986; Pinnell *et al.* 1991; Hulen *et al.* 1993). The precise relationship is unclear; it may be that igneous-derived volatiles and/or hydrothermal fluids are effective at maturing and entraining organic material from the intruded sediments. Alternatively, the hydrocarbons may be a by-product of an extremely prolific metal-producing system, such as a mid-oceanic rift system. Study of the relationship may improve both hydrocarbon and metal exploration.

Source

Igneous rocks and hydrocarbon source rocks are generally not considered together. Although most of the hydrocarbons found in igneous rocks come from sedimentary rocks, some volcanic rocks may be primary source rocks, and organic-rich sediments directly associated with volcanic environments may be significant hydrocarbon sources.

Ignimbrites may be local source rocks, due to the woody material incorporated into them as they entrain the local vegetation (Czochanska *et al.* 1986; Clifton *et al.* 1990). Murchison & Raymond (1989) noted that the organic material in tuff generally had similar vitrinite values to the surrounding sediments; they suggested that the contained water in the organic debris generally protected it from the transient heat of emplacement.

Subaerial volcanics often develop lakes and swamps, which contain hydrocarbon-rich sediments. Kirkham (1935) attributed the non-associated gas in the Rattlesnake Hills field of Washington to lacustrine deposits within the flood basalts; he noted that the gas contained considerable N_2. Liu *et al.* (1989) noted that basaltic volcanism in the Bohai Basin was pene-contemporaneous with source rock deposition. They suggested that in the lacustrine basins volcanically produced warm waters enhanced the production of oil-prone organic material. Khadkikar *et al.* (1999) suggested a similar phenomenon in a lake in the Deccan Traps of western India.

Zimmerle (1995) commented on the common association of volcanics, particularly tuffs, and organic-rich sediments. Although he did not advocate a cause-and-effect relationship in every case, he suggested that volcanism might contribute to temporary anoxia. Most of his examples are more probably associated with overall reduced sedimentation in marine condensed sections, where volcanic ashes are commonly expressed in organic-rich sediments (Loutit *et al.* 1988); but the concept may be more applicable in lacustrine environments, where seeping volcanic volatiles (such as CO_2), as well as volcanic debris, can produce anoxic conditions. Fu *et al.* (1988) noted distinctive geochemical characteristics from oils sourced in tuffs, volcaniclastics and interbedded mudstones of the Junggar Basin of western China, but did not link them specifically to the volcanic activity.

Felts (1954) noted that tar-filled vesicles and voids of the Columbia Plateau basalts were sometimes found above diatom- and algal-rich lacustrine deposits between flows. He suggested that the flows entering the lakes were highly vesiculated and disrupted from the steam, providing space for hydrocarbons from the lake sediments. However, this model has not been rigorously documented by geochemistry. If it is a valid model, it suggests the possibility of a 'stratigraphic' trap within volcanic-filled basins (Fig. 1); the basin axis would be the presumed site for lakes and organic-rich sediments. The lakes would produce their own reservoir rock from the disruption of entering flows; hyaloclastites and pillow lavas from subaqueous eruptions could also contribute porous reservoir rock. The lakes would produce their own hydrocarbons and would be sealed by more lacustrine sediments, altered volcanic ash, or non-disrupted flows.

The igneous activity along mid-oceanic ridge systems may also produce hydrocarbon source

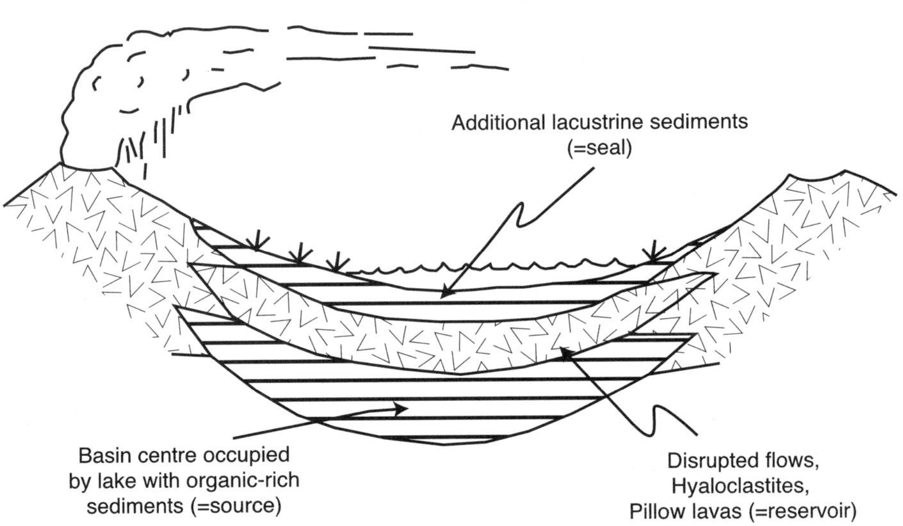

Fig. 1. 'Stratigraphic' trap in a volcanic-filled basin. The basin-centre lacustrine facies provide the source rock and also the seal; hot, mineral-rich groundwater may enhance biological productivity in the lakes. The reservoir facies consist of disrupted flows, hyaloclastites and pillow lavas, with the lateral seal provided by the transition to subaerial facies: massive flows and clay-rich weathered zones. The model is theoretical, based on elements commonly found in rift basins.

material. The hydrothermal vents support very productive thermophilic communities. The organic-rich sediments are often intruded by shallow sills and exposed to very hot fluids, leading to early maturation and hydrocarbon generation (Simoniet 1985; Kvenvolden & Simoniet 1990). Although the preservation potential and possible trapping mechanisms for the resulting hydrocarbons have not been established, they may be significant. In the Guaymas Basin of the Gulf of California, seismic anomalies indicate the escape of hydrocarbons. Simoniet (1985) estimated that in the Guaymas Basin area known to be actively producing hydrocarbons (3×9 km, greater than 120 m thick), given a 2% TOC and a 50% expulsion efficiency, at least 30 MMbbl of oil could be generated. Further, if the hydrocarbons associated with mercury mineralization in the serpentines of the Franciscan melange of California were originally from organics associated with mid-oceanic volcanic activity, it could indicate a considerable volume and preservation potential for the derived hydrocarbons. Rasmussen & Buick (2000) reported on an association of oil and hydrothermal sulphides from an Archaean deep marine assemblage in Western Australia, demonstrating that this type of system has been around a long time.

Abiotic hydrocarbons

While most hydrocarbons found associated with igneous rocks are derived from maturation of organic-rich sediments, there is some possibility of other origins. Abrajano *et al.* (1988) discussed one such possible origin in conjunction with natural gas seeps in an ophiolite in the Philippines. Under some circumstances, the serpentinization of ultramafic rocks may produce hydrogen from the reaction of olivine with water; if carbon is also present, methane may be the product. The reaction resembles the Fischer–Tropsch reaction for generating synthetic hydrocarbons (Szatmari 1989):

$$nCO_2 + (2n+1)H_2 \xrightarrow[\text{Fe, Co catalyst}]{} C_nH_{2n+2} + nH_2O$$

Hawkes (1980) noted that such a reaction could take place with any igneous rock containing reduced iron. Molchanov (1968) produced hydrogen gas by grinding olivine, hedenbergite and dunite in water. Stevens & McKinley (1995) conducted experiments with crushed basalt in water and found that hydrogen was produced; even crushed granite produced a minimal amount of hydrogen, apparently from the ferromagnesian minerals present. Szatmari (1989) stated that serpentinization in a CO_2-rich fluid produced hydrocarbons, particularly methane; he noted that the process produces abundant waxes, which parallels the Fischer–Tropsch process.

Abiotic hydrocarbons from serpentinization or from the mantle may be identified by the anomalous distributions of carbon isotopes and helium isotope ratios (Abrajano *et al.* 1988). Giardini & Melton (1981) stated that hydrocarbons with a $\delta^{13}C$ value more depleted than -18‰ may be abiogenic in origin. Sakata *et al.* (1984) noted that Lancet & Anders (1970) had found that the Fischer–Tropsch reaction strongly partitioned ^{13}C in heavier hydrocarbons (defined as non-volatile at 400 K or 127 °C). Sakata *et al.* (1984) concluded that such hydrocarbons should have $\delta^{13}C$ values of -39 to -42‰.

Sherwood *et al.* (1988) discussed the origin of CH_4 found in the Precambrian crystalline rocks of the Canadian Shield. They noted that the CH_4 lacked the characteristic isotopic signature of either organic matter or a mantle source. Some of the CH_4 was strongly depleted in deuterium, and some was accompanied by H_2; Sherwood *et al.* (1888) noted that strong deuterium depletion is characteristic of serpentinization, when depleted H_2 is a reactant in producing CH_4. One reported occurrence was in a hardrock boring near the ultramafic body at Sudbury, Ontario. There was a small flow of gas, up to 26% H_2 and 55% CH_4, with most of the remainder heavier hydrocarbons.

Gerlach (1980) discussed the origin of CH_4 from cooling alkaline magmas. If the original magma contained sufficient H_2O and CO_2 as dissolved volatiles, CH_4 became an abundant species at lower temperatures (below the subsolidus) when oxygen fugacities dropped rapidly. (Another necessary condition is low sulphur fugacity, or H_2S becomes the favoured gas.)

Another possibility is mantle-derived methane. Its abundance probably does not justify the type of exploration Gold & Soter (1980) suggested and which led to the drilling of the Siljan exploratory hole in Sweden (Jeffrey & Kaplan 1988), but it may be locally significant.

Maturation

Magoon & Dow (1994) described as atypical the petroleum systems where maturation was the result of igneous intrusion rather than burial. Maturation is one of the most difficult variables to interpret in hydrocarbon exploration near igneous rocks. Igneous activity does not condemn an area, but rather provides new complexities and opportunities. Numerous studies have shown hydrocarbons depleted from near small intrusions and condensed further away (e.g., Perregaard & Schiener 1979; Saxby & Stephenson 1987). On a larger scale, in the Solimões Basin of Brazil, the role of intrusions can be shown by comparing the Juruá gas field with the Urucu oil field (Mullin 1988; Castro & da Silva 1990; Kingston & Matzko 1995). At Juruá, a dolerite sill up to 250 m thick is intruded into the evaporitic section immediately above the reservoir interval, resulting in overmaturation; at Urucu, where the intruded interval is farther above the reservoir, the oil is preserved.

Thermal effects of igneous activity vary widely. Volcanics have very little direct impact on maturation, because they cool so quickly. Even with flood basalts, the principal thermal effect is from burial beneath the thickness of the flows (Skogseid 2001). Intrusive rocks show considerable variation. Goulart & Jardim (1982) cited estimates of the thermal aureole of an intrusion extending from one half to five times its thickness; most estimates (e.g., Dow 1977; Mullins 1988) are about twice the thickness.

The number of intrusions complicates the problem. Zalán *et al.* (1990) modelled maturation in the Iratí source rock (Paraná Basin, Brazil) by the aggregate thickness of sills. The Iratí averages 130 m thick; when the sills within the interval exceeded an aggregate thickness of 30 m, the Iratí was usually overmature. If the sills totalled 10–30 m in aggregate thickness, the Iratí was mature.

Souther & Jessøp (1990) found a similar pattern, estimating that areally each 1% of dyke by volume will raise the temperature in the area by 10 °C (for basaltic dykes). In the dyke swarms they studied in the Queen Charlotte Islands of British Columbia, they estimated extension of 1–10%, yielding temperature increases of 10° to 100 °C in the vicinity of the dyke swarms. Gordoyeva *et al.* (2001) also modelled the thermal influence of sills, and found that unless multiple sills were intruded simultaneously, their effects were minimized.

Water in the system has extremely variable effects. In some cases, hydrothermal systems carry heat away effectively to heat the surrounding country rock, while in others the heating of groundwater disperses the heat with no effect. Einsele *et al.* (1990) found that basalts intruding highly porous water-saturated sediments in the Gulf of California developed extensive hydrothermal systems; the sediments contained biogenic CH_4, overprinted by thermogenic CH_4 to C_5H_{12} near the sills (Simoniet 1994). Reeckman & Mebberson (1984) observed similar effects near intruded porous sediments in the Canning Basin off Western Australia. Hydrothermal systems associated with ash-flow tuffs also matured hydrocarbons (Czochanska *et al.* 1986; Clifton *et al.* 1990). Summer & Verosub (1992) found that heated groundwater produced uniform maturation beneath the Columbia Plateau basalts; Krehbiel (1993) found other areas where maturation decreased downward.

Simoniet (1994) noted that a principal difference with a hydrothermal system is much more rapid maturation at higher temperatures. While normal burial-driven maturation takes place at about 60 °C to 150 °C, hydrothermal maturation takes place at about 60 °C to greater than 400 °C, and maturation takes years to thousands of years. However, because of the active hydrothermal system, released hydrocarbons can be entrained and removed from the heated region, preserving them from overmaturation. Simoniet also noted that supercritical water near an intrusion could be very effective at mobilizing and removing hydrocarbons, since the water loses the hydrogen bonds that make hydrocarbons immiscible.

Raymond & Murchison (1988) and England *et al.* (1993) had a different assessment. They suggested that the conversion of water into steam by the intrusions limited the thermal effects to near the intrusions; only water-poor consolidated sediments showed significant aureoles. These contrasting interpretations on the role of water may possibly be reconciled, depending on whether steam production is possible due to pressure conditions.

While field-sized intrusions, a few kilometres across, cool in a geologically brief time, very large intrusions may be a different situation. Nodop (1971) seismically studied the very large Bramsche mafic laccolith in the Lower Saxony Basin of northwestern Germany and found it to be up to 4 km thick and 25 km across; the thermal effects also increased seismic velocities above it. Bartenstein *et al.* (1971) found the laccolith had an outer halo of oil fields in the Mesozoic sediments, with an inner halo of dry gas from Westphalian coals near the intrusion. Leythaeuser *et al.* (1987) studied the nearby Vlotho Massif, and found that the vitrinite reflectance increased from 0.48 (immature) to 1.45 (wet gas) over 47 km as the massif was approached. Kettel (1983) identified a similar large intrusion beneath the East Groningen gas field on the Dutch–German border, possibly contributing to the development of that gas field. French (1964) observed that the kerogen-to-graphite transition was 3–5 km from the Duluth gabbro complex of Minnesota, one of the largest ultramafic bodies in the world. Hollister (1980) noted that the lower Duluth gabbro contained low- to high-pressure methane and graphite; he concluded that hydrocarbons were released from the underlying organic-rich sediments, migrated upward, and were further heated within the cooling gabbro.

Thrasher (1992) studied the thermal effects of the Tertiary Cuillins intrusive complex in the Hebrides of Scotland. She found that the oil maturation aureole extended only 2 km from the complex, affecting an area of no more than 213 km^2. In contrast, Lewis *et al.* (1992) reported on the Skye granite intrusions of the same region. They reported evidence of palaeotemperatures over 100 °C up to 10 km from the intrusions, and attributed the effect to heated groundwater.

Maturation in an area with igneous activity may be due more to an elevated regional heat flow than to the intrusions themselves (e.g. the Taranaki Basin of New Zealand, discussed by Pilaar & Wakefield 1984, and the western Delaware Basin of Texas and New Mexico, discussed by Barker & Pawlewicz 1987). Hurter & Pollack (1995) studied the Paraná Basin of Brazil, and concluded that the intrusions and flood basalts significantly affected the surrounding sediments for 10^6 years or less, in most cases 2×10^5 years or less. In contrast, the underplated magma was a significant thermal influence for about 10^7 years; Skogseid (2001) found similar values for the effect of underplating. Even so, the thermal effects of the igneous activity were small (due to the short cooling time involved) compared to burial by 1–2 km of basalt.

Maturation modelling should be part of the analysis of a basin with igneous activity, although it is difficult. Maturation effects may be difficult to measure near igneous rocks, affecting interpretation and assessment. Summer & Verosub (1992) noted that in some cases vitrinite reflectance is higher than T_{max}. In contrast, Altebäumer *et al.* (1983) reported that higher temperatures were required to reach a given vitrinite reflectance if less time was involved. Ujiié (1986) and Raymond & Murchison (1992) found that optical maturation measures, such as vitrinite reflectance, responded much more quickly to heating than did molecular measures (which more closely reflect hydrocarbon maturation). As a result, the 'oil window' near intrusions is often at a higher R_0 range than that due to normal burial maturation. An assessment based solely on vitrinite reflectance data from the vicinity of intrusions might incorrectly condemn a prospective area.

Heat flow values for maturation models associated with igneous activity are difficult to find and quite variable. Some representative values from various settings are given in Table 1.

Rapid maturation associated with igneous activity may produce a distinctive suite chemical signature in the organics. The range of temperatures approaching an intrusion may produce hydrocarbons with a range of maturation signatures, including natural fractionation. Dow (1977), Bostick & Pawlewicz (1984), and Raymond & Murchison (1988) found that the temperature and maturation level already present before the intrusion were important variables in the ultimate maturation; this implies that maturation from the intrusion did not reach equilibrium due to rapid cooling.

Simoniet *et al.* (1981) and Püttmann *et al.* (1989) analysed the effects of intrusions on organic-rich shales, and found distinctive changes in the distribution of alkanes and alteration of organic markers. George & Jardine (1994) found ketones (relatively rare in oil) in a Precambrian dolerite sill, and suggested that they

Table 1. *Geothermal gradients and heat flow in basins with igneous activity*

Location and setting	Heat flow	Reference
Most active spreading site, Gulf of California	20 HFU	Einsele *et al.* 1980
Miocene volcanoes, Pannonian Basin, Hungary	>7 HFU	Sachsenhofer 1994
Volcanic arcs (general)	>2.8 HFU	Souther & Jessup 1992
Kamchatka volcanic arc, Russia	2.2 HFU	Adam 1978
Niigata Basin, Japan	>2.0 HFU	Fukuta 1986
Bohai Basin, China	>2.0 HFU	Lee 1989
Rio Grande Rift, New Mexico	1.8 to 3.2 HFU	Reiter 1986
Cape Verde plume	plus 1 HFU over background	Courtney & White 1986
Peak igneous event, Jameson Land Basin, Greenland	1.5 HFU	Mathiesen *et al.* 1995
Taranaki Basin, New Zealand	1.4 HFU, 1.8 HFU near volcanoes	Armstrong *et al.* 1997

HFU = heat flow unit (10^{-6} cal/cm^2/sec)

might have been produced by rapid pyrolysis of the source rock. Murchison & Raymond (1989) found high levels of polycyclic aromatic hydrocarbons (PAH) near intrusions; these compounds are generated by combustion or pyrolysis at high temperatures. Simoniet & Fetzer (1996) reported PAHs in petroleums from submarine hydrothermal vents. Mello *et al.* (2000) found that the distribution characteristics of highly stable diamondoids near intrusions could be used for a number of purposes, particularly calibrating oil-to-gas conversion models and estimating expulsion efficiencies. Simoniet (1994) noted that hydrothermal hydrocarbons tend to have more aromatics, polar compounds and associated non-hydrocarbons than normal hydrocarbons generated by burial of sediments. Hydrothermal hydrocarbons may also be relatively depleted in light aliphatic hydrocarbons and soluble aromatics, which are more efficiently removed by the hydrothermal system. T. J. Weismann *et al.* (Anon. 1971) examined stable isotopes in natural gases and concluded that many gases were influenced by igneous-related maturation. Neto *et al.* (2001) examined stable carbon isotope distribution in some natural gases, and found evidence of multiple levels of maturation, with some from preexisting hydrocarbons cracked by intrusive activity. However, Simoniet & Didyk (1978) found an unusual non-igneous modification: natural gas escaping near diorite intrusions provided the substrate for bacteria, which in turn produced hopanoid-rich ‘paraffins’ lacking alkanes.

Yükler & Dow (1990) noted that rapid heating might increase expulsion efficiency from the source rock by producing higher pressures. The higher pressures may also increase stress fracturing within the source rock, also contributing to expulsion efficiency. Barker (1994) calculated that approximately 64 m^3 of CH_4 are produced when a barrel of oil (about 159 litres) is cracked, producing sufficient pressure to fracture the enclosing rock. Hutchinson (1994) noted that around a Texas ‘serpentine plug’, the Austin Chalk reservoir was more fractured and porous than normal, as well as being thermally more mature. Hutchinson interpreted the hydrocarbons present as being locally sourced and trapped beneath the altered volcanics of the submarine volcano; the early hydrocarbon charge additionally maintained porosity against later burial and diagenesis.

Araújo *et al.* (2000) estimated the amount of hydrocarbons expelled from the Iratí source rock in the Paraná Basin of Brazil due to Cretaceous intrusions. Their values were not based on theoretical models but on empirical observations in a large number of wells. While adequate data sets may not be available for the analysis of other basins, this is a useful example of calculating volumes of expelled hydrocarbons.

Migration

Hydrocarbons can be found in igneous rocks (excluding weathered basement) for several reasons: (1) hydrocarbons matured in sedimentary rocks can migrate vertically or laterally into structurally higher igneous rocks; (2) hydrocarbons may be forced from compacting sedimentary rocks into more porous igneous rocks; (3) cooling igneous rocks may achieve a lower vapour pressure, with hydrocarbons forced in; (4) hydrothermal fluids may dissolve hydrocarbons and precipitate them in igneous rocks; or (5) the hydrocarbons may originate within the igneous rocks. With the last possibility, there are several variants: (5a) volcanic rocks, such as ignimbrites, may have entrained a significant volume of organic material when they were emplaced; (5b) the hydrocarbons may have been produced by the Fischer–Tropsch reaction, when hydrogen is produced from water in the presence of reduced iron, and joins with available carbon; or (5c) the hydrocarbons may have been produced by reactions within the low-oxygen volatiles at the end of magmatic crystallization.

Igneous activity can influence the effectiveness of migration by converting groundwater into a supercritical state. In this state, it loses its hydrogen bonds and becomes an excellent solvent for hydrocarbons (Simoniet 1994). Therefore, supercritical water is good at scavenging and removing hydrocarbons, which are dropped in cooler regions when the water cools.

Possible products of this process are froth veins, reported from mercury deposits with serpentine bodies in California. Froth veins apparently form when hydrocarbons separate from cooling hydrothermal fluids, producing multitudinous globules. The hydrocarbon-fluid interface may become mineralized, resulting in a ‘froth’ of globule shells.

Volatiles associated with the magma as part of a hydrothermal system may also play a role in petroleum generation and migration. In the Otway Basin of Australia, CO_2 associated with a maar volcano reacted with Type III and Type IV woody organics, removing aromatics and the sparse saturated hydrocarbons present, thus producing a modest amount of a peculiar oil associated with the CO_2 (McKirdy & Chivas 1992). Kvenvolden & Claypool (1980) studied a hydrocarbon-bearing CO_2 seep in the Norton

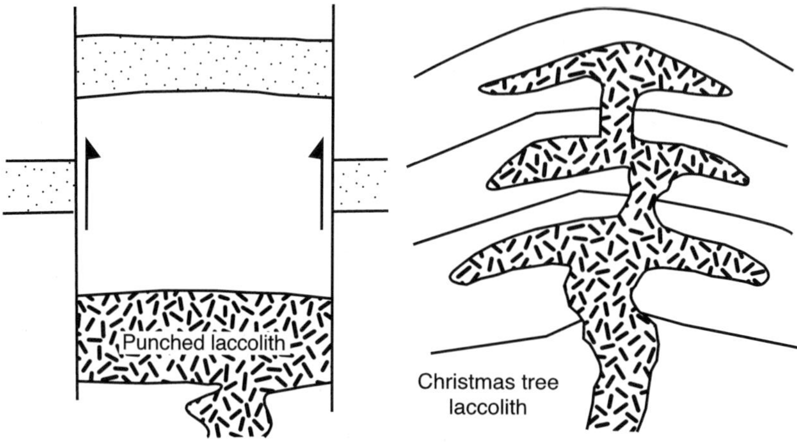

Fig. 2. Laccolith end members. A punched laccolith moves its overburden vertically along bounding faults; a Christmas-tree laccolith intrudes a series of weak layers, progressively deforming the overburden. Most laccoliths have characteristics of both end members; both provide structural closure. Concept from Corry (1988). The Omaha Dome of the Illinois Basin is a well-documented example of a Christmas tree laccolith.

Basin off Alaska, and noted that light hydrocarbons are readily soluble in CO_2, while heavy hydrocarbons, particularly those with N, S and O, are not. Gize & Macdonald (1993) attributed a bitumen occurrence in a lava flow at the Suswa volcano of Kenya to mobilization by CO_2, and noted that some CO_2-rich systems with hydrocarbons also contain mercury. Kvenvolden & Simoniet (1990) reported hydrothermally derived petroleum from sediments rich in terrigenous organics as well as those with marine organics.

CO_2 may affect hydrocarbon migration in another way. In the northern Kaiparowitz Basin of southern Utah, CO_2 associated with the Marysvale volcanic centre may have been associated with a natural CO_2 flood for the hydrocarbon system of the area (Shirley 1998; Anonymous (Utah Geol. Surv.) 1999). The structures nearest to the volcanic centre may have been swept and are full of CO_2, while the more distant structures may have oil displaced off-structure by the strong regional hydrodynamic system; most such fields have associated CO_2 as a gas cap, rather than light hydrocarbons.

Traps

As with sedimentary rocks, hydrocarbon traps with igneous rocks may be stratigraphic or structural. However, like salt structures, igneous activity can produce traps independent of regional tectonics. At shallow depths, igneous intrusions are rarely emplaced by stoping and almost never by melting; usually, magma wedges into the country rock, adding volume and producing deformation. Sills and laccoliths frequently result in closed structures in the intruded sediments.

Corry (1988) recognized two end types of laccolith. Punched laccoliths are characterized by vertical peripheral faults, with the roof lifted like a piston by magma flowing into the underlying chamber. Christmas-tree laccoliths are a series of lens-shaped intrusions along bedding planes, stacked in succession along a central feeder (Fig. 2). Most laccoliths are somewhere between the end members; all can produce trapping closures.

The Omaha Dome in the Illinois Basin is one of the best-documented examples of oil production associated with a Christmas-tree laccolith produced by an ultramafic intrusive; there are several of these features of Permian age in the Illinois Basin. Discovered in 1940, it has a cumulative production of 6.5 million bbl, with a productive area of 450 acres on a structure of 15 000 acres (English & Grogan 1948; Seyler & Cluff 1990). The stratigraphic section is similar in both areas. The Lower Palaeozoic section, composed mostly of massive carbonates, is penetrated cleanly. But when the intrusions reached the Upper Mississippian and Pennsylvanian sections, with abundant interbedded shales, the section was intruded with many sills, producing Christmas-tree laccoliths. The structural closure, some 10–15 km in diameter (Nicolaysen & Ferguson 1990) is restricted to the intruded zone and above; the punched carbonate section may show gentle closure developed before piercement

(English & Grogan 1948; Brown *et al.* 1954; Wojcik & Knapp 1990). There may be a secondary graben around the piercement due to withdrawal of magma at depth. These features, coupled with extreme brecciation if the magma encounters groundwater at relatively shallow depths or as volatile-rich magma depressurizes, has prompted some workers to identify these domes as astroblemes (Rampino & Volk 1996). Nicolaysen & Ferguson (1990) and Luczaj (1998) noted that the association of ultramafic rocks with these 'cryptoexplosion' features indicated that igneous activity could produce shocked quartz and shatter cones, phenomena considered diagnostic of astroblemes. Nicolaysen & Ferguson (1990) related the petrology of alkalic and alkaline ultramafic rocks (including kimberlites, lamproites and carbonatites) to very high initial volatile contents—up to 27 wt% of C–O–H fluid. These magmas originate at great depths and ascend very rapidly (McGetchin *et al.* 1973, estimate 1–10 hours to reach the surface, possibly at velocities of 350–400 m/sec); devolatilization can be explosive. Apparently, if the relative volume of magma is small, a collapse crater without a dome may result. Notably, Omaha Dome is one such feature that did not reach the surface. These features often have associated hydrocarbons, either migrated from the surrounding sediments or perhaps resulting from inorganic processes. If the high temperatures produce unusual hydrocarbons, such as polycyclic aromatics (found with some igneous-related hydrocarbons), the diatremes would also have another feature frequently assigned to impact features.

Corry (1988) affirmed Gilbert's (1877) observation that there are no small (<1 km diameter) laccoliths. Amaral (1967, cited by Bigarella 1971) stated that laccoliths in the Paraná Basin of Brazil ranged up to 10 km in diameter, with 4° to 5° dips on the flanks. Mesner & Wooldridge (1964) stated that laccoliths in the Maranhão (Parnaíba) Basin of Brazil could theoretically create more than 500 m of closure. Leyptsig (1971) noted that Siberian laccoliths had radial and concentric crestal faults, similar to salt domes.

Some lithologies are preferentially intruded. Evaporites are particularly prone to intrusion; for example, in the Solimões and Amazonas basins of Brazil, the widespread doleritic sills are almost exclusively in the Permo-Carboniferous Itaituba and Nova Olinda evaporites (Mosmann *et al.* 1986; Mullins 1988); little or none of the igneous activity reached the surface. In a similar situation in the Lena–Tunguska province of Siberia, Kontorovich *et al.* (1990) suggested that intrusion of the dolomite-anhydrite interval resulted in high CO_2 levels as well as high sulphur content in nearby oil. Oil shales and similar source rocks are also preferentially intruded. In the Paraná Basin of Brazil, the Iratí oil shale is preferentially intruded by doleritic sills and laccoliths associated with the Serra Geral flood basalts, enhancing maturation (Zalán *et al.* 1990); the common intrusions make the Iratí one of the few seismically mappable units beneath the basalts. In South Africa, the Karoo dolerites have a similar affinity for the 'White Band' shale (Hawthorne 1968), which correlates with the Iratí. The reason for this correlation is unclear. It could be due to the weakness of the intruded rocks, the increased strength of the overlying rocks, or the reactivity of the intruded rocks in response to magma (evaporites may melt or dissolve; organic-rich rocks may generate hydrocarbons, reducing the lithostatic pressure). Better understanding of this relationship would greatly improve the predictability and modelling of hydrocarbon systems associated with igneous rocks.

Fractured sills or laccoliths themselves are also common igneous traps. Cooling may produce fracturing; some sills are also fractured by later tectonism. A good example is Dineh-bi-Keyah oil field in northeastern Arizona. It is a fractured syenite sill on an anticline. The sill intruded black shale of the Hermosa Group, the source rock in the nearby Paradox Basin. Dineh-bi-Keyah has produced more than 18 million barrels of oil (Kornfeld & Travis 1967; Pye 1967; McKenny & Masters 1968; Biederman 1986; Ray 1989; Masters 2000). Wichian Buri field of the Phetchabun Basin, Thailand (Fig. 3), is another example of an oil field related to laccolithic intrusion of source rock facies.

Buried volcanoes are another common trap for hydrocarbons (Fig. 4). In addition to the volcanic cone, uplift around the conduit and fracturing of the country rock may provide additional traps and reservoirs. Volcanoes are known traps in Japan and New Zealand, but the best known are the volcanoes and associated laccoliths, plugs and dykes of the Texas 'serpentine plug' trend. These were small Late Cretaceous volcanoes, composed of silica-poor alkalic basalt, active during deposition of the Austin Chalk (Fig. 5). The first oil field hosted by a volcano was discovered in 1915; because they are excellent gravity and magnetic anomalies, they provided the early impetus of geophysical exploration for hydrocarbons. In addition to hydrocarbons in the altered basalts and pyroclastic rocks, oil is also found in associated shoal facies, fractured carbonates beneath the

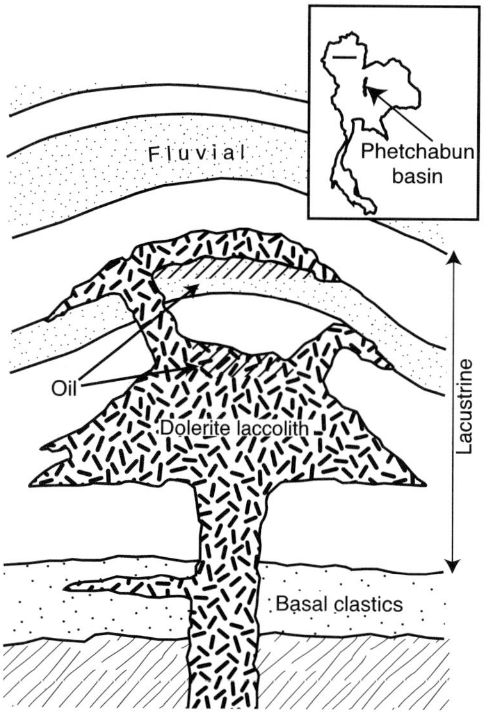

Fig. 3. The oil field at Wichian Buri, Phetchabun Basin, Thailand (*see* inset map), is an excellent example of hydrocarbon reservoirs associated with igneous intrusions. It formed as a wrench basin with high heat flow, but later doleritic intrusions into the Oligocene–Miocene lacustrine shales locally matured the hydrocarbons. Reservoirs are deltaic sands within the lacustrine source, domed above the laccolith. Oil is also recovered from the dolerite intrusions themselves. Depth to the laccolith is about 1200 m; depth to basement is about 2600 m. Wichian Buri was originally estimated to contain 10 MMbbl of waxy oil; recently that has been increased to 30 MMbbl (Remus *et al.* 1993; Williams *et al.* 1995; Anon. 2002a, b).

volcanoes and sands draped over the plugs. The plugs occur in a band about 250 miles (400 km) long (Ewing & Caran 1982; Matthews 1986). Approximately 225 surface and subsurface igneous bodies have almost 90 associated oil fields, producing 54 million barrels of oil; 32 fields are larger than 100 000 barrels, while the largest, Lytton Springs, has produced 11 million barrels (Table 2). Trap density averages 3.6 plugs per 100 mi^2 (1.4 plugs per 100 km^2); in the densest area, it reaches 5.5 plugs per 100 mi^2 (2.1 plugs per 100 km^2). The individual plugs are usually 1.5–2.5 km^2 in size; the volcanic necks are usually less than 0.8 km in diameter (Lewis 1984). In exploration terms, the Texas 'serpentine plug' trend would be comparable to a pinnacle reef trend; the individual fields are usually not large, but the quantity of fields accounts for a large total volume of hydrocarbons. [In contrast, the oil-bearing Kora volcano in the Taranaki Basin of New Zealand is 10–12 km in diameter and about 1 km thick (Russell, R. O., pers. comm. 1997, cited in Batchelor 2000).]

Although most widespread in Texas, volcanic centres around the northern Gulf of Mexico produced hydrocarbon traps. The Jackson Dome in Mississippi and the Monroe Uplift in Louisiana were large shoal areas developed around a cluster of volcanoes (Fig. 6); while the principal reservoir rock is the shoal-water carbonate facies, volcanics are intermixed. A similar platform, the Anacacho Platform, is exposed near San Antonio in Texas, where the shoal-water carbonate is locally saturated with tar.

Reservoirs and seals

There is a wide variety of porosity and permeability types associated with igneous rocks. (The reservoir characteristics of associated sediments and metasedimentary rocks, such as atoll carbonates and turbiditic volcaniclastic sands, are more appropriately discussed elsewhere.) Igneous rocks may have primary porosity (associated with extrusive rocks); secondary porosity from late-stage retrograde metamorphism or hydrothermal alteration; and fracturing, from cooling or weathering. An important aspect of porosity in igneous rocks is that, except in tuffs, it is lost only slowly through compaction; porous lava flows in the deeper parts of a basin may be more likely to have porosity than the surrounding sediments.

Primary porosity in igneous rocks may be intergranular (as in agglomerates and tuffs) or vesicular (as in vesicular flow tops and bases). The Conejo oil field of the Ventura Basin in southern California has oil in a basalt agglomerate; the seal is the surficial asphalt mat (Taliaferro *et al.* 1924; Powers 1932; Nagle & Parker 1971, pp. 269, 273). Chen *et al.* (1999) reported porosities in vesicular basalt and andesite in the Bohai Basin, northeastern China, of 30%, sometimes as high as 50%; the vesicles were 0.5–5 mm in diameter. Luo *et al.* (1999) did a detailed study of porosity and permeability in the Bohai igneous reservoirs, and showed that vesicles were usually the leading source of porosity. The tuffs and breccias of the Kora volcano of the Taranaki Basin have porosities up to 30% and permeabilities up to 300 millidarcies (mD) (Hart 2001).

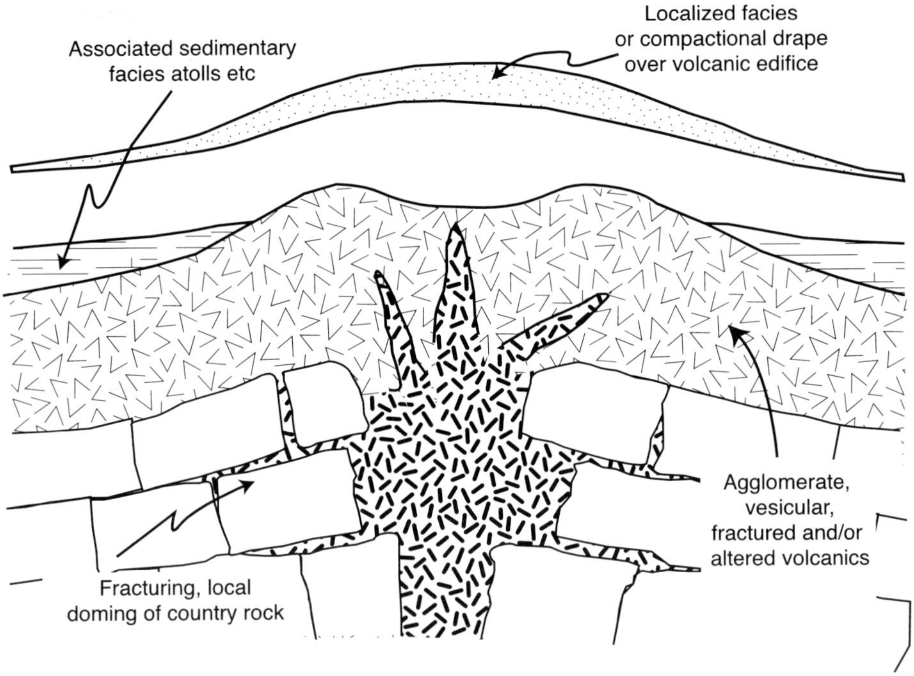

Fig. 4. Possible hydrocarbon reservoirs associated with buried volcanoes.

Another type of reservoir rock is péperite, a mixture of sedimentary and igneous rock that fills maar craters. Péperites are characterized by a very high ratio of country rock to juvenile igneous rock, with country rocks usually 60% to 80% or more of the debris. Maars are often filled with lacustrine or swamp deposits, which may provide a seal, or even a hydrocarbon source if there is later activity. Hydrothermal activity may also mature the country rock;

Fig. 5. Pilot Knob, one of the Texas 'serpentine plugs' exposed immediately SE of Austin; similar volcanoes produce oil in the subsurface. The 'plug' is about 1 km in diameter. It is surrounded by a moat developed on the McKown Formation, a shoal-water facies of the upper Austin Chalk Group. The unaltered igneous rock is described as a nepheline basanite (Young *et al.* 1982); the activity took place during the earliest Campanian (Young & Woodruff, 1985). Pilot Knob is about 25 km up dip from Lytton Springs, the largest of the Texas 'serpentine plug' oil fields (Ewing & Caran, 1982).

Table 2. *Lytton Springs, Texas 'serpentine plug' oil field reservoir data*

Estimated ultimate recovery	11 million barrels
Original oil in place	90 million barrels
Recovery efficiency	12%
Average porosity of producing igneous rocks	6%
Average permeability of producing igneous rocks	7 millidarcies

(From Galloway *et al.* 1983)

fracturing may release trapped hydrocarbons. Barrabé (1932) described oil shows in péperite in the Limogne Graben of central France, an area with abundant maars. Ridd (1983) noted péperite in the lower volcanics of the Faroe–Shetland Basin.

Secondary porosity in many igneous hydrocarbon reservoirs is very important. Frequently, this is due to alteration by the latest stages of the igneous activity, which may alter the earlier-formed minerals and result in intracrystalline or vuggy porosity.

Many of the hydrocarbon fields of Japan are in altered volcanics, in the 'Green Tuff Belt' of western Japan. Katahira & Ukai (1976) compared volcanic reservoirs to those in carbonates, characterized by vugs connected by fractures, and sometimes with similar shapes and log responses as well. In Japanese oil and gas fields, volcanic rock porosities range up to 40% (Uchida 1992). In the Samgori and Teleti oil fields of eastern Georgia, laumontite tuff reservoirs may have porosities greater than 27% and permeabilities exceeding 400 mD (Vernik 1990; Patton 1993; Grynberg *et al.* 1993).

An unusual reservoir derived from volcanic debris was described by Aoyogi (1985) in the Fukubezawa oil field in the Akita Basin of northern Honshu, Japan. Bioclastic limestone was deposited with volcanic debris (mostly tuff). The siliceous volcanic debris was altered to fine-grained dolomite, with lenses of fossiliferous limestone and dolomites. The resulting reservoir rock ranges in porosity from 5% to 30%, and in permeability from 0.1 mD to 12.5 mD.

Fracturing may enhance primary or secondary porosity, or it may provide the only pore space present. Igneous rocks commonly have fractures due to cooling (such as the well-known columnar fractures in basalts) and sometimes from unloading. Fracturing due to cooling is important in the West Rozel heavy oil field of Utah, where wells in basalt produced up to 1000 BOPD (Nelson 1985). Igneous rocks (particularly intrusive rocks) are usually quite brittle, and may be subject to fracturing during tectonism. In the Thrace Basin of northwestern Turkey, Ozkanli & Kumsal (1993) reported that silicified rhyolitic tuff was fractured by tectonism and a reservoir, while dacitic tuff was not fractured and was tight. Levin (1995) proposed a rule of thumb: acidic igneous rocks are generally more fractured than basic igneous rocks, and are thus better reservoirs; also, lava flows tend to have better reservoir characteristics than pyroclastic rocks. Intrusive rocks, particularly sills and laccoliths, frequently owe their porosity to fracturing.

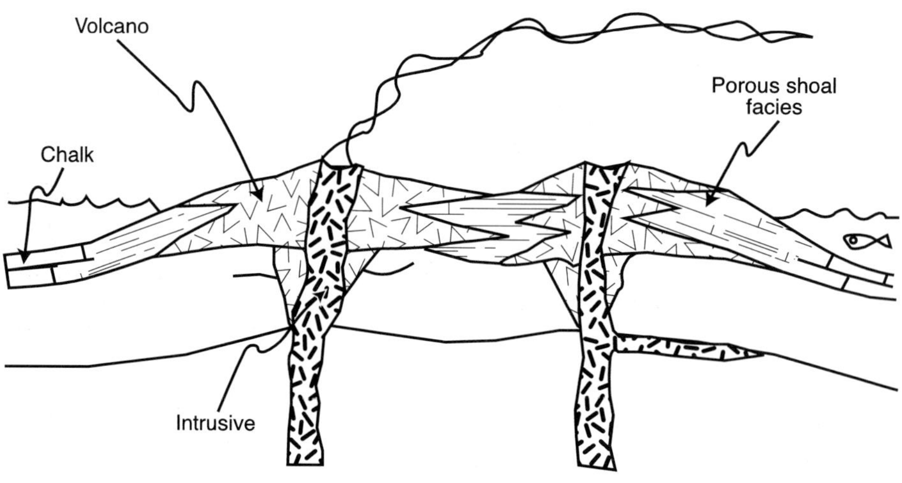

Fig. 6. Upper Cretaceous 'domes' (composite volcanic-carbonate platforms). On the Jackson Dome (Mississippi) and the Monroe Uplift (Louisiana), the shoal facies has produced large volumes of gas. Near San Antonio, a similar platform (the Anacacho Platform) is exposed, but some of the reservoir facies is saturated with asphalt.

Fracturing may also be present on the flanks of intrusions; gas-filled fractures are reported along the edges of dolerites in the Karoo Basin of South Africa (Petroleum Agency SA 2000).

Igneous rocks, particularly extrusive rocks, may have both porous zones and tight sealing zones. In ignimbrites, the upper tuff may be rapidly altered to clay, while the lower welded portion may only be fractured. Similar relationships may occur in basalts; in the Rattlesnake Hills gas field in Washington State, the gas gathered in reservoirs in the vesicular zones at the tops of the basalts, while the interflow clays (bentonites or soil zones) provided the seals (Kirkham 1935). In the Kipper field of the Gippsland Basin off southeastern Australia, the top seal is highly altered basaltic volcanics (Sloan *et al.* 1992).

Exploration

Geological methods, mapping, imagery, seeps

Exploration of hydrocarbons in and around igneous rocks can involve a wide range of techniques, once the decision is made to look for the hydrocarbons. Simple surface mapping may be useful. Layered igneous rocks, particularly volcanics, are deformed in regional structures so mapping may indicate deeper structures. Komatsu *et al.* (1984) noted that many of the oil and gas fields in the Niigata Basin of northwestern Honshu were found by mapping surface structures, since there the thick volcanic cover rendered geophysical methods useless. Local, igneous-related structures may also be mapped; Collingwood & Rettger (1926) noted that Lytton Springs, the largest of the Texas 'serpentine plug' oil fields, was identifiable on the surface due to doming, apparently due to compactional doming over the volcano. Photogeological and satellite imagery may also help; feeder dyke swarms may show up as lineaments and post-emplacement structuring would be apparent, while pre-emplacement features (such as those preceding flood basalts) would be visible only beyond the margins of the igneous cover. Fritts & Fisk (1985) used photogeology and satellite imagery to help assess the basalt-covered Columbia Basin in Washington and Oregon. Stanley *et al.* (1985) discussed rivers in the Paraná Basin of Brazil that follow lineaments and parallel the Ponta Grossa feeder dyke swarm.

The presence of surface seeps also supplements exploration data. Link (1952) showed examples of oil seeps associated with igneous rocks from the Cuban serpentine fields and the Golden Lane region of Mexico. The contacts between igneous rocks and the surrounding country rocks are often migration pathways, producing surface seeps. Such seeps have led to the opening of major hydrocarbon provinces.

Geochemical methods may be valuable exploration tools. Johnson *et al.* (1993) reported on a study of methane in Columbia flood basalt aquifers. The methane was apparently concentrated near faults and fractures where it could leak up from the buried sediments beneath the basalts. Through isotope analysis, they identified a biogenic and a thermogenic component to the methane, with the thermogenic portion apparently derived from deeply buried coals. Bortz (1994) reported that a soil gas survey was useful in delineating an oil field in welded tuff in Nevada; the field apparently showed up because of the leaky bounding fault.

Gravity and magnetic methods

The various geophysical methods are highly variable in their effectiveness. Geophysical exploration programmes must take this into account, and reliance on a single technique is hazardous.

Gravity and magnetic methods immediately suggest themselves. Mafic igneous rocks are more amenable; they offer sufficient contrast to the regional sediments that shows up well on gravity and magnetic surveys. These methods were among the earliest geophysics used in hydrocarbon exploration when they were applied to the 'serpentine plugs' of Texas (Collingwood 1930; Jenny 1951) and Louisiana (Spooner 1928). In comparison, felsic igneous rocks have relatively low density contrasts with the country rock and are generally not exceptionally magnetic.

Gravity and magnetic methods depend on local conditions. Williams & Finn (1985) found that the intrusions beneath volcanoes and small calderas (<15 km in diameter) usually produced positive gravity anomalies, due to the contrast of the intrusions with the older extrusive rocks. Larger calderas usually have negative gravity anomalies, due to the contrast between silicic intrusive rocks (with variable amounts of tuff) and the surrounding metamorphic rocks. In the Taranaki Basin off New Zealand, Bergman *et al.* (1992) noted that some buried volcanoes had strong gravity and magnetic anomalies, while others had virtually none.

Gravity and magnetic methods may be useful at the regional scale or the prospect scale. Gunn (1998) reported on an aeromagnetic survey of the Otway Basin, off southern Australia. Irregularities in a broad magnetic sheet were interpreted as topography on a flood basalt; intense high-amplitude anomalies were interpreted as volcanic centres. The magnetic survey was used in conjunction with a marine seismic survey. On a prospect scale, in the Durham Basin of North Carolina, Daniels (1988) used high-resolution ground-based magnetic and gravity surveys to model a dolerite sheet 120–250 m thick. He noted that locally the hornfels of the contact aureole was sufficiently magnetic to have a signature like the dolerite. Recently (Anon. 2001), Conoco announced a proprietary method for inverting gravity and magnetic data for exploration beneath volcanics as well as beneath salt.

Seismic methods

Sonic velocities in unaltered igneous rocks can be quite high (Table 3); they are also high in some extrusive rocks, such as unaltered flows, but pyroclastic rocks and altered igneous rocks can be very variable. Intrusive rocks generally are well expressed in lower-velocity sediments, although near-vertical dykes may be obscure (Jansa & Pe-Piper 1988). Flood basalts and other volcanics can be problematic. If they have little weathering, minimal topographic irregularities and no interbedded sediments, internal and external seismic reflectors can be good. However, this is frequently not the case, and thick volcanics are often seismically opaque. Planke & Eldholm (1994) noted that reflections within flood basalt intervals are usually the result of interference or tuning effects, although thick flows or thick sediment/weathered zone intervals may be laterally traceable. In some cases, seismic exploration in and around igneous rocks may be possible, but in others it may be useless.

A broad range of techniques can be used to improve the interpretation of seismic data in and around igneous rocks. Henkel (1989) reported on seismic surveying in the San Juan Sag of Colorado, an area largely covered by volcanics. He reported that the dominant variable was outcrop lithology: andesites and volcaniclastics produced good data, ash-flow tuffs produced poor data and basalts yielded uniformly very poor data. Seismic source appeared to have only a minor effect on quality. Jenyon (1990) stated that buried flood basalts usually wipe out seismic data because of the strong impedance contrast with the overlying sediments. Intrusive rocks (such as dolerite) badly attenuate low-frequency seismic energy, particularly when the seismic input is into an outcropping intrusive body (Fatti 1972).

A number of techniques have been tried to achieve better seismic data. Krehbiel (1993) reported on a seismic survey of a sub-basin beneath the Columbia Plateau basalts, originally located by magnetotelluric data. He found Vibroseis with high fold (125 to 200) to be helpful, but structural outlines were still mostly based on packages of reflectors rather than single events. Campbell & Reidel (1994) found diving waves to be useful in determining the thickness of the basalt; they also found that the brecciation along fault zones caused a pronounced velocity anomaly, allowing the faults to be well defined. Silva & de Brito (1973), working on the Paraná Basin of Brazil, found that shaped charges and Vibroseis improved the amount of energy penetrating the rocks, resulting in better data. Zalán *et al.* (1990) noted that Vibroseis and dynamite were the best sources in the basin; dynamite was often necessary as rough terrain required

Table 3. *Sonic velocities of igneous rocks*

Igneous rocks	Velocity	Reference
Near-surface unweathered intrusives	5.0–6.2 km/sec	Smithson & Shive 1975
Dolerite (South Africa)	6.1 km/sec	Fatti 1972
Dolerite (Proterozoic, Ontario)	6.7 km/sec	Zaleski *et al.* 1997
Basalt flows, top to bottom variation	2.9–6.1 km/sec	Symonds *et al.* 1998
Basalt flows, average	4.2 km/sec	Symonds *et al.* 1998
Unaltered mafics, ultramafics (Texas 'serpentine plugs')	5.5–7.3 km/sec	Ewing & Caran 1982; Matthews 1986
Altered palagonite tuff (Texas 'serpentine plugs')	2.9 km/sec	Ewing & Caran 1982
Plateau basalts (Columbia Basin)	5.8 km/sec	Pujol *et al.* 1989
Interbedded clay layers in basalt (Columbia Basin)	1.7 km/sec	Pujol *et al.* 1989
Unaltered andesite tuff (Georgia)	5.0 km/sec	Grynberg *et al.* 1993
Altered laumontite tuff (Georgia)	3.3 km/sec	Grynberg *et al.* 1993
Rhyolitic lavas, welded tuffs (Nevada)	>5.5 km/sec	Carroll 1968
Ash-flow, ash-fall tuffs (Nevada)	<2.1 km/sec	Carroll 1968

crooked lines. They described geological problems including diffractions related to sills and dykes and loss of high frequencies in the flood basalts. Jarchow *et al.* (1990) noted that large explosive sources provided a very good signal-to-noise ratio and produced a surprisingly high content of high-frequency data. They suggested that such sources would be applicable in areas such as the Columbia Plateau, the Paraná Basin and the Permo-Triassic basins of Northern Ireland and Britain (part of the North Atlantic Igneous Province).

Jarchow *et al.* (1991, 1994) found that very long offsets (in their case, greater than 18 km) and large explosive charges minimized the effects of reverberations by using only the first arrivals to analyse the base of basalt and the depth to basement. Richardson *et al.* (1999) found that very long offsets (up to 36 km) helped to identify sediments beneath flood basalts in the Faeroe–Shetland Basin; they also found very large air-guns producing low-frequency waves to be useful. Campbell & Reidel (1994) commented that the high-explosive, long-offset technique offered a great improvement to conventional seismic data, which often had errors exceeding 10%. Even so, it is effectively limited to modelling the basalt-sediment interface.

Working in the basalt-covered Paraná Basin, Zalán *et al.* (1990) found that some of the problems could be minimized by appropriate processing procedures. Silva & Vianna (1982) concluded that an adequate velocity model is critical, and that statics analysis also greatly improves the data. Miller & Steeples (1990) conducted a near-surface seismic survey on interbedded basalt flows and sediments on the Snake River plain of Idaho. They found very rapid lateral variations in the near-surface section, and suggested that such a shallow survey would be very helpful for statics corrections in a conventional seismic survey. Montgomery (1997) reported on recent efforts to improve seismic data quality beneath ash-flow tuffs in Nevada. The analysis of statics was a major problem, so while collecting three-dimensional seismic data a coincident high-resolution gravity survey was conducted and used to interpret near-surface lateral variations. Seismic data quality was significantly improved.

Some igneous rocks do show seismic features that can be useful in interpretation. Mathisen & McPherson (1981) discussed seismic exploration in volcaniclastics (principally tuffs, pyroclastic rocks and epiclastic sediments). Such volcaniclastics may have higher impedance due to welding or early cementation; large-volume ignimbrites and pyroclastic fall beds may be good seismic markers, while pyroclastic flows and lahars are discontinuous.

Basalt-covered areas also do not uniformly have poor seismic data. Shulman & May (1989) reported excellent data beneath the basalts of the Golan Heights, which is part of the large Harrat Ash-Shamah volcanic field in northeastern Saudi Arabia, Jordan and Syria (Saif & Shah 1988). Mahfoud & Beck (1995) stated that the plateau basalts in southern Syria are up to 1150 m thick. Shulman & May (1989) suggested that the good data quality might be due to acquisition during the rainy season, which might saturate weathered basalt and reduce the velocity variations. Recently (Saunders 1997), an exploration programme beneath the basalt in Jordan was proposed, applying subsalt technology from the Gulf of Mexico. Apparently, the intent is to apply techniques used around salt sills, such as pre-stack depth migration, to get better seismic data below the basalts.

Ogilvie *et al.* (2001) used seismic velocity analysis to outline larger-scale packages of volcanics west of the Shetlands. They could identify the areas with significant sub-volcanic sediments as well as the internal structure of the volcanic intervals.

In some cases, improvements in seismic technology can help. Nurmi *et al.* (1991) illustrated the changing interpretation of the Beykan oil field, underneath flood basalts in southeastern Turkey. The original interpretation, done with two-dimensional seismic technology, was an anticline with a series of cross faults. With three-dimensional seismic technology, the interpretation was rotated 90° and became a thrust fault.

Mjelde *et al.* (1993) reported on using refraction seismic data to image below buried flood basalts in the North Atlantic, using a sea bottom receiver system. They found that they could identify the presence and thickness of the sub-basalt sediments, but no internal features.

Planke *et al.* (2000) found that volcanic passive margins and other large-volume extrusive volcanic constructions frequently have good internal and subvolcanic reflectors. They have developed the concept of seismic volcanostratigraphy, analogous to the concept of seismic stratigraphy. They identified a set of distinct seismic facies, and in conjunction with observations from dredge and well samples, correlated the seismic facies with volcanic facies within the evolving basaltic province. In turn, these facies can be used to interpret the history of the igneous activity. While many of these events are only peripherally relevant to hydrocarbon exploration, the changing position of the palaeoshoreline and the margin subsidence history are both important. Notably, this seismic volcanostratigraphy has been applied only to the large volumes of basaltic volcanics at passive margins and related events. Smaller-scale volcanic features, as well as large-scale felsic provinces (such as ignimbrites) have not been systematically analysed.

Magnetotelluric methods

Vozoff (1972) and Christopherson (1988) suggested that magnetotelluric (MT) methods might be helpful in areas covered with near-surface volcanics. While MT surveys do not provide much resolution, they may help with the gross structure of the basin, particularly in conjunction with geological data and when integrated with other geophysical methods. Ádám *et al.* (1989) found MT data useful in modelling high-resistivity volcanics in a sedimentary section, and particularly useful below volcanics where seismic data were often poor. Calvert *et al.* (1987) found MT data useful for outlining a subsurface volcanic pile, when combined with seismic and well data. Mitsuhata *et al.* (1999) used MT methods to model the different igneous lithologies and reservoir characteristics within a basaltic volcanic reservoir.

Stanley *et al.* (1985) conducted a large-scale MT survey in the Paraná Basin of Brazil. They used the results to pinpoint favourable areas for specific targets and higher resolution surveys. Generally, they found they could determine the outlines of the section, such as the thickness of the flood basalts, the depth to basement and areas of extensive dyke development. They could also define intervals with low resistivity (shale-prone) or high resistivity (sand-prone or sills). They presented an example of how resistivity logs could be linked to MT models by Bostick's (1977) method.

Ilkisik & Jones (1984) studied the ability of an MT survey to resolve the geology beneath 100–200 m of basalts in southeastern Turkey. They concluded that the most significant variable was the resistivity of the basalt itself, which could vary by two orders of magnitude, depending on fracturing and water saturation. However, they thought broad features of the deeper section could be resolved and applied to hydrocarbon exploration.

More recently, Matsuo & Negi (1999) conducted a three-dimensional MT survey in the Akita Basin of northern Japan over a reservoir in sandy tuffs. They found that they could reasonably define structures in an area with mixed volcanics (basalt, acidic tuffs) and volcanic-rich sediments.

Young & Lucas (1988) reported on an experimental survey across the boundary of the volcanic-covered Snake River plain in eastern Idaho. The overall survey included coincident gravity, MT, and seismic refraction and reflection surveys. They concluded that the coincident surveys substantially increased the reliability of the interpretation. Some subordinate observations were also included; one was that closely spaced survey stations (especially for the MT survey) were important. They also concluded that surveys perpendicular to an edge of the volcanics were particularly useful. Another observation was that the saturation of the volcanics was quite prominent. The shallow, dry volcanics were very porous, with some interflow sediments; they were resistive and slow (2–3 km/sec), while the slightly different deeper volcanics (with more welded tuffs) below the water table were much less resistive and considerably faster (5.3 km/sec).

Beamish & Travassos (1993) recognized statics as a major problem for MT surveys. They reinterpreted older surveys from the Paraná and Solimões basins of Brazil, using surveys focused on hydrocarbon prospects. They obtained good results in the Paraná Basin, where the resistive basalts were at the surface, so that they did not mask the resistivity profile of the underlying Palaeozoic sediments. In contrast, in the Solimões Basin, the surficial sediments are highly conductive and no vertical resolution was possible; they had some success in modelling lateral variation due to structure, however.

Withers *et al.* (1994) reported on an exploration project beneath the Columbia Plateau basalts in north-central Oregon that employed several geophysical techniques. Magnetotelluric data was used to constrain the seismic interpretation; in addition gravity data was used on the broader interpretation. They found that the MT data needed a near-surface statics correction to account for near-surface variations in resistivity. They found that transient electromagnetic (TEM) methods (as suggested by Pellerin & Hohmann 1990) were effective in defining near-surface variations. Also, in the course of the MT survey, they found that the upper Columbia River basalts were more than twice as resistive as the lower basalts (200 ohm-m vs 70 ohm-m), so that internal stratigraphy could be resolved within the basalt pile.

Geological modelling

Developing a usable basin history can have an influence on play concepts. For example, an area particularly prone to dyke development might be more closely examined if dyke-related traps were likely. Knowledge of palaeoslopes and their impact on topography might be considered in exploration for buried topographic traps. Isopachs of net sill thickness (Bellieni *et al.* 1984; Peate *et al.* 1990) would be useful in looking for sill-related traps and for maturation considerations. If sills are dependent on the characteristics of the overburden, it may be possible to model the sill-prone areas of the basin. Another important line of evidence is geohistory modelling, such as that illustrated by França & Potter (1991) for an area of the Paraná Basin with a subcommercial gas discovery. Mathiesen *et al.* (1995) applied basin modeling to exploration in a heavily intruded basin partially covered by flood basalts in East Greenland. Significantly, flood basalts are apparently emplaced in 1–2 Ma (Peate *et al.* 1990), which has a significant impact on the geohistory model.

Well log analysis

Since exploration does not end until a field is developed and in production, early log analysis, drilling and even initial well testing may be considered to be part of the exploration process, since they contribute to the decision on commerciality. Thus, assessment of the rock characteristics in a well may be considered part of exploration.

There is no systematic assessment of the best way to evaluate igneous reservoirs, since igneous reservoirs have rarely been considered systematically. The first problem is in simply recognizing igneous rocks. Jansa & Pe-Piper (1988) cited an example from an exploratory well on the Grand Banks off Newfoundland where diorite in dykes was originally identified as arkose and sandstone. The accompanying resistivity and density logs varied with compositional changes, rather than porosity as originally assumed. Jansa & Pe-piper advocated greater care in examining cuttings and logs. Clegg & Bradbury (1956) noted that the mica peridotites of Illinois (associated with the Omaha oil field) had an electric log pattern similar to some limestones; the cuttings could also be extensively altered, with abundant calcite making them effervesce when tested with acid.

Interpretation of log data varies widely, depending on the type of igneous rocks involved (Table 4). For example, K-feldspar content can affect the gamma ray logs; porosity logs can be influenced by the presence of micas or clay alteration products. Fracturing of igneous reservoirs is generally important, both to provide and to connect pore space; thus, much of the log analysis is directed toward fracture analysis.

Flow units may be identified in log patterns. Grabb (1994) concluded that reservoir units and E-log responses in ignimbrites were expressions of post-emplacement cooling history, weathering and tectonic activity. Welding decreases porosity and increases fracturing and resistivity. Snyder (1968) noted that caliper logs were also helpful. However, at least in rhyolitic ignimbrite suites the familiar siliciclastic pattern (weathering produces wide, washed-out holes) is reversed: brittle, little-altered flows tend to cave, while altered tuffs with zeolites or clays have more cement and are competent.

Planke (1994) reviewed the interpretation of a number of wireline logs in flood basalt. He noted that even self-potential (SP) logs were useful, since the weathered and permeable zones contrasted with the unweathered flow interiors.

Calvert *et al.* (1987) found that in the igneous rocks of the San Juan Sag of southern Colorado, the gamma ray log could be used as a qualitative indicator of silica percentage. Also, lahars have a framework-clast relationship identifiable from cuttings, so that laharic cycles within a volcanic apron can be determined from logs and cuttings.

Zalán *et al.* (1990, figs 33, 34) illustrate logs related to a sill or laccolith overlying a gas sand in the Paraná Basin of Brazil. They note that the gamma ray profile shown has a 'crystallization profile' characteristic of a sill, with lower values toward the margin and a more radioactive zone near the last-cooled interior.

Flanigan (1989) concluded that in Nevada wireline logs were of only marginal value (because of the active freshwater aquifers) and that drillstem testing was the best single open-hole evaluation method. Sembodo (1973) reached a similar conclusion about evaluation of the reservoir in the Jatibarang oil field in Java. The most useful procedure was to evaluate all of the logs together, although the SP and resistivity logs were most helpful. The best reservoir evaluations come from careful observations of the cuttings, the zones of mud loss (indicating fracturing) and the drilling rate.

Drilling and production

Because fracturing is generally important in hydrocarbon reservoirs in or associated with igneous rocks, fracture analysis is often very important in their development. Directional drilling is frequently an important strategy. In some cases, methods of enhancing the fracture system, such as hydrofracturing, may improve

Table 4. *Log evaluation of igneous rocks*

Log(s)	Observations	Reference
Resistivity	Igneous rocks generally resistive (Argentina)	Khatchikian 1983
Resistivity, caliper	Weathered, altered tuff with much lower resistivity, weaker than unaltered tuff (Nevada)	Snyder 1968; French & Freeman 1979; Grabb 1994
Resistivity, gamma ray, caliper, self potential, compensated neutron, density	Different extrusives (lavas, pyroclastic breccias, tuffs) have different characteristics (NE China); includes several log examples	Luo *et al.* 1999
Gamma ray	Igneous rocks variable, depending on K-feldspar content (Argentina)	Khatchikian 1983
Gamma ray, rate of penetration	Igneous rocks variable, depending on K-feldspar content (Thailand); ROP 'shoulder' in baked sediments next to intrusion	Remus *et al.* 1993
Gamma ray, neutron	Cross-plot used to distinguish different igneous rocks	Sanyal *et al.* 1980
Spectral gamma ray	Useful for distinguishing different igneous rocks	Keys 1979
Sonic	Generally low transit times	Passey *et al.* 1990
Sonic	Zeolitized tuffs with low density, high transit times	Khatchikian 1983
Sonic	Transit time in tuff correlates with porosity (Nevada)	Carroll 1968
Sonic, resistivity	Useful in evaluating fractured igneous rocks	Daniel & Hvala 1982
Caliper, sonic, neutron	Useful in evaluating laumontite tuffs	Vernik 1990
Caliper, sonic, neutron	Distinguished between fractured rhyolitic tuff reservoir and unfractured dacitic tuffs Thrace Basin, Turkey)	Ozkanli & Kumsal 1993
Sonic, density, neutron	Determine tuff and clay content	Khatchikian & Lesta 1973
Sonic, density, self potential, resistivity, porosity, gamma ray, vertical seismic profile	Useful in evaluation of flood basalt stratigraphy, especially identification of porous zones/flow tops	Planke 1994
Density, porosity, resistivity	Useful in evaluating fractured igneous rocks	Kumar *et al.* 1985
Dipmeter	Useful in evaluating fractured sill reservoirs (Argentina)	Perea & Giordano 1988

yields. Because of the importance of the fracture systems, understanding the origin and orientation of the fracture systems can have a significant impact.

Drilling and production practices in reservoirs associated with igneous rocks sometimes require special consideration. Sensitive clays often develop due to weathering or alteration of igneous rocks; particularly with tuffaceous rocks, air-foam is used as a drilling fluid (Flanigan 1989; O'Sullivan 1992). Hunter & Davies (1979) noted that in addition to clays, goethite, ankerite and zeolites might also cause problems in altered igneous rocks.

Reservoirs in or near igneous rocks are generally free from notable pressure or gas problems, apparently because the hydrocarbons mature and migrate during the late stages of the igneous event or later during normal maturation. However, in a few cases problems may be present. Parker (1974) attributed high geopressures to thermal cracking of pre-existing oil, with high H_2S levels because the oil was sulphur-rich. Alternatively, Castro & da Silva (1990) suggested another model for the high H_2S levels found in the Juruá gas field in the Solimões Basin of western Brazil; the gas field is immediately below a thick dolerite sill intruded into a sulphate-rich evaporitic section. High CO_2 levels in hydrocarbons near igneous rocks have long been attributed to the reaction of magma with carbonates. Kontorovich *et al.* (1990) suggested that both processes were at work in the Lena–Tunguska superprovince of Siberia, where the intrusive rocks were frequently in the Cambrian dolomite–evaporite interval, directly above the principal reservoir horizons. In other cases, the CO_2 may be from devolatilization of the magma itself, and may produce a natural CO_2 flood phenomenon, displacing hydrocarbons updip.

In some cases, oil produced by igneous activity may have unusual characteristics. Such oils may be heavy or contain unusual amounts of aromatics or other non-chain hydrocarbons. This is due to unusual migration pathways, especially those involving derivation from terrestrial organics. If hydrocarbons are produced inorganically by the Fischer–Tropsch reaction, they may be high in paraffins.

Grynberg *et al.* (1993) discussed production problems with the Samgori oil field in Georgia. The reservoir facies, laumontite tuff, is enclosed in unaltered andesite tuff and connected by fractures. The unaltered tuff protects the laumontite tuff from collapse due to regional and overburden stresses, but the connecting fractures are kept open by fluid pressure. Without careful planning, water breakthrough can occur and pockets of oil may be isolated.

Drilling through igneous rocks is generally assumed to be a tedious process, due to their crystalline structure and density. Basalts usually are slow drilling, but Gries (1985) observed that drilling through intermediate to silicic flows and tuffs is comparable to drilling through quartz arenites, siltstones and shales.

Completion descriptions in igneous rocks are rare. The Dineh-bi-Keyah oil field in Arizona had relatively simple completion procedures, since it is a fractured reservoir with little alteration. The producing interval was perforated, and then fractured with sand before being put on production (Kornfeld & Travis 1967). Sembodo (1973) noted that the Jatibarang oil field of Java originally had perforated completions, but later holes were completed naturally. More recently, deviated holes have been used to improve production in fractured volcanics (O'Sullivan 1992). Modern fracturing techniques may also increase production.

Conclusions

Hydrocarbons can occur within and in association with igneous rocks, sometimes in commercially significant quantities. Exploration for such hydrocarbons requires consideration of unique features of igneous rocks and the hydrocarbon system. For example, igneous or closely associated rocks can be hydrocarbon sources in the conventional sense (biotic) as well as possibly through abiotic processes. Maturation is extremely variable, depending on the extrusive/intrusive nature of the activity and the relative importance of a deep heat source. Igneous volatiles and hydrothermal fluids may also be important in mobilizing and moving hydrocarbons. Igneous rocks can have good reservoir qualities, and they can produce their own trapping structures as well as being part of a larger feature.

Many exploration methods are individually unreliable in and around igneous rocks due to the unique properties of the rocks. An integrated approach is probably more effective. Seismic, magnetotelluric, gravity and magnetic surveys all provide helpful information. Geological techniques, including mapping, geochemistry and remote imagery, may also be helpful. Pinpointing promising areas for exploration may be helped by geological models.

Evaluation of potentially commercial hydrocarbon accumulations requires interpretation of well logs, which may have unusual characteristics due to the igneous rocks. Drill stem and production tests may also be needed for evaluation before exploration ends and development begins.

I would like to thank J. Brenneke and H. Mueller, who provided valuable criticism and suggestions on an earlier version of this manuscript. I would also like to thank the many people who provided examples and leads on hydrocarbons in and around igneous rocks; I hope to receive many more. I would list the people who provided information, but the names and notes were lost in the floods of Tropical Storm Allison.

References

Abrajano, T. A., Sturchio, N. C., Bohlke, J. K., Lyon, G. L., Poreda, R. J. & Stevens, C. M. 1988. Methane-hydrogen gas seeps, Zambales ophiolite, Philippines: deep or shallow origin? *Chemical Geology*, **71**, 211–222.

Ádám, A. 1978. Geothermal effects in the formation of electrically conducting zones and temperature distribution in the earth. *Physics of the Earth and Planetary Interiors*, **17**, P21–P28.

Ádám, A., Nagy, Z. & Varga, G. 1989. Magnetotelluric (MT) research and exploration in Hungary. *Geophysics*, **54**, 795–797.

Altebäumer, F. J., Leythaeuser, D. & Schaefer, R. G. 1983. Effect of geologically rapid heating on maturation and hydrocarbon generation in Lower Jurassic shales from NW-Germany. *In*: Bjorøy, M. et al. (eds) *Advances in organic geochemistry 1981*. John Wiley and Sons, Chichester, 80–86.

Amaral, S. E. 1967. *Contribução ao conhecimento geológico, petrográfico e sedimentológico de Formação Iratíno Estado do São Paulo*. Thesis, Universidade São Paulo.

Anon 1971. Igneous rocks seen promising for gas. *Oil & Gas Journal*, **69**, 5 April (n. 14), 33–34.

Anon 1999. New study suggests flanks of anticlines may be better targets for oil and gas deposits in GSENM. *Utah Geological Survey, Petroleum News*, April, 2.

Anon 2001. Conoco exploration technique patented. *Oil & Gas Journal*, **99**, 26 November (n. 48), 40.

Anon 2002a. Pacific Tiger roars. *Upstream*, **7**, 15 February (n. 7), 28.

Anon 2002b. Wichian's riches look set to treble. *Upstream*, **7**, 28 March (n. 13), 23.

AOYOGI, K. 1985. Origin of the Miocene carbonate reservoir rocks, Fukubezawa oil field, Akita Province, northeast Honshu, Japan. *In*: ROEHL, P. O. & CHOQUETTE, P. W. (eds) *Carbonate petroleum reservoirs*. Springer-Verlag, New York, 561–569.

ARAÚJO, L. M., TRIGÜIS, J. A., CERQUEIRA, J. R. & FREITAS, L. C. DE S. 2000. The atypical Permian petroleum system of the Paraná Basin, Brazil. *In*: MELLO, M. R. & KATZ, B. J. (eds) *Petroleum systems of South Atlantic margins*. American Association of Petroleum Geologists, Memoirs, **73**, 377–402.

ARMSTRONG, P. A., KEMP, P. J. J., ALLIS, R. G. & CHAPMAN, D. S. 1997. Thermal effects of intrusion below the Taranaki Basin (New Zealand): evidence from combined apatite fission track age and vitrinite reflectance data. *Basin Research*, **9**, 151–169.

BAILEY, E. H. 1959. Froth veins, formed by immiscible hydrothermal fluids, in mercury deposits, California. *Geological Society of America, Bulletin*, **70**, 661–664.

BARKER, C. 1990. Calculated volume and pressure changes during the thermal cracking of oil to gas in reservoirs. *American Association of Petroleum Geologists, Bulletin*, **74**, 1254–1261.

BARKER, C. E. & PAWLEWICZ, M. J. 1987, The effects of igneous intrusions and higher heat flow on the thermal maturity of Leonardian and younger rocks, western Delaware Basin, Texas. *In*: CROMWELL, D. W. & MAZZULLO, L. (eds) *The Leonardian facies in West Texas and Southeast New Mexico*, Society of Economic Paleontologists and Mineralogists—Permian Basin Section, Publications, **87–27**, 69–81.

BARRABÉ, L. 1932, Oil in Limagne area, France. *American Association of Petroleum Geologists, Bulletin*, **16**, 825–832.

BARTENSTEIN, H., TEICHMÜLLER, M. & TEICHMÜLLER, R. 1971. Die Umwandlung der organischen Substanz im Dach des Bramscher Massivs. *Fortschritte Geologische Rheinland und Westfalen*, **18**, 501–538.

BATCHELOR, A. S. (compiler) 2000. *Hydrocarbon production from fractured basement formations*. Geoscience Ltd., Falmouth, Cornwall, online publication, Version 6, November, 2000, unpaginated.

BEAMISH, D. & TRAVASSOS, J. M. 1993. Magnetotelluric studies from two contrasting Brazilian basins: a reassessment of old data. *Physics of the Earth and Planetary Interiors*, **81**, 261–278.

BELLIENI, G., COMIN-CHIARAMONTI, P., MARQUES, L. S., MELFI, A. J., PICCIRILLO, E. M. & STOLFA, D. 1984. Low pressure evolution of basalt sills from boreholes in the Paraná Basin. *TMPM Tschermaks Mineralogische und Petrographische Mitteilungen*, **33**, 25–47.

BERGMAN, S. C., TALBOT, J. P. & THOMPSON, P. R. 1992. The Kora Miocene submarine andesite stratovolcano hydrocarbon reservoir, northern Taranaki Basin, New Zealand. *New Zealand Oil Exploration Conference, Proceedings*. Ministry of Commerce, Wellington, New Zealand, 178–206.

BIEDERMAN, E. W., JR. 1986. *Atlas of selected oil and gas reservoir rocks from North America*. John Wiley and Sons, New York, 1–399.

BIGARELLA, J. J. 1971. Geology of the Iratí Formation. *Conferência do Simpósio sobre Ciência e Taechnologia do Xistos, Anais, Dec. 12–17, 1971. Curitiba, Paraná, Brazil, 1–113.*

BORTZ, L. C. 1994. Petroleum geology of the Eagle Springs oil field, Nye County, Nevada. *In*: SCHALLA, R. A. & JOHNSON, E. H. (eds) *Oil fields of the Great Basin*. Nevada Petroleum Society, Reno, 285–294.

BOSTICK, F. X., JR. 1977. A simple almost exact method of MT analysis. *In*: WARD, S. H. (ed.) *Workshop of electrical methods in geothermal exploration*. United States Geological Survey, Contract 14-8-001G-359, 174–183.

BOSTICK, N. H. & PAWLEWICZ, M. J. 1984. Paleotemperatures based on vitrinite reflectance of shales and limestones in igneous dike aureoles in the Upper Cretaceous Pierre Shale, Walsenburg, Colorado. *In*: WOODWARD, J., ET AL., *Hydrocarbon source rocks of the greater Rocky Mountain region*. Rocky Mountain Association of Geologists, Denver, 387–392.

BRAY, R., LAWRENCE, S. & SWART, R. 1998. Source rock, maturity data indicate potential off Namibia. *Oil & Gas Journal*, **96**, 10 August (n. 32), 84–89.

BROWN, J. S., EMERY, J. A. & MEYER, P. A., JR. 1954. Explosion pipe in test well on Hicks Dome, Hardin County, Illinois. *Economic Geology*, **49**, 891–902.

CALVERT, R. H., GRIES, R. R. & TERBEST, H., JR. 1987. San Juan volcanics and oil exploration. *Geological Society of America, Abstracts with Programs*, **19**, n. 5, 264.

CAMPBELL, N. P. & REIDEL, S. P. 1994. Further exploration for gas warranted in Columbia Basin. *Oil & Gas Journal*, **92**, 2 May (n. 18), 127–131.

CARROLL, R. D. 1968. Velocity-porosity logging in volcanic rocks. *Journal of Petroleum Technology*, **12**, 1371–1374.

CASTRO, L. O. C. & DA SILVA, O. B. 1990. Ocurrência de gás sulfidrico na bacia do Solimões—modelo geológico e implicações. *Boletim Geociências da Petrobrás*, **4**, 285–298.

CHEN, Z., YAN, H., LI, J., ZHANG, G., ZHANG, Z. & LIU, B. 1999. Relationship between Tertiary volcanic rocks and hydrocarbons in the Liaohe Basin, People's Republic of China. *American Association of Petroleum Geologists, Bulletin*, **83**, 1004–1014.

CHRISTOPHERSON, K. R. 1998. MT gauges Earth's electric fields. *AAPG Explorer*, **19**, December (n. 12), 22, 24–25.

CLEGG, K. E. & BRADBURY, J. C. 1956. *Igneous intrusive rocks in Illinois and their economic significance*. Illinois State Geological Survey, Report of Investigations, **197**, 1–19.

CLIFTON, C. G., WALTERS, C. C. & SIMONIET, B. R. T. 1990. Hydrothermal petroleums from Yellowstone National Park, U.S.A. *Applied Geochemistry*, **5**, 169–191.

COLLINGWOOD, D. M. 1930. Magnetics and geology of Yoast Field, Bastrop County, Texas. *American Association of Petroleum Geologists, Bulletin*, **14**, 1191–1197.

COLLINGWOOD, D. M. & RETTGER, R. E. 1926. The Lytton Springs oil field, Caldwell County, Texas. *American Association of Petroleum Geologists, Bulletin*, **10**, 953–975.

CORRY, C. E. 1988. *Laccoliths: mechanics of emplacement and growth.* Geological Society of America, Special Papers, **220**, 1–110.

COURTNEY, R. G. & WHITE, R. S. 1986. Anomalous heat flow and geoid across the Cape Verde Rise: evidence for dynamic support from a thermal plume in the mantle. *Journal of the Royal Astronomical Society*, **87**, 815–867.

CZOCHANSKA, Z., SHEPPARD, C. M., WESTON, R. J., WOODHOUSE, A. D. & COOK, R. A. 1986. Organic geochemistry of sediments in New Zealand, Pt. 1. A biomarker study of the petroleum seepage in the geothermal region of Waitapu. *Geochimica et Cosmochimica Acta*, **50**, 507–515.

DANIEL, G. A. & HVALA, O. 1982. Analisis de rocas tobaceas y volcanicas a partir de los perfiles sonicos y de resistividad. *Primer Congreso Nacional de Hidrocarburos, Petroleo y Gas*. Trabajos Tecnicos, Explotacion, **4**, 99–126.

DANIELS, D. L. 1988. Magnetic and gravity fields of early Mesozoic diabase sheets in the northern Durham Basin, North Carolina. *In*: FROELICH, A. J. & ROBINSON, G. R., JR. (eds) *Studies of the early Mesozoic basins of the eastern United States.* United States Geological Survey, Bulletin, **1776**, 200–210.

DOW, W. G. 1977. Kerogen studies and geological interpretations. *Journal of Geochemical Exploration*, **7**, 79–99.

EINSELE, G., GIESKES, J. M. ET AL. 1990. Intrusion of basaltic sills into highly porous sediments, and resulting hydrothermal activity. *Nature*, **283**, 441–445.

ENGLAND, R. W., BUTLER, R. W. H. & HUTTON, D. H. W. 1993. The role of Paleocene magmatism in the Tertiary evolution of basins on the NW seaboard. *In*: PARKER, J. R. (ed.) *Petroleum geology of Northwest Europe*. Proceedings 4th Conference, Geological Society London, **1**, 97–105.

ENGLISH, R. M. & GROGAN, R. M. 1948. Omaha pool and mica-peridotite intrusives, Gallatin County, Illinois. *In*: HOWELL, J. V. (ed.) *Structure of Typical American Oil Fields*. American Association of Petroleum Geologists, Tulsa, **3**, 189–212.

EWING, T. E. & CARAN, S. C. 1982. Late Cretaceous volcanism in south and central Texas—stratigraphic, structural, and seismic models. *Gulf Coast Association of Geological Societies, Transactions*, **32**, 137–145.

FATTI, J. L. 1972. The influence of dolerite sheets on reflection seismic profiling in the Karroo Basin. *Geological Society of South Africa, Transactions*, **75**, pt. 2, 71–75.

FELTS, W. M. 1954. Occurrence of oil and gas and its relation to possible source beds in continental Tertiary of Intermountain region. *American Association of Petroleum Geologists, Bulletin*. **38**, 1661–1670.

FLANIGAN, T. 1989. Borehole evaluation and completion: carbonate, volcanic, clastic reservoirs of Nevada. *Oil & Gas Journal*, **87**, 23 October (n. 43), 92, 94–96.

FRANÇA, A. B. & POTTER, P. E. 1991. Stratigraphy and reservoir potential of glacial deposits of the Itararé Group (Carboniferous-Permian), Paraná Basin, Brazil. *American Association of Petroleum Geologists, Bulletin*, **75**, 62–85.

FRENCH, B. M. 1964. Graphitization of organic material in a progressively metamorphosed Precambrian iron formation. *Science*, **146**, 917–918.

FRENCH, D. E. & FREEMAN, K. J. 1979. Tertiary volcanic stratigraphy and reservoir characteristics of Trap Spring Field, Nye County, Nevada. *In*: NEWMAN, G. W. & GOODE, H. D. (eds) *Basin and Range Symposium*. Rocky Mountain Association of Geologists–Utah Geological Association, 487–502.

FRITTS, S. G. & FISK, L. H. 1985. Tectonic model for formation of Columbia Basin: implications for oil, gas potential of north central Oregon. *Oil & Gas Journal*, **83**, 26 August (n. 34), 84–86; 2 September (n. 35), 85–89.

FU, J., SHENG, G. & LIU, D. 1988. Organic geochemical characteristics of major types of terrestrial petroleum source rocks in China. *In*: FLEET, A. J., KELTS, K. & TALBOT, M. R. (eds) *Lacustrine petroleum source rocks*, Geological Society, London, Special Publications, **40**, 279–289.

FUKUTA, O. 1986. Natural gas expected in the lakes originating in the Rift Valley system of East Africa, and analogous gas in Japan. *In*: HALBOUTY, M. T. (ed.) *Future petroleum provinces of the world.* American Association of Petroleum Geologists, Memoir, **40**, 445–456.

GALLOWAY, W. E., EWING, T. E., GARRETT, C. M., TYLER, N. & BEBOUT, D. G. 1983. *Atlas of major Texas oil reservoirs*. University of Texas, Bureau of Economic Geology, 1–139.

GEORGE, S. C. & JARDINE, D. R. 1994. Ketones in a Proterozoic dolerite sill. *Organic Geochemistry*, **21**, 829–839.

GERLACH, T. M. 1980. Chemical characteristics of the volcanic gases from Nyiragongo lava lake and the generation of CH_4-rich fluid inclusions in alkaline rocks. *Journal of Volcanology and Geothermal Research*, **8**, 177–189.

GIARDINI, A. A. & MELTON, C. E. 1981. Experimentally-based arguments supporting large crustal accumulations of non-biogenic petroleum. *Journal of Petroleum Geology*, **4**, 187–190.

GILBERT, G. K. 1877. *Geology of the Henry Mountains, Utah*. United States Geographical and Geological Survey of the Rocky Mountain Region, 1–170.

GIZE, A. P. 1986. The development of a thermal mesophase in bitumens from high temperature ore deposits. *In*: DEAN, W. E. (ed.) *Organics and Ore Deposits*. Denver Region Exploration Geologists, Symposium, Proceedings, 137–150.

GIZE, A. P. & MACDONALD, R. 1993. Generation of compositionally atypical hydrocarbons in CO_2-rich geologic environments. *Geology*, **21**, 129–132.

GOLD, T. & SOTER, S. 1980. The deep earth-gas hypothesis. *Scientific American*, **242**, n. 6, 154–161.

GORDOYEVA, A. O., ZHIDKOVA, L. V. & KHOMENKO, A. V. 2001. Effect of dolerite intrusions on oil-gas potential of South Tunguska oil-gas region. *Petroleum Geology*, **35**, n. 4, 347–353.

GOULART, E. P. & JARDIM, N. S. 1982. Avaliação geoquimica das formaçãos Ponta Grossa e Iratí—Bacia do Paraná. *In*: *Geologia da Bacia do Paraná—Reavaliação de Potentcialidade e Prospectividade em Hidrocarbonetos*. Instituto Pesquisas Tecnica, Estado São Paulo, Publicação Especial, **12**, 41–74.

GRABB, R. 1994. Sand Spring Oil Field, Railroad Valley, Nye County, Nevada. *In*: SCHALLA, R. A. & JOHNSON, E. H. (eds) *Oil fields of the Great Basin*. Nevada Petroleum Society, Reno, 241–252.

GRIES, R. R. 1985. San Juan Sag: Cretaceous rocks in a volcanic-covered basin, south-central Colorado. *Mountain Geologist*, **22**, n. 4, 167–179.

GRYNBERG, M. E., PAPAVA, D., SHENGELIA, M., TAKAISHVILI, A., NANADZE, A. & PATTON, D. K. 1993. Petrophysical characteristics of the Middle Eocene laumontite tuff reservoir, Samgori Field, Republic of Georgia. *Journal of Petroleum Geology*, **16**, 312–322.

GUNN, P. 1998. Aeromagnetics locates prospective areas and prospects. *Geophysics*: *The Leading Edge of Exploration*, **17**, 67–69.

HART, A. 2001. New Zealand potential. *Oil & Gas Journal*, **99**, 16 July (n. 29), 38–44; 23 July (n. 30), 40–43.

HAWKES, H. E. 1980. Geothermal hydrogen. *Mining Engineering*, **32**, 671–675.

HAWTHORNE, J. B. 1968. Kimberlite sills. *Geological Society of South Africa, Transactions*, **71**, 291–311.

HENKEL, R. P. 1989. Seismic data acquisition over San Juan volcanic field, San Juan Sag, south-central Colorado. *American Association of Petroleum Geologists, Bulletin*, **73**, 1159.

HOLLISTER, V. F. 1980. Origin of graphite in the Duluth Complex. *Economic Geology* **75**, 764–766.

HULEN, J. B., COX, J. W., NIELSON, D. L. & LAMBORN, J. C. 1993. Oil in Carlin-type sediment-hosted gold deposits. *American Association of Petroleum Geologists, 1993 Annual Convention, Program*, 122.

HUNTER, B. E. & DAVIES, D. K. 1979. Distribution of volcanic sediments in the Gulf Coastal Province—significance to petroleum geology. *Gulf Coast Association of Geological Societies, Transactions*, **29**, 147–155.

HURTER, S. J. & POLLACK, H. N. 1995. Effect of the Serra Geral igneous event on the temperature and heat flow of the Paraná Basin, southern Brazil. *Basin Research*, **7**, 215–220.

HUTCHINSON, P. J. 1994. Upper Cretaceous (Austin Group) volcanic deposits as a hydrocarbon trap. *Gulf Coast Association of Geological Societies, Transactions*, **44**, 293–303.

ILCHIK, R. P., BRIMHALL, G. H. & SCHULL, H. W. 1986. Hydrothermal maturation of indigenous organic matter at the Alligator Ridge gold deposits, Nevada. *Economic Geology*, **81**, 113–130.

ILKISIK, O. M. & JONES, A. G. 1984. Statistical evaluation of MT and AMT methods applied to a basalt-covered area in southeastern Anatolia, Turkey. *Geophysical Prospecting*, **32**, 706–724.

JANSA, L. F. & PE-PIPER, G. 1988. Middle Jurassic to Early Cretaceous igneous rocks along eastern North American continental margin. *American Association of Petroleum Geologists, Bulletin*, **72**, 347–366.

JARCHOW, C. M., CATCHINGS, R. D. & LUTTER, W. J. 1991. How Washington crew got good, thrifty seismic in bad data area. *Oil & Gas Journal*, **89**, 17 June (n. 24), 54–55.

JARCHOW, C. M., CATCHINGS, R. D. & LUTTER, W. J. 1994. Large-explosive source, wide-recording aperture, seismic profiling on the Columbia Plateau, Washington. *Geophysics*, **59**, 259–271.

JARCHOW, C. M., GOODWIN, E. B. & CATCHINGS, R. D. 1990. Are large explosive sources applicable to resource exploration? *Geophysics*: *The Leading Edge of Exploration*, **9**, 12–17.

JEFFREY, A. W. A. & KAPLAN, I. R. 1988. Hydrocarbons and inorganic gases in the Gravberg-1 well, Siljan Ring, Sweden. *Chemical Geology*, **71**, 237–255.

JENNY, W. P. 1951. Aerial magnetic oil discoveries. *World Oil*, **133**, November (n. 6), 85–92.

JENYON, M. K. 1990. *Oil and gas traps*. John Wiley and Sons, Chichester, 1–396.

JOHNSON, V. G., GRAHAM, D. L. & REIDEL, S. P. 1993. Methane in Columbia River basalt aquifers: isotope and geohydrologic evidence of a deep coal-bed gas source in the Columbia Basin, Washington. *American Association of Petroleum Geologists, Bulletin*, **77**, 1192–1207.

KARTANEGARA, A. L., BAIK, R. N. & IBRAHIM, M. A. 1996. Volcanics oil bearing in Indonesia. *American Association of Petroleum Geologists, 1996 Annual Convention, Meeting Abstracts*, **5**, A73.

KATAHIRA, T. & UKAI, M. 1976. Petroleum fields of Japan with volcanic-rock reservoirs—summary. *In*: HALBOUTY, M. T., MAHER, J. C. & LIAN, H. M. (eds) *Circum-Pacific Energy and Mineral Resources*. American Association of Petroleum Geologists, Memoir **25**, 276–279.

KETTEL, D. 1983. The East Groningen Massif—detection of an intrusive body by means of coalification. *Geologie en Mijnbouw*, **62** 203–210.

KEYS, W. S. 1979. Borehole geophysics in igneous and metamorphic rocks. *Log Analyst* 20, n. 4, 14–28.

KHADKIKAR, A. S., SANT, D. A., GOGTE, V. & KARANTH, R. V. 1999. The influence of Deccan volcanism on climate: insights from lacustrine intertrappean deposits, Anjar, western India. *Palaeogeography, Palaeoclimatology, Palaeoecology*, **147**, 141–149.

KHARKIV, A. D., BORIS, E. I., SHABO, Z. V., MAMCHUR, G. P. & SHEREMEEV, V. F. 1988 [Oil shows in diatremes of the Siberian Platform]. *Sovetskaya Geologiya*, **1988**, n. 4, 60–70 (English abstract in: *Petroleum Abstracts*, **28**, 3014, Abstract No. 451996, 1988).

KHATCHIKAN, A. 1983. Log evaluation of oil-bearing igneous rocks. *World Oil*, **197**, December (n. 7), 79–82, 87, 91, 95, 98.

KHATCHIKAN, A. & LESTA, P. 1973. Log evaluation of tuffites and tuffaceous sandstones in southern Argentina. *Society of Professional Well Log Analysts, 14th Annual Logging Symposium, Transactions*, K1–K24.

KINGSTON, J. & MATZKO, J. R. 1995. Undiscovered petroleum of the Brazilian interior sag basins. *International Geology Review*, **37**, 959–980.

KIRKHAM, V. R. D. 1935. Natural gas in Washington, Idaho, eastern Oregon, and northern Utah. *In*: LEY, H. A. (ed.) *Geology of natural gas*. American Association of Petroleum Geologists, Tulsa, 221–243.

KOMATSU, N., FUJITA, Y. & SATO, O. 1984. Cenozoic volcanic rocks as potential hydrocarbon reservoirs. *World Petroleum Congress, 11th, Proceedings*, **2**, 411–422.

KONTOROVICH, A. E., MANDEL'BAUM, M. M., SURKOV, V. S., TROFIMUK, A. A. & ZOLOTOV, A. N. 1990. Lena-Tunguska Upper Proterozoic—Palaeozoic petroleum super province. *In*: BROOKS, J. (ed.) *Classic Petroleum Provinces*. Geological Society, Special Publication **50**, 473–489.

KORNFELD, J. A. & TRAVIS, M. M. 1967. Arizona's spectacular oil strike tops Rocky Mountain field interest. *World Oil*, **164**, May (n. 6), 180–190.

KREHBIEL, S. C. 1993. Seismic, maturity data point to SW Washington potential. *Oil & Gas Journal*, **91**, 22 March (n. 12), 107–112.

KUMAR, C. P., CHANDRA, D. & GILL, T. S. 1985. Identification of fractured reservoirs in igneous formations using well logs. *11th Annual Conference and Seminar on Exploration Geophysics, Nov. 22–24, 1985*, Oil India Ltd.–Dept. Geology, Uktal University–Association of Exploration Geophysicists, A31.

KVENVOLDEN, K. A. & CLAYPOOL, G. E. 1980. Origin of gasoline-range hydrocarbons and their migration by solution in carbon dioxide in Norton Basin, Alaska. *American Association of Petroleum Geologists, Bulletin*, **64**, 1078–1086.

KVENVOLDEN, K. A. & SIMONIET, B. R. T. 1990. Hydrothermally derived petroleum: examples from Guaymas Basin, Gulf of California, and Escanaba Trough, northeast Pacific Ocean. *American Association of Petroleum Geologists, Bulletin*, **74**, 223–237.

LANCET, M. S. & ANDERS, E. 1970. Carbon isotope fractionation in the Fischer–Tropsch synthesis and in meteorites. *Science*, **170**, 980–982.

LEE, K. Y. 1989. *Geology of petroleum and coal deposits in the North China Basin, eastern China*. United States Geological Survey, Bulletin, **1871**, 1–36.

LEVIN, L. E. 1995. Volcanogenic and volcaniclastic reservoir rocks in Mesozoic-Cenozoic island arcs: examples from the Caucasus and the NW Pacific. *Journal of Petroleum Geology*, **18**, 267–288.

LEWIS, C. L. E., CARTER, A. & HURFORD, A. J. 1992. Low-temperature effects of the Skye Tertiary intrusions on Mesozoic sediments in the Sea of Hebrides Basin. *In*: PARNELL, J. (ed.) *Basins of the Atlantic seaboard: petroleum geology, sedimentology and basin evolution*. Geological Society, London, Special Publications, **62**, 175–188.

LEWIS, J. O. 1984. Seismic signature of serpentine plugs in Maverick Basin. *Gulf Coast Association of Geological Societies*, **34**, 109–122.

LEYPTSIG, A. G. 1971 [Folds formed by intrusive activity in the Tunguska syneclise]. *Akademii Nauk SSSR, Doklady*, **201**, n. 2, 427–430 (English translation, 1971, 67–69).

LEYTHAEUSER, D., LITTKE, R., RADKE, M. & SCHAEFER, R. G. 1987. Geochemical effects of petroleum migration and expulsion from Toarcian source rocks in the Hils syncline area, NW Germany. *Organic Geochemistry*, **13**, 489–502.

LINK, W. K. 1952. Significance of oil and gas seeps in world oil exploration. *American Association of Petroleum Geologists, Bulletin*, **36**, 1505–1540.

LIU, Z., XIN, Q., WANG, Y., XU, P. & ZHANG, X. 1989. Formation conditions and distribution regularities of oil-gas pools in Tertiary volcanic rocks in western Huimin Depression, Shandong Province. *Chinese Journal of Geochemistry*, **8**, n. 2, 146–154.

LOUTIT, T. S., HARDENBOL, J., VAIL, P. R. & BAUM, G. R. 1988. Condensed sections: the key to age dating and correlation of continental margin sequences. *In*: WILGUS, C. K., HASTINGS, B. S., POSAMENTIER, H., VAN WAGONER, J., ROSS, C. A. & KENDALL, C. G. ST. C. (eds) *Sea-level changes: an integrated approach*. Society of Economic Paleontologists and Mineralogists, Special Publication, **42**, 183–213.

LUCZAJ, J. 1998. Argument supporting explosive igneous activity for the origin of "cryptoexplosion" structures in the Midcontinent, United States. *Geology*, **26**, 295–298.

LUO, J., ZHANG, C. & QU, Z. 1999. Volcanic reservoir rocks: a case study of the Cretaceous Fenghuadian Suite, Huanghua Basin, eastern China. *Journal of Petroleum Geology*, **22**, 397–415.

MAGOON, L. B. & DOW, W. G. 1994. The petroleum system. *In*: MAGOON, L. B. & DOW, W. G. (eds) *The petroleum system—from source to trap*. American Association of Petroleum Geologists, Memoir, **60**, 3–24.

MAHFOUD, R. F. & BECK, J. N. 1995. Composition, origin and classification of extrusive carbonatites in rifted southern Syria. *International Geology Review*, **37**, 361–378.

MASTERS, J. A. 2000. The slow discovery of Dineh-bi-Keyah. *Mountain Geologist*, **37**, 91–100.

MATHIESEN, A., CHRISTIANSEN, F. G., BIDSTRUP, T., MARCUSSEN, C., DAM, G., PIASECKI, S. & STEMMERIK, L. 1995. Modeling of hydrocarbon generation in the Jameson Land Basin, East Greenland. *First Break*, **13**, 329–341.

MATHISEN, M. E. & MCPHERSON, J. G. 1981. Volcaniclastic deposits: implications for hydrocarbon exploration. *In*: FISHER, R. V. & SMITH, G. A. (eds) *Sedimentation in volcanic settings*. Society of Economic Paleontologists and Mineralogists, Special Publications, **45**, 27–36.

MATSUO, K. & NEGI, T. 1999. Oil exploration in the difficult Minami-Noshiro area; Pt. 2: magnetotelluric survey. *Geophysics—the Leading Edge of Technology*, **18**, 1411–1413.

MATTHEWS, T. F. 1986. The petroleum potential of "serpentine plugs" and associated rocks, Central and South Texas. *Baylor Geological Studies, Bulletin*, **44**, 1–43.

McDowell, F. W. & Clabaugh, S. E. 1979. Ignimbrites of the Sierra Madre Occidental and their relation to the tectonic history of western Mexico. *In*: Chapin, C. E. & Elston, W. E. (ed.) *Ash-Flow Tuffs*. Geological Society of America, Special Paper **180**, 113–124.

McGetchin, T. R., Nikhanj, Y. S. & Chodos, A. A. 1973. Carbonatite-kimberlite relations in the Cane Valley diatreme, San Juan County, Utah. *Journal of Geophysical Research*, **78**, 1854–1869.

McKenny, J. W. & Masters, J. A. 1968. Dineh-bi-Keyah field, Apache County, Arizona. *American Association of Petroleum Geologists, Bulletin*, **52** 2045–2057.

McKirdy, D. M. & Chivas, A. R. 1992. Nonbiodegraded aromatic condensate associated with volcanic supercritical carbon dioxide, Otway Basin: implications for primary migration from terrestrial organic matter. *Organic Geochemistry*, **18**, 611–627.

Mello, M. R., Dahl, J., Maldowan, M. & Requejo, A. J. 2000. Petroleum geochemistry applied to petroleum system investigation. *In*: Mello, M. R. & Katz, B. J. (eds) *Petroleum systems of South Atlantic margins*. American Association of Petroleum Geologists, Memoirs, **73**, 41–52.

Mencher, E., Fichter, H. J., Renz, H. H., Wallis, W. E., Patterson, J. M. & Robie, R. H. 1953. Geology of Venezuela and its oil fields. *American Association of Petroleum Geologists, Bulletin*, **37**, 690–777.

Mesner, J. C. & Wooldridge, L. C. P. 1964. Maranhão Paleozoic basin and Cretaceous coastal basins, North Brazil. *American Association of Petroleum Geologists, Bulletin*, **48**, 1475–1512.

Miller, R. D. & Steeples, D. W. 1990. A shallow seismic reflection survey in basalts on the Snake River Plain, Idaho. *Geophysics*, **55**, 761–768.

Mitsuhata, Y., Matsuo, K. & Minegishi, M. 1999. Magnetotelluric survey for exploration of a volcanic-rock reservoir in the Yurihara oil and gas field, Japan. *Geophysical Prospecting*, **47**, 195–218.

Mjelde, R., Sellevoll, M. A., Shinamura, H., Iwasaki, T. & Kanazawa, T. 1993. Ocean bottom seismographs used in a crustal study of an area covered with flood-basalt off Lofoten, N. Norway. *Terra Nova*, **5**, 76–84.

Molchanov, V. I. 1968 [Sedimentation and free hydrogen]. *Akademii Nauk SSSR, Doklady*, **182**, 445–448 (English translation, **182**, 192–194).

Montgomery, S. L. 1997. Lone Tree prospect area, Railroad Valley, Nevada. *American Association of Petroleum Geologists, Bulletin*, **81**, 175–186.

Mosmann, R., Falkenhein, F. U. H., Gonçalves, A. & Nepumuceno Filho, F. 1986. Oil and gas potential of the Amazon Paleozoic basins. *In*: Halbouty, M. T. (ed.) *Future petroleum provinces of the world*. American Association of Petroleum Geologists, Memoirs, **40**, 207–241.

Mullin, R. P. 1988. Maturation model for Middle Amazon Basin, Brazil. *Congreso Brasiliera Geologica, 35e, Anais*, **6**, 2457–2471.

Murchison, D. G. & Raymond, A. C. 1989. Igneous activity and organic maturation in the Midland Valley of Scotland. *International Journal of Coal Geology*, **14**, 47–82.

Nagle, H. E. & Parker, E. S. 1971. Future oil and gas potential of the onshore Ventura Basin, California. *In*: Cram, I. A. (ed.) *Future petroleum provinces of the United States—their geology and potential*. American Association of Petroleum Geologists, Memoirs, **15**(1), 254–297.

Nelson, R. A. 1985. *Geologic analysis of naturally fractured reservoirs*. Gulf Publishing, Houston, 1–320.

Neto, E. S., Cerqueira, J., Takaki, T. & Prinzhofer, A. 2001. Ponta Grossa—Itararé: an unusual gas system in the Paraná Basin, Brazil. *American Association of Petroleum Geologists, Annual Meeting*, online abstract, paper_7981.htm at aapg.confex.com.

Nicolaysen, L. O. & Ferguson, J. 1990. Cryptoexplosion structures, shock deformation and siderophile concentration related to explosive venting of fluids associated with alkaline ultramafic magmas. *Tectonophysics*, **171**, 303–335.

Nodop, I. 1971. Tiefenrefraktionsseismischer Befund im Profil Versmold—Lübbecke—Nienburg. *Fortschritte Geologische Rheinland und Westfalen*, **18**, 411–422.

Nurmi, R., Abdine, S. et al. 1991. Super structures. *In*: Charnock, G. & Gamble, S. (eds) *Middle East Well Evaluation Review*. Special Supplement—Structural Geology, Schlumberger Technical Services, Dubai, 66–81.

Ogilvie, J. S., Crompton, R. & Hardy, N. M. 2001. Characterization of volcanic units using detailed velocity analysis in the Atlantic margin, West of Shetlands, United Kingdom. *Geophysics—the Leading Edge of Exploration*, **20**, 34–50.

O'Sullivan, T. 1992. Scotia—gas resource in a fractured andesite reservoir. *American Association of Petroleum Geologists, Bulletin*, **76**, 1120–1121.

Ozkanli, M. & Kumsal, K. 1993. Petrophysical evaluation of volcanic gas reservoirs of Upper Eocene formations in Thrace Basin, Turkey. *American Association of Petroleum Geologists, 1993 Annual Convention, Program*, 162.

Parker, C. A. 1974. Geopressures and secondary porosity in the deep Jurassic of Mississippi. *Gulf Coast Association of Geological Societies, Transactions*, **24**, 69–80.

Passey, Q. R., Creaney, S., Kulla, J. B., Moretti, F. J. & Stroud, J. D. 1990. A practical model for organic richness from porosity and resistivity logs. *American Association of Petroleum Geologists, Bulletin*, **74**, 1777–1794.

Patton, D. K. 1993. Samgori Field, Republic of Georgia: critical review of island-arc oil and gas. *Journal of Petroleum Geology*, **16**, 153–168.

Peabody, C. E. & Einaudi, M. T. 1992. Origin of petroleum and mercury in the Culver-Baer cinnabar deposit, Macayamas District, California. *Economic Geology*, **87**, 1078–1103.

Peate, D. W., Hawkesworth, C. J., Mantovani, M. S. M. & Shukowsky, W. 1990. Mantle plumes and flood-basalt stratigraphy in the Paraná Basin, South America. *Geology*, **18**, 1223–1226.

PELLERIN, L. & HOHMANN, G. W. 1990. Transient electromagnetic inversion: a remedy for magnetotelluric static shifts. *Geophysics*, **55**, 1242–1250.

PEREA, D. A. & GIORDANO, J. D. 1988. Igneous rocks as hydrocarbon reservoirs in Malargue, Argentina. *Society of Professional Well Log Analysts, 29th Annual Logging Symposium, Transactions*, Z1–Z24.

PERREGAARD, J. & SCHIENER, E. J. 1979. Thermal alteration of sedimentary organic matter by a basalt intrusive (Kimmeridgian shales, Milne Land, East Greenland). *Chemical Geology*, **26**, 331–343.

PETROLEUM AGENCY SA 2000. *South Africa—petroleum exploration opportunities*. 1–24.

PILAAR, W. F. H. & WAKEFIELD, L. L. 1984. Hydrocarbon generation in the Taranaki Basin, New Zealand. *In*: DEMAISON, G. & MORRIS, R. J. (eds) *Petroleum geochemistry and basin evolution*. American Association of Petroleum Geologists, Memoirs, **35**, 405–423.

PINNELL, M. L., BLAKE, J. G. & HULEN, J. B. 1991. Active oil seep at Nevada gold mine holds intrigue for more exploration. *Oil & Gas Journal*, **89**, 15 July (n. 28), 74–79.

PLANKE, S. 1994. Geophysical response of flood basalts from analysis of wire line logs—Ocean Drilling Program Site 642, Vøring volcanic margin. *Journal of Geophysical Research*, **99**, n. B5, 9279–9296.

PLANKE, S. & ELDHOLM, O. 1994. Seismic response and construction of seaward dipping wedges of flood basalts, Vøring volcanic margin. *Journal of Geophysical Research*, **99**, n. B5, 9263–9278.

PLANKE, S., SYMONDS, P. A., ALVESTAD, E. & SKOGSEID, J. 2000. Seismic volcanostratigraphy of large-volume basaltic extrusive complexes on rifted margins. *Journal of Geophysical Research*, **105**, 19335–19351.

POWERS, S. 1932. Notes on minor occurrences of oil, gas, and bitumen with igneous and metamorphic rocks. *American Association of Petroleum Geologists, Bulletin*, **16**, 837–858.

PUJOL, J., FULLER, B. N. & SMITHSON, S. B. 1989. Interpretation of a vertical seismic profile conducted in the Columbia Plateau basalts. *Geophysics*, **54**, 1258–1266.

PÜTTMANN, W., ECKHARDT, C. B. & SCHWARK, L. 1989. Use of biological marker distributions to study thermal history of the Permian Kupferschiefer of the Lower Rhine Basin. *Geologische Rundschau*, **78**, 411–426.

PYE, W. D. 1967. How unique Arizona syenite oil reservoir formed. *World Oil*, **165**, July (n. 1), 81–83.

RAMPINO, M. R. & VOLK, T. 1996. Multiple impact event in the Paleozoic: collision with a string of comets or asteroids? *Geophysical Research Letters*, **23**, 49–52.

RASMUSSEN, B. & BUICK, R. 2000. Oily old ores: evidence for hydrothermal petroleum generation in an Archean volcanogenic massive sulfide deposit. *Geology*, **28**, 731–734.

RAY, R. R. 1989. Seismic expressions of igneous sills in San Juan Sag, south-central Colorado. *American Association of Petroleum Geologists, Bulletin*, **73**, 1159.

RAYMOND, A. C. & MURCHISON, D. G. 1988. Development of organic maturation in the thermal aureoles of sills and its relation to sediment compaction. *Fuel*, **67**, 1599–1608.

RAYMOND, A. C. & MURCHISON, D. G. 1992. Effect of igneous activity on molecular-maturation indices in different types of organic matter. *Organic Geochemistry*, **18**, 725–735.

REECKMANN, S. A. & MEBBERSON, A. J. 1984. Igneous intrusions in the north-west Canning Basin and their impact on oil exploration. *In*: PURCELL, P. G. (ed.) *The Canning Basin*. Geological Society of Australia–Petroleum Exploration Society of Australia, 389–399.

REITER, M. A. 1986. Estimates of terrestrial heat flow from deep petroleum tests along the Rio Grande Rift in central and southern New Mexico. *Journal of Geophysical Research*, **91**, 6225–6245.

REMUS, D., WEBSTER, M. & KEAWKAN, K. 1993. Rift architecture and sedimentology of the Phetchabun intermontane basin, central Thailand. *Journal of Southeast Asian Earth Sciences*, **8**, 421–432.

RICHARDSON, K. R., WHITE, R. S., ENGLAND, R. W. & FRUEHN, J. 1999. Crustal structure east of the Faeroe Islands: mapping sub-basalt sediments using wide-angle seismic data. *Petroleum Geoscience*, **5**, 161–172.

RIDD, M. F. 1983. Aspects of the Tertiary geology of the Faeroe-Shetland Channel. *In*: BOTT, M. H. P., SAXOV, S., TALWANI, M. & THIEDE, J. (eds) *Structure and development of the Greenland-Scotland Ridge*. Plenum Press, New York, 91–108.

SACHSENHOFER, R. F. 1994. Petroleum generation and migration in the Styrian Basin (Pannonian Basin system, Austria): an integrated geochemical and numerical modeling study. *Marine and Petroleum Geology*, **11**, 684–701.

SAIF, S. I. & SHAH, S. M. A. 1988. Field and petrographic characteristics of Cenozoic basaltic rocks, northwestern Saudi Arabia. *Journal of African Earth Sciences*, **7**, 805–809.

SAKATA, S., MATSUSHISA, Y. & IGARI, S.-I. 1994. Organic geochemistry of petroleum from the Green Tuff basin, Japan. *Organic Geochemistry*, **21**, 27–34.

SALAS, G. P. 1968. Petroleum evaluation of north-central Mexico. *American Association of Petroleum Geologists, Bulletin*, **52**, 665–674.

SANYAL, S. K., JUPRASERT, S. & JUSBASCHE, M. 1980. An evaluation of a rhyolite-basalt-volcanic ash sequence from well logs. *Log Analyst*, **21**, n. 1, 3–9.

SAUNDERS, B. 1997. Subsalt exploration interest spreading outside Gulf of Mexico. *Oil & Gas Journal*, **95**, July 14 (n. 28), 22.

SAXBY, J. D. & STEPHENSON, L. G. 1987. Effect of an igneous intrusion on oil shale at Rundle (Australia). *Chemical Geology*, **63**, 1–16.

SEMBODO 1973. Notes on formation evaluation in the Jatibarang volcanic reservoir. *Indonesian Petroleum Association, 2nd Annual Convention, Proceedings*, 131–147.

SEYLER, B. & CLUFF, R. M. 1990. Petroleum traps in the Illinois Basin. *In*: LEIGHTON, M. W., KOLATA, D. R., OLTZ, D. F. & EIDEL, J. J. (eds) *Interior Cratonic Basins*, American Association of Petroleum Geologists, Memoirs, **51**, 361–401.

SHERWOOD, B., FRITZ, P., FRAPE, S. K., MACKO, S. A., WEISE, S. M. & WHELAN, J. A. 1988. Methane occurrences in the Canadian Shield. *Chemical Geology*, **71**, 223–236.

SHIRLEY, K. 1998. Controversy grows for monument. *AAPG Explorer*, **19**, July (n. 7), 8–11, 31.

SHULMAN, H. & MAY, P. R. 1989. Seismic survey in the Golan Heights and its resultsr *Israel Geological Society, Annual Meeting, Abstracts* (Israel Geological Survey), 155–156.

SILVA, W. F. & DE BRITO, C. G. 1973. Novas possibilidades de resolução estrutural com o método sísmico na Bacia do Paraná. *Congreso Brasiliera Geologica, 27e, Anais*. **3**, 357–371.

SILVA, G. R. & VIANNA, R. B. 1982. Aplicabilidade do método de reflexão sismica en Bacia do Paraná. *Geologia da Bacia do Paraná—Reavaliação de Potentcialidade e Prospectividade em Hidrocarbonetos*. Instituto Pesquisas Tecnica, Estado São Paulo, Publicacão Especial, **12**, 117–142.

SIMONEIT, B. R. T. 1985. Hydrothermal petroleum: genesis, migration and deposition in the Guaymas Basin, Gulf of California. *Canadian Journal of Earth Sciences*, **22**, 1919–1929.

SIMONEIT, B. R. T. 1994. Organic matter alteration and fluid migration in hydrothermal systems. *In*: PARNELL, J. (ed.) *Geofluids: origin, migration and evolution of fluids in sedimentary basins*. Geological Society, Special Publication **78**, 261–274.

SIMONEIT, B. R. T. & DIDYK, B.M. 1978. Organic geochemistry of a Chilean paraffin dirt. *Chemical Geology*, **23**, 21–40.

SIMONEIT, B. R. T. & FETZER, J. C. 1996. High molecular weight polycyclic aromatic hydrocarbons in hydrothermal petroleums from the Gulf of California and northeast Pacific Ocean. *Organic Geochemistry*, **24**, 1065–1077.

SIMONEIT, B. R. T., BRENNER, S., PETERS, K. E. & KAPLAN, I. R. 1981. Thermal alteration of Cretaceous black shale by diabase intrusions in the Eastern Atlantic—II. Effects on bitumen and kerogen. *Geochimica et Cosmochimica Acta*, **45**, 1581–1602.

SKOGSEID, J. 2001. Volcanic margins: geodynamic and exploration aspects. *Marine and Petroleum Geology*, **18**, 457–461.

SLOAN, M. W., MOORE, P. W. & MCCUTCHEON, A. 1992. Kipper—a unique oil and gas discovery. *APEA Journal*, **32**, pt. 1, 1–8.

SMITHSON, S. B. & SHIVE, P. N. 1975. Field measurements of compressional wave velocities in common crystalline rocks. *Earth & Planetary Science Letters*, **27**, 170–176.

SNYDER, R. P. 1968. Electric and caliper logs as lithologic indicators in volcanic rocks, Nevada Test Site. *In*: ECKEL, E. B. (ed.) *Nevada Test Site*. Geological Society of America, Memoir, **110**, 117–124.

SOUTHER, J. G. & JESSOP, A. M. 1990. Dyke swarms in the Queen Charlotte Islands, British Columbia, and implications for hydrocarbon exploration. *In*: WOODSWORTH, G. (ed.) *Evolution and hydrocarbon potential of the Queen Charlotte Basin*. Geological Survey of Canada, Paper **90–10**, 465–487.

SPARLIN, M. A. & LEWIS, R. D. 1994. Interpretation of the magnetic anomaly over the Omaha oil field, Gallatin County, Illinois. *Geophysics*, **59**, 1092–1099.

SPOONER, W. C. 1928. Production in East Carroll Parish, Louisiana. *American Association of Petroleum Geologists, Bulletin*, **12**, 765.

STANLEY, W. D., SAAD, A. R. & OHOFUGI, W. 1985. Regional magnetotelluric surveys in hydrocarbon exploration, Paraná Basin, Brazil. *American Association of Petroleum Geologists, Bulletin*, **69**, 346–360.

STEVENS, T. D. & MCKINLEY, J. P. 1995. Lithautotrophic microbial ecosystems in deep basalt aquifers. *Science*, **270**, 450–454.

STOFFERS, P., HANNINGTON, M., WRIGHT, I., HERZIG, P. & DE RONDE, C. 1999. Elemental mercury at submarine hydrothermal vents in the Bay of Plenty, Taupo volcanic zone, New Zealand. *Geology*, **27**, 931–934.

SUMMER, N. S. & VEROSUB, K. L. 1992. Diagenesis and organic maturation of sedimentary rocks under volcanic strata, Oregon. *American Association of Petroleum Geologists, Bulletin*, **76**, 1190–1199.

SYLVESTER-BRADLEY, P. C. & KING, R. J. 1963. Evidence for abiogenic hydrocarbons. *Nature*, **198**, 728–731.

SYMONDS, P. A., PLANKE, S., FREY, O. & SKOGSEID, J. 1998. Volcanic evolution of the Western Australian continental margin and its implication for basin development. *In*: PURCELL, P. G. & PURCELL, R. R. (eds) *The sedimentary basins of Western Australia 2, Symposium, Proceedings*. Petroleum Exploration Society of Australia, Perth, 33–54.

SZATMARI, P. 1989. Petroleum formation by Fischer–Tropsch synthesis in plate tectonics. *American Association of Petroleum Geologists, Bulletin*, **73**, 989–998.

TALIAFERRO, N. L., HUDSON, F. S. & CRADDOCK, W. N. 1924. The oil fields of Ventura County, California. *American Association of Petroleum Geologists, Bulletin*, **8**, 789–829.

THOMPSON, G. A. 1998. Deep mantle plumes and geoscience vision. *GSA Today*, **8,** April (n. 4), 17–25.

THRASHER, J. 1992. Thermal effect of the Tertiary Cuillins intrusive complex in the Jurassic of the Hebrides: an organic geochemical study. *In*: PARNELL, J. (ed.) *Basins of the Atlantic seaboard: petroleum geology, sedimentology and basin evolution*. Geological Society, London, Special Publication, **62**, 35–49.

UCHIDA, T. 1992. Mode of occurrences of secondary porosities in volcanic rock reservoirs distributed along the west coast of the Northeastern Honshu Arc. *International Geological Congress, 29th, Abstracts*, **3**, 810.

UJIIÉ, Y. 1986. Contact-metamorphic effect on parameters for kerogen maturation. *Organic Geochemistry*, **9**, 375–378.

VERNIK, L. 1990. A new type of reservoir rock in volcaniclastic sequences. *American Association of Petroleum Geologists, Bulletin*, **74**, 830–836.

VOZOFF, K. 1972. The magnetotelluric method in exploration of sedimentary basins. *Geophysics*, **37**, 98–141.

WILLIAMS, D. L. & FINN, C. 1985. Analysis of gravity data in volcanic terrain and gravity anomalies and subvolcanic intrusions in the Cascade Range, U.S.A., and at other selected volcanoes. *In*: HINZE, W. J. (ed.) *The utility of regional gravity and magnetic anomaly maps*, Society of Exploration Geophysicists, Tulsa, 361–374.

WILLIAMS, H. H., FOWLER, M. & EUBANK, R. T. 1995. Characteristics of selected Palaeogene and Cretaceous lacustrine source basins of Southeast Asia. *In*: LAMBAISE, J. J. (ed.) *Hydrocarbon habitat in rift basins*. Geological Society, London, Special Papers, **80**, 241–282.

WITHERS, R., EGGERS, D., FOX, T. & CREBS, T. 1994. A case study of integrated hydrocarbon exploration through basalt. *Geophysics*, **59**, 1666–1679.

WOJCIK, K. M. & KNAPP, R. W. 1990. Stratigraphic control of the Hills Pond lamproite, Silver City Dome, southeastern Kansas. *Geology*, **18**, 251–254.

YOUNG, K. & WOODRUFF, C. M., JR. 1985. *Austin Chalk in its type area—stratigraphy and structure*. Austin Geological Society, Guidebook **7**, 1–88.

YOUNG, K., CARAN, S. C. & EWING, T. C. 1982. *Cretaceous volcanism in the Austin area, Texas*. Austin Geological Society, Guidebook **4** (revised), 1–66.

YOUNG, R. A. & LUCAS, J. E. 1988. Exploration beneath volcanics: Snake River Plain, Idaho. *Geophysics*, **53**, 444–452.

YÜKLER, M. A. & DOW, W. D. 1990. The effect of heating rate on kerogen maturation and oil generation. *American Association of Petroleum Geologists, Bulletin*, **74**, 796.

ZALÁN, P. V., WOLFF, S. ET AL. 1990. The Paraná Basin. *In*: LEIGHTON, M. W., KOLATA, D. R., OLTZ, D. F. & EIDEL, J. J. (eds) *Interior Cratonic Basins*. American Association of Petroleum Geologists, Memoirs, **51**, 681–708.

ZALESKI, E., EATON, D. W., MIKEREIT, B., ROBERTS, B., SALISBURY, M. & PETRIE, L. 1997. Seismic reflections from subvertical diabase dikes in an Archean terrane. *Geology*, **25**, 707–710.

ZIMMERLE, W. 1995. New aspects on the formation of hydrocarbon source rocks. *Geologische Rundschau*, **74**, 385–416.

ZOLOTUKHIN, V. V. & AL'MUKHAMEDOV, A. I. 1988. Traps of the Siberian Platform. *In*: MACDOUGALL, J. D. (ed.) *Continental Flood Basalts*, Kluwer Academic Publishers, 273–310.

Occurrences of hydrocarbons in and around igneous rocks

STEPHEN R. SCHUTTER

Subsurface Consultants & Associates, LLC, 2500 Tanglewilde, Suite 120, Houston, Texas 77063, USA (e-mail:104034.454@compuserve.com)
Current address: Murphy Exploration and Production Company, 550 Westlake Park Boulevard, Suite 1000, Houston, Texas 77079, USA (e-mail: steve_schutter@murphyoilcorp.com)

Abstract: Data on the occurrence of hydrocarbons in and around igneous rocks show them to be global in extent, occurring in over 100 countries worldwide. While this list is not exhaustive, it is possibly the first of its kind to be published, and will serve as a useful source of reference for those wishing to embark on further study.

This is not intended to be a comprehensive list of hydrocarbon occurrences in and associated with igneous rocks. Because such hydrocarbons have not been systematically studied, references to them are anecdotal, and are principally mentioned in passing in the context of regional studies or exploration techniques. The primary purpose here is to provide a data base for more systematic study, and secondarily to encourage reexamination of old, neglected occurrences and the reporting of new occurrences. This list also has practical applications, indicating poorly explored areas where hydrocarbons may be present.

The tables are broadly arranged by continents, which emphasizes the range of hydrocarbon occurrences and their abundance. Many of the occurrences are poorly documented. Hydrocarbon trapping mechanisms are listed when known, but do not apply in the cases of seeps (hydrocarbons leaking to the surface) or shows (hydrocarbons leaking into quarries, mines and boreholes).

From: PETFORD, N. & MCCAFFREY, K. J. W. (eds) 2003. *Hydrocarbons in Crystalline Rocks*. Geological Society, London, Special Publications, **214**, 35–68. 0305-8719/03/$15 © The Geological Society of London.

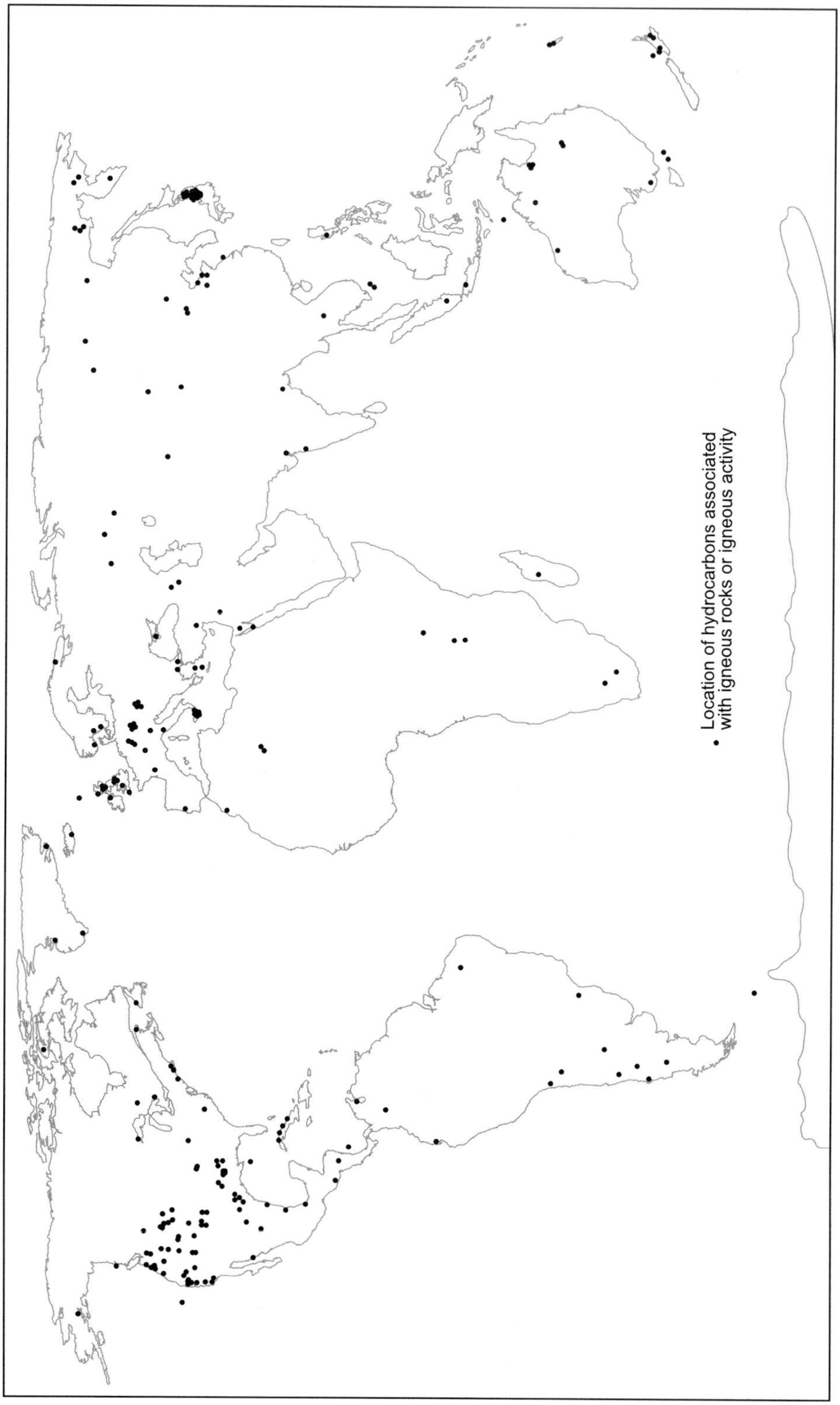

Fig 1. Map showing the worldwide geographical distributions of hydrocarbons associated with igneous rocks, as listed in Table 1.

Table 1. *Hydrocarbons associated with igneous rocks or igneous activity*

Field or show	Basin	Type	Size	Reservoir rock	Trap	Reference
United States						
Alaska						
Novarupta		gas	Seep (fumaroles, 4–14% CH_4)		NA	White & Waring 1963, p. K-2
Arizona						
Dineh-bi-Keyah	Paradox/Defiance Uplift	oil, gas	18.1 MM bbl, 4.7 BCF	syenite	anticline (fractured sill)	Kornfeld & Travis 1967; Pye 1967; McKenny & Masters 1968; McKenny 1969; Biederman 1986; Ray 1989; Masters 2000; Rauzi 2001
Arkansas						
Murfreesboro		oil	seep	peridotite	NA	Powers 1932
California						
Abbott Mine (Wilbur Springs district, Lake County)	Coast Ranges	froth vein	show		NA	Bailey 1959; Moiseyev 1968
Bogess and Harrington creeks	Coast Ranges	oil	seep	dolerite	NA	Powers 1932
Cinnabar King Mine (Sonoma County)	Coast Ranges	froth vein	show	serpentinite	NA	Bailey 1959
Colusa/Moon Bend	Central Valley	gas	60 BCF EUR	sandstone	dome over intrusion	Hopkins *et al.* 1956; Poreda *et al.* 1986; MacKevett 1998
Conejo	Ventura	oil	<1 MM bbl	basalt agglomerate	tar at surface (fractured monocline)	Taliaferro *et al.* 1924; Powers 1932; Nagle & Parker 1971, pp. 269, 273
The Geysers	Coast Ranges	gas	show (superheated ground water, 15% CH_4, some heavier hydrocarbons)			White & Waring 1963, p. K-2; Des Marais *et al.* 1981
Jumbo Mine (Sonoma County)	Coast Ranges	froth vein	show	serpentinite		Bailey 1959
Manhattan Mine (Napa County)	Coast Ranges	froth vein	show	silicified tuff	NA	Bailey 1959
Marysville or Sutter Buttes	Central Valley	gas	205 BCF EUR	sandstone	faults, anticline with intrusion; intrusion flanks	Powers 1932; Hunter 1955a; Hopkins *et al.* 1956; Weismann 1971; Gries 1985; Poreda *et al.* 1986; MacKevett 1998
Mono Lake	Great Basin	oil, gas	seep	'Recent volcanics'	NA	Powers 1932; Oremland *et al.* 1987

Table 1. *Continued*

Field or show	Basin	Type	Size	Reservoir rock	Trap	Reference
United States						
New Almaden (Santa Clara County)	Coast Ranges	oil, gas, froth veins	seeps	andesite, serpentinite, metasediments	faulted anticline	Powers 1932; Bailey 1959; Peabody 1993
Parkfield	Coast Ranges	oil	seep	metasediments near peridotite	NA	Powers 1932
Petaluma	Coast Ranges	oil	seep	basalt	NA	Powers 1932
Santa Clara Avenue	Ventura	oil	7 MM bbl cumulative	sandstone	mafic dyke blocking migration	Richards 1985
Santa Ynez River canyon	Coast Ranges	oil	seep	serpentinite	NA	Powers 1932
Terhel-1 (Wilbur Springs district)	Coast Ranges	oil (35.7° API, Terhel-1 test)	show			Peabody 1993
West Butte/Mapco-Kylling	Central Valley	gas				Poreda *et al.* 1986
Wilbur Springs	Coast Ranges	oil	seep	serpentinite	anticline	Powers 1932
Wild Goose	Central Valley	gas	103 BCF EUR	sandstone	dome over intrusion	Matjasic 1954; Hunter 1955b; Hawley 1968; Poreda *et al.* 1986; MacKevett 1998
Offshore California/East Pacific Ocean						
	Escanaba Trough	oil	seep	turbidites, sulphides	NA	Kvenvolden *et al.* 1986
Colorado						
Campbell Ranch	Raton	oil	show	'felsite'	anticline (fractured sill)	Creely & Saterdal 1956
Del Norte	San Juan Sag	oil	1500 bbl (1 well at 30 BOPD)	fractured sill	sill in shale	Gries 1985; Ray 1989
Summer Coon	San Juan Sag (near Del Norte)	oil	seep	volcanic complex	NA	Gries 1985
Tow Creek and Tow Creek North	(Rouett County, 20 mi west of Steamboat Springs)	oil	3 MM bbl ultimate recovery	fractured shale	anticline over laccolith	Heaton 1929; Saterdal 1955; Mallory 1977
Connecticut						
Connecticut River valley	Hartford	bitumen	seep	dolerite, basalt	NA	Percival 1878, *cited in* Hodge 1927; Powers 1932
Idaho						
Grandview-Marsing	Snake River Downwarp	oil, gas		sandstone	associated with plugs, sills	Kirkham 1935
Payette-Weiser	Snake River Downwarp	oil, gas	IP up to 75 MMCFGPD	sandstone	associated with plugs, sills	Kirkham 1935

Illinois						
Omaha	Illinois	oil	6.5 MM bll cumulative production	sandstone	laccolith	English & Grogan 1948; Seyler & Cluff 1990; Sparlin & Lewis 1994
Saline County	Illinois	oil	show	'igneous dyke'	NA	Powers 1932
Louisiana						
Baskinton	Monroe Uplift	oil		sandstone	up-dip migration blocked by dyke	F. Spooner, pers. comm. 1999
Door Point (offshore)	Gulf of Mexico	gas	show	sandstone	drape over intrusion	Braunstein & McMichael 1976
Epps (Floyd)	Monroe Uplift	gas	2.1 BCF	sandstone	drape over intrusion	Spooner 1928; Gordon 1931; Easton 1935; Moody 1949; Shreveport Geol. Soc. 1968
Monroe	Monroe Uplift	gas	7.3 TCF cumulative; 365 mi^2	limestone	structure formed by igneous intrusions	Zimmerman & Sasser 1993
Richland	Monroe Uplift	gas	141 BCF by 1/31	tuff and tuffaceous sandstone	faulted anticline	Gordon 1931; Easton 1935
South Epps	Monroe Uplift	gas	1.4 BCF	sandstone	drape over intrusion	Shreveport Geol. Soc. 1968
Massachusetts						
Hampden	Hartford	bitumen		basalt	NA	Pratt & Burruss 1988
Minnesota						
(northeastern)	Lake Superior	gas, graphite	shows	gabbro	NA	Hollister 1980
Mississippi						
Carey	Monroe/Sharkey Uplift	oil	53 M bbl by by 1/46; 25.5° API	limestone	drape over intrusion	Shreveport Geol. Soc. 1946; Moody 1949; Harrelson & Bicker 1979
Jackson	Mississippi Salt Basin	gas	118.8 BCF, minor oil	limestone	drape over intrusion	Moody 1949; Beebe 1968
Montana						
Flat Coulee	Western Canada	oil, gas	5.1 MM bbl, 4.2 BCF	sandstone	intrusion flanks (partial)	Lopez 1995
Utopia	Western Canada	oil, gas	0.7 MM bbl, 20.6 BCF	sandstone, dolomite	intrusion flanks (partial)	Lopez 1995
Whitlash	Western Canada	oil, gas	4.5 MM bbl, 45.4 BCF	sandstone	intrusion flanks (partial)	Lopez 1995
15 smaller fields	Western Canada	oil, gas	1.0 MM bbl, 90.4 BCF (total)	sandstone	intrusion flanks (whole or in part)	Lopez 1995

Table 1. *Continued*

Field or show	Basin	Type	Size	Reservoir rock	Trap	Reference
United States						
Nevada						
Bacon Flat	Sheep Pass (Great Basin)	oil	65,000 bbl (1981–1982) 28° API	hydrothermally fractured dolomite		Bortz 1983; Duey 1983
Blackburn	Great Basin	oil	6 MM bbl EUR 27° API	hydrothermally fractured dolomite, tuff	horst	Bortz 1983; Flanigan 1989; Hulen *et al.* 1990
Carlin	Great Basin	bitumen (mesophase)	show	hydrothermal deposit in limestone and dolomite	NA	Gize 1986
Duckwater Creek	Sheep Pass (Great Basin)	oil	12 M bbl cumulative (1993); 31° API, 0.3% S	lava, tuff, welded tuff, tuffaceous sandstone		French 1994a
Eagle Springs	Sheep Pass (Great Basin)	oil	5.5 MM bbl EUR; 26°–29° API, 1.7% S	rhyolitic ash-flow tuff, sediments	down-dropped fault block	Bortz 1983; Duey 1983; Flanigan 1989; Hazlett & Hersch 1993; Bortz 1994
Grant Canyon	Great Basin	oil	17 MM bbl by 1993	hydrothermally fractured dolomite		Hulen *et al.* 1990; Hazlett & Hersch 1993
Kate Springs	Great Basin	oil		hydrothermally fractured carbonates		Hulen *et al.* 1990
Sand Spring	Sheep Pass (Great Basin)	oil	100 M bbl EUR; 160 acres IP discovery 1253 BOPD; 28° API	ignimbrites	fault	Grabb 1994a
Standard-Amoco No. 1 S. P. Land	Great Basin	oil	show	vuggy basalt breccia	NA	Bortz 1983
Three Bar	Great Basin	oil	23 M bbl cumulative (1993)	tuffs, interbedded sediments		Grabb 1994a
Tomera Ranch	Great Basin	oil		ash-flow tuff		Flanigan 1989
Trap Spring	Sheep Pass (Great Basin)	oil	11 to 15 MM bbl EUR; wells range 600–1200 BOPD; 21°–25° API	rhyolitic ash-flow tuff	down-dropped fault block	French & Freeman 1979a, b; Bortz 1983; Duey 1983; Flanigan 1989; Hazlett & Hersch 1993; French 1994b
Northwest Exploration No. 6 White River Valley	Sheep Pass (Great Basin)	heavy oil	show	volcanics		Bortz 1983
Willow Creek	Great Basin	oil		ash-flow tuff		Flanigan 1989

New Jersey						
(unspecified locality)	Newark	bitumen	seep	dolerite	NA	Russell 1878, *cited in* Hodge 1927; Powers 1932; Hedberg 1964
New Mexico						
Aztec (10 mi east of)		oil	seep	'dyke'	NA	Powers 1932
Raton	Raton	oil	seep	basalt	NA	Powers 1932
New York						
(unspecified locality)	Newark	bitumen	seep	basalt	NA	Hodge 1927, p. 404
North Carolina						
SEPCO Butler No. 1	Deep River (Sanford)	oil, gas	show (free oil recovered)	fractured intrusive		Ziegler 1983
Oregon						
Clarno		asphalt	seep	basalt	NA	Powers 1932
Clatsop County	Astoria	bitumen	seep	basalt	NA	Hodge 1927; Powers 1932
Note: Clatsop seep across Columbia River from a similar seep in Ilwaco, WA						
Florence (Lane Co.)		oil	seep	basalt	NA	Hodge 1927; Powers 1932
Mist	Nehalem Arch (between Willamette and Astoria basins)	gas	40 to 50 BCF	sandstone	anticline	Armentrout & Suek 1985; Summer & Verosub 1987
Note: Mist partially under edge of Columbia basalts; Summer & Verosub 1987, stated that local intrusions provided maturation						
Post (Crook County)		asphalt	seep	rhyolite	NA	Hodge 1927; Powers 1932
Wheeler County		bitumen	seep	tuff	NA	Hodge 1927
Texas						
Note: There are nearly 90 fields producing from volcanics or associated facies in the 'serpentine plug' trend (Ewing & Caran 1982; Matthews 1986). For lists, data on fields, see Sellards 1938; Simmons 1967; Rives 1968; Matthews 1986. Some are given here for comparison.						
Hilbig		oil	6.12 MM bbl	'serpentine'	buried volcano	Matthews 1986
Lytton Springs		oil	>11 MM bbl	'serpentine'	buried volcano	Matthews 1986
Sutil		oil	1.55 MM bbl	facies associated with buried volcano	buried volcano	Matthews 1986
Terlingua	Rio Grande Rift	oil, gas	show (sufficient free oil for furnace fuel)	(various lithologies), breccia	breccia pipe	Powers 1932; Sylvester-Bradley & King 1963; Yates & Thompson 1976; Fallin 1989
Texarkana West	East Texas	oil		sandstone	buried topography from dolerite sill	Tieman 1992
Thrall		oil	2.39 MM bbl	'serpentine'	buried volcano	Matthews 1986
Torch		oil	2.54 MM bbl	sandstone	drape over buried volcano	Rives 1968; Matthews 1986

Table 1. *Continued*

Field or show	Basin	Type	Size	Reservoir rock	Trap	Reference
United States						
Utah						
Rozel Point	North Basin, Great Salt Lake	oil	3000+ bbl since 1956; 9.4° API	fractured vesicular basalt	horst	Bortz 1983; Kendell 1994a
San Rafael Desert	SE Utah	oil, bitumen	Seep	lamproite, sediments	dyke following fault	Hulen *et al.* 1998
West Rozel	North Basin, Great Salt Lake	oil	>100 MM bbl OIP; wells up to 2160 BOPD; 28,000 bbl produced; 4°–6° API, 13% S	fractured vuggy basalt, agglomerate	faulted anticline	Bortz 1983, 1987; Kendell 1994b
Washington						
Ilwaco	Astoria	bitumen	Seep	basalt	NA	Hodge 1927; Powers 1932
Note: Across the Columbia River from a similar seep in Clatsop County, OR						
Rattlesnake Hills	Columbia	gas	1.3 BCF cumulative production; best well up to 1.5 MMCFGPD	basalt	faulted anticline	Powers 1932; Hammer 1934; Kirkham 1935; McFarland 1979; Kuuskraa & Schmoker 1998
(unnamed localities)	Columbia	bitumen	seep	basalt	NA	Felts 1954
Wyoming						
Calcite Springs	Yellowstone	oil	seep	rhyolitic volcanics	NA	Love & Good 1970; Clifton *et al.* 1990
Castle Rocks	Absaroka	oil, gas	seep	andesitic volcanics	NA	Sundell & Love 1986
Chimney Rock	Absaroka	bitumen	seep	andesitic volcanics	NA	Sundell & Love 1986
Hot Springs County	Absaroka	heavy oil/ bitumen	>10 MM bbl	andesitic volcanics	tar seal above anticline	Bailey & Sundell 1986a, b; Keener 1987
Middle Fork	Absaroka	oil	seep	andesitic volcanics	NA	Sundell & Love 1986
Rainbow springs	Yellowstone	oil	seep	rhyolitic volcanics	NA	Love & Good 1970; Clifton *et al.* 1990
(unnamed localities)	Bighorn	oil	seeps	andesitic volcanics	NA	Schmoker *et al.* 1996
Canada						
British Columbia						
Queen Charlotte Islands (many localities)	Queen Charlotte	oil, gas, bitumen	seeps	basalt, andesite flows, agglomerates, tuffs, sills	NA	Hamilton & Cameron 1989
Tian Head	Queen Charlotte	oil, gas	shows	basalt flows, breccias; rhyolite flows	NA	McCaslin 1971

Newfoundland						
(unspecified locality)		bitumen	seep	dolerite	NA	Powers 1932
Northwest Territories						
Nanisvik	NW Baffin Island	bitumen	show	hydrothermal deposit in dolomite	NA	Gize 1986
Ontario						
Parry Sound district		oil, bitumen	show	feldspar-quartz pegmatite	NA	Spence 1930; Parnell 1990
Timmins		bitumen (mesophase)	show	ultramafic volcanics, interflow sediments	NA	Goodzari *et al.* 1992
Quebec						
Tar Point	Gaspe	oil	seep	dolerite	NA	Powers 1932; Dott & Reynolds 1969; Sikander 1975, p. 283
Costa Rica						
Cocoles No. 2	Bocas del Toro	oil	8000 bbl	fractured andesite	anticline	Morris *et al.* 1990
(unspecified locality)		oil	seep	andesite, serpentinite, metasediments	NA	Hedberg 1964
Cuba						
Bacuanao-Cruz Verde	North Cuba	oil	343,000 bbl cumulative production 1955–1959; IP = 25 to 200 BOPD	fractured serpentine		Lewis 1932; Wassall 1957; Anon. 1959; Echevarria-Rodriguez *et al.* 1991
Camarioca	North Cuba	oil		fractured serpentine		Echevarria-Rodriguez *et al.* 1991
Cantel	North Cuba	oil		fractured serpentine		Echevarria-Rodriguez *et al.* 1991
Cristales	South Cuba	oil	2500 BOPD from field in 1971	basaltic tuff		Sawyer 1975; Petersilie 1973
Guanabo	North Cuba	oil	76,000 bbl cumulative production 1956–1959	fractured serpentine		Echevarria-Rodriguez *et al.* 1991
Jarahueca	North Cuba	oil	1.25 MM bbl cumulative production 1943–1959; wells up to 1000 BOPD; field produced 12,000 bbl/mo in 1/46	fractured serpentine		Noguera 1946; Wassall 1957; Anon. 1959; Echevarria-Rodriguez *et al.* 1991
Jatibonico	South Cuba	oil	1.25 MM bbl cumulative production 1954–1959; IP = 175 BOPD +75 BWPD	basaltic tuff		Irizarry 1954; Anon. 1959; Dott & Reynolds 1969; Sawyer 1975

Table 1. *Hydrocarbons associated with igneous rocks or igneous activity*

Field or show	Basin	Type	Size	Reservoir rock	Trap	Reference
Cuba						
Motembo	North Cuba	oil	1.8 MM bbl cumulative production 1885–1959	fractured serpentine		Lewis 1932; Wassall 1957; Anon. 1959; Echevarria-Rodriguez *et al.* 1991
Penas Altas	North Cuba	oil		fractured serpentine		Echevarria-Rodriguez *et al.* 1991
Pina		oil		fractured tuffs and agglomerate		Rodriquez 1996; IHS Energy Group 2001
Santa Maria del Mar	North Cuba	oil	88,000 bbl cumulative production 1955–1959	fractured serpentine		Wassall 1957; Anon. 1959; Echevarria-Rodriguez *et al.* 1991
El Salvador						
Carolina district	Dpto. San Miguel	oil	seep	andesite	NA	Powers 1932; Hazzard 1957
Greenland						
Disco Island	Nussauq	oil	seep	basalt	NA	Powers 1932
Ilimaussaq Peninsula	south	oil, gas, bitumen	show	syenite	NA	Petersilie & Sorensen 1970; Konnerup-Madsen *et al.* 1979; Gize & Macdonald,1993
Marrait	Nussauq	oil	seep (>50 MM bbl OOIP in 8 × 5 km area)	basalt (in vesicular flow tops)	NA	Pedersen 1981; Anon. 1993; Christiansen *et al.* 1994, 1996
(unspecified locality)	Nussauq	oil	show (90 m saturation), 3 MM bbl estimated	vesicular basalt		G. Vigh, oral comm. 1994, *cited in* Patton 1995
Romer Fjord	Jameson Land	oil	seep	basalt	NA	Watt & Wrang 1984
Scoresby Sund	Jameson Land	oil	seep	basalt	fault zone	Watt & Wrang 1984
Mexico						
Casiano No. 7	75 mi south of Tampico	oil, gas	20,000 BOPD and 10 to 12 MMCFGPD; cumulative production in 4 years was 33.58 MM bbl	fractured shale	flank of stock	I. C. White *in* Ordoñez 1915
Furbero	Vera Cruz	oil		fractured shale	gabbro sill or laccolith	DeGoyler 1915, 1932
Naica	Chihuahua Trough	oil	seep	sulphide mineralization with igneous intrusion	NA	Salas 1968
Ojinaga	Chihuahua Trough	oil	seep	sulphide mineralization with igneous intrusion	NA	Salas 1968
Topila	Vera Cruz	oil		tuff		DeGoyler 1932

Gulf of California						
	Guaymas	oil	seeps 27.6° API, 0.99% S	organic-rich sediments, sulphides	NA	Simoniet 1985; Didyk & Simoniet 1989; Kaka & Simoniet 1990
Nicaragua						
Chontales	10 mi NE of Sto. Tomas	asphalt	seep	tuff	NA	Powers 1932
Argentina						
Los Cavaos	Neuquen	oil	cumulative (1988) 3.5×10^6 m^3 IP = 30,000 l/hr	andesite	faulted anticline/sill	Perea *et al.* 1984; Perea & Giordano 1988; Belotti *et al.* 1995
Chihuido de la Sierra Negra	Neuquen	oil, gas?	show	andesite	sill or laccolith	Pucci 1996
Malal de Medio	Neuquen	oil		andesite	faulted anticline/sill	Perea *et al.* 1984; Belotti *et al.* 1995
El Manzano	Neuquen	oil, gas		andesite	faulted anticline/sill	Perea *et al.* 1984
YPF Palmar Largo es-1	Noroeste (Oran)	oil, gas	550 m^3/d oil, 33,800 m^3/d gas	vuggy basalt		Turic *et al.* 1985
Rio Grande	Neuquen	oil		andesite	faulted anticline/sill	Perea *et al.* 1984; Belotti *et al.* 1995
El Sosneado	Neuquen	oil		sediments	andesite plugs and sills	Lahee 1932
Tupungato	Cuyo	oil	8.7 MM bbl as of 1/43	fractured tuff	faulted anticline	Baldwin 1944
Vega Grande	Neuquen	oil, gas	IP up to 1635 BOPD, 380 MCFGPD	fractured andesite		Turic *et al.* 1986; Wiman 1987
25 de Mayo-Medanito	Neuquen	oil	23.9 MM bbl EUR	altered rhyolite, rhyolite tuff, agglomerate		Daniel & Hvala 1982
Los Volcanes	Neuquen	oil, gas		andesite	faulted anticline/sill	Perea *et al.* 1984
Brazil						
2-CM-1-MA	Parnaiba/Maranhao	gas	show	dolerite	sill	Agencia Nacional do Petroleo 2002
(multiple localities)	Parana	asphalt	seeps	dolerite	NA	Powers 1932
Chile						
25 mi south of Concepcion		bitumen, asphalt	seep	sediments near granodiorite	NA	Mueller 1964
Siglia	Salar de Atacama	gas	seeps	flank of diorite intrusion	NA	Simoniet & Didyk 1978
(unspecified locality)	Tamarugal	oil	seep	anticline around intrusive	NA	Roos 1996

Table 1. *Continued*

Field or show	Basin	Type	Size	Reservoir rock	Trap	Reference
Colombia						
Magdalena Valley, Dpto. Tolima	Upper Magdalena	oil	seep	volcanic ash	NA	Powers 1932
Ecuador						
Santa Elena	Santa Elena	oil		sandstone, igneous dykes		Garfias 1923; Powers 1932
Venezuela						
Totumo	Maracaibo	oil	150,000 bbl ultimately recovered; best well produced 2100 BOPD for short time; 22° API	Pre-Mesozoic igneous rocks and Jurassic volcanics		Mencher *et al.* 1953; Smith 1956; Guariguata & Richardson 1960; Dott & Reynolds 1969
Azerbaijan						
Muradkhanly	western	oil	30° API	andesite and basalt, interflow sediments	anticline	Rustamov 1982; Buryakovsky 1993
Czech Republic						
Mitov		bitumen (mesophase)	seep	pillow basalt, interflow stromatolites	NA	Kribek *et al.* 1993
Radotin		bitumen	seep	dolerite	NA	Powers 1932
Rybnik		bitumen	seep	dolerite	NA	Powers 1932
Semil		bitumen	seep	dolerite	NA	Powers 1932
Faeroe Islands						
Lopra	Faeroe-Shetland	oil, gas	show	basalt		Waagstein 1988; Laier *et al.* 1997
France						
Near Clermont-Ferrand	Limagne Graben	oil	show	peperite, basalt		Barrabe 1932
Georgia						
Manavi		oil		laumontite tuff		Grynberg *et al.* 1993
Ninotsminda		oil		laumontite tuff		Grynberg *et al.* 1993
Rustavi		oil		laumontite tuff		Grynberg *et al.* 1993
Samgori		oil	>165 MM bbl (as of 1993) IP up to 3000 BOPD	laumontite tuff		Vernik 1990; Grynberg *et al.* 1993; Patton 1993
Teleti		oil	>3 MM bbl (as of 1990)	laumontite tuff		Vernik 1990; Grynberg *et al.* 1993; Patton 1993

Germany						
Kaiserstuhl	Baden	bitumen	seep	phonolite		Powers 1932
Oberstein	Palatinate	bitumen	seep	metaphyre		Powers 1932
Stahlberg		bitumen	show	rhyolite?	faulted anticline	Peabody 1993
Werra district		oil	seep	basalt		Powers 1932
Greece						
Faredjik	Thrace	oil	seep	andesite	NA	Powers 1932
Hagiostrati Island	Aegean	oil	seep	'volcanic'	NA	Powers 1932
Santorin Volcano	Aegean	H_2	seep		NA	Baker 1928
Hungary						
Nagy Balony		bitumen	seep	andesite	NA	Powers 1932
Parad		bitumen	seep	andesite	NA	Powers 1932
Pasztori	Little Plain	oil	shows	agglomerate	buried volcano	Mattick *et al.* 1996
Recsk		bitumen	seep	rhyolitic tuff	NA	Powers 1932
Iceland						
Skyndidalur, Lon, SE Iceland		oil	seep	basalt	NA	Jakobbsson & Fridleifsson 1990, *cited in* Parnell *et al.* 1992
Italy						
Etna		oil	seep	dolerite, basalt	NA	Powers 1932
Francavilla	Etna	oil	seep	basalt	NA	Reeves 1953
Iblei Mountains	SE Sicily	asphalt	seep	basalt, tuff	NA	Powers 1932
Monte Amiata		bitumen, gas	show	andesite?	NA	Peabody 1993
Pachino	Sicily	oil	seep	dolerite	NA	Reeves 1953
Palogonia	Sicily	oil	seep	tuff	NA	Reeves 1953
Paterno	Etna	oil	seep	basalt	NA	Reeves 1953
Rea 1 dir	Po	oil	show	volcanics		IHS Energy Group 2002
Norway						
Dypvika (Arendal)	Oslo Graben	oil, bitumen	seeps	dolerite, associated veins	NA	Dons 1956; Evans *et al.* 1964; Gize 1986
Portugal						
Near Saccario, N of Cintra, NW of Lisbon		oil	seep	basalt	NA	Powers 1932
Russia						
(unspecified locality)	Anadyr	oil	show	tuff	anticline	Burlin *et al.* 1975
Cape Parthenit	Crimea	bitumen	seep	keratophyre	NA	Powers 1932
Khibiny pluton	Kola Peninsula	oil, bitumen	show	pegmatite	NA	Zezin & Sokolova 1967
Kolyma headwaters area	E Siberia	bitumen	seeps	sandstone, flows	NA	Meyerhoff & Meyer 1987

Table 1. *Continued*

Field or show	Basin	Type	Size	Reservoir rock	Trap	Reference
Russia						
Novoyelkhovskaya	Tatarstan	gas?	shows	'crystalline basement'		A. A. Kitchka, pers. comm. 1998, *cited in* Batchelor 2000
Pasmurnyi, Tat'ianka and Zheldon diatremes	Vilyui	asphalt	show	kimberlite agglomerate	volcanic pipes	Beskrovnyi 1958
Plammenoye	Kolyma region	bitumen	show	rhyolite	fault	Peabody 1993
Tamvatnei	Kamchatka (Koryak region)	biutmen	show	altered serpentinite	NA	Peabody 1993
(unspecified locality)	Tunguska	bitumen (mesophase)	show	sediment adjacent to dolerite	NA	Bogdanova *et al.* 1977
Ural-1	Southern Urals	bitumen	show	basalt, serpentine	fault	Peabody 1993
Uzon Caldera	Kamchatka	oil, bitumen	seep	mafic to felsic volcanics	NA	Beskrovnyy & Lebedev 1971; Meyerhoff & Meyer 1987; Dmitriyevskiy *et al.* 1993; Peabody 1993
Sweden						
Hunneberg	near Wenersberg	oil	seep	dolerite	NA	Powers 1932
Nyhamn	NW of Helsingborg	oil	seep	dolerite	NA	Powers 1932
Ukraine						
Butovo		bitumen	show	andesite	fault	Peabody 1993
Kamennyi Kar'er		bitumen	show	basalt	fault	Peabody 1993
United Kingdom						
England						
Cornwall		oil	shows	mineralized zones in granite		Parnell 1988
Hartequin	E Midlands	oil	show (100s of gallons of free oil)	dolerite dyke	NA	Kent 1954; Sylvester-Bradley & King 1963
Mountsorrel	E Midlands (Leicestershire, Derbyshire)	bitumen	seep	dolerite	NA	Ponnamperuma & Pering 1966; King & Ford 1968; Parnell 1988
Windy Knoll	E Midland (Castleton district, Derbyshire)	oil, bitumen	show	hydrothermal veins	NA	Mueller 1964; Khavari-Khorosani & Murchison 1978
Northern Ireland						
Giants Causeway		bitumen	seep	basalt	NA	Powers 1932; Parnell *et al.* 1992

Scotland						
Alva	Midland Valley	bitumen	seep	Late Devonian lava	NA	Robinson *et al.* 1986
Mid-Calder	Midland Valley	bitumen	seep	dolerite dyke in oil shale		Murchison & Raymond 1989
(multiple localities)	Midland Valley	oil, gas, bitumen	seeps	dolerite dykes, sills and plugs, and equivalent extrusives	NA	Powers 1932; Parnell 1984
Inver Tote, Elgol	Minch	bitumen	seep	sandstone, basalt	NA	Parnell 1983
Wales						
Hollybush	Llanelwedd, Welsh Borderlands	oil, bitumen	seep	Builth volcanics (lavas, pyroclastics) and dolerite dykes	NA	Sylvester-Bradley & King 1963; Parnell 1983
Algeria						
Mereksene	Illizi	oil, gas	280 MMBOE	sandstone	anticline over laccolith	MacGregor 1998
Stah	Illizi	oil, gas	1200 MMBOE	sandstone	anticline over laccolith	MacGregor 1998
(unspecified locality)		oil?	seep	plateau basalt	NA	Batchelor 2000
Republic of the Congo						
Cape Benza	North Tanganyika Trough	gas	seep	fractured schist	NA	Tiercelin *et al.* 1989, 1993
Cape Kalamba	North Tanganyika Trough	bitumen, gas	seep	fractured schist, lake sediments	NA	Tiercelin *et al.* 1989, 1993
Lake Kivu		gas	estimated 1.76 TCF dissolved in lake		NA	Gerlach 1980
Pemba	North Tanganyika Trough	gas	seep	fractured schist	NA	Tiercelin *et al.* 1989, 1993
Egypt						
(unspecified locality)	Sinai	oil	seep	'dykes'	NA	Powers 1932
Northern Red Sea						
	Kobrit, Shaban deeps	oil	seeps	shaly carbonates, sulphides		Michaelis *et al.* 1990
Kenya						
Suswa Volcano	south-central Kenya	bitumen	seep	fractured trachyte	NA	Gize & Macdonald 1993
Madagascar						
(unspecified localities)	western	oil	seeps	sediments near 'dykes'	NA	Powers 1932

Table 1. *Continued*

Field or show	Basin	Type	Size	Reservoir rock	Trap	Reference
South Africa						
(unspecified localities)	Karoo	oil	shows	dolerite, lavas, adjacent sediments	NA	Powers 1932; Haughton *et al.* 1953; Rilett 1956
(unspecified localities)	Karoo	gas	seeps, shows	sediments near intrusives	NA	Powers 1932; Petroleum Agency SA 2000
Beerkliist		asphalt	show	kimberlite	volcanic pipe	Beskrovnyi 1958
Bultfontein		gas	show	kimberlite	volcanic pipe	Beskrovnyi 1958
Kimberley		gas	show	kimberlite	volcanic pipe	Beskrovnyi 1958
China						
Abei	Erlian	oil		andesite, basalt	fault block	Dou 1997
Bachu Arch	Tarim	oil	5000 BOPD	sediments near dolerite	anticline	Zhou *et al.* 1984; Meyerhoff & Meyer 1987
Note: Hydrocarbons matured by doleritic intrusions; Qunkuqiake oil field may be example						
Binnan	Bohai	oil		basalt		Zhou 1988; Liu *et al.* 1989
Bintian	Bohai?	oil		basaltic volcanics?		Liu *et al.* 1989
Caoqiao/Caojiawu	Bohai	oil		basaltic volcanics		Liu *et al.* 1989; Hu *et al.* 1999
	Erlian	asphalt		tuff, sandstone		Meyerhoff & Meyer 1987
(several unnamed fields)	Erlian	oil	1 to 30 MM bbl	andesites, basalts and tuffs		Du *et al.* 1984; Yu 1989, *cited in* Traynor & Sladen 1995
Fenghuadian	Bohai	oil		andesite	horst	Zhang 1992
(unnamed field, NW Junggar)	Junggar	gas	produces up to 1×10^5 m^3/d	volcanics		Hu *et al.* 1999
Linpan	Bohai	oil	128.82×10^6 tons	basaltic volcanics, deltaic sediments	buried volcano	Liu *et al.* 1989; Qiang & McCabe 1998
(unnamed fields)	North Jiangsu	oil, gas		fractured basalt flows	enclosed in lacustrine source rock	Zhang *et al.* 1989
Qijia	Bohai	oil		fractured andesite, tuff		Lee 1989
Rehetai	Bohai	oil	612 BOPD from volcanics	andesite?	faulted anticline	Chen *et al.* 1999
Shanghe	Bohai	oil		basaltic volcanics	buried volcano	Liu *et al.* 1989
Shijutuo	Bohai	oil, gas	up to 2000 BOPD from basalts and andesite	fractured limestone, basalt, andesite		Lee 1989
Xinglongtai	Bohai	oil	760 BOPD	andesite, basalt, basement granite	fault, drape over buried hill	P'an 1982; Chen *et al.* 1999

Yibei	Bohai	oil		lamprophyre, sandstone		Zhou 1987
Yuhungmiao	Bohai	oil		basaltic volcanics	buried volcano	Liu *et al.* 1989
India						
Bokaro Coalfield		oil	show	sandstone near dolerite, lamprophyre intrusions	NA	B. Biswas *in* Mueller 1964
Bombay Island		bitumen	seep	dolerite	NA	Fox 1922, *in* Meyer 1987
(unspecified localities)	Cambay	oil	shows	basalt		Nurmi *et al.* 1991
Japan						
Fujikawa	Niigata	gas		andesite	anticline	Magara 1968
Fukubezawa	Akita	oil, gas	2.6 MM bbl; 1.8 BCF (cumulative 1964–1978); 35° API	dolomite (altered tuff)		Aoyagi 1985
Higashi-Sanjo	Niigata	gas		dacite, tuff	anticline	Kujiraoka 1967; Magara 1968
Honjoji	Niigata	gas		tuff		Kujiraoka 1967
Honjoji	Niigata	gas	1.8 BCF cumulative; 350 MCFGPD (1 well)	fractured andesite	anticline	Kujiraoka 1967; Magara 1968
Kumoide	Niigata	gas		andesite agglomerate	anticline	Kujiraoka 1967; Magara 1968
Kurokawa	Akita	oil		andesite		Powers 1932; Katahira & Ukai 1976
Kurosaka	Niigata	oil		tuff		Kujiraoka 1967
Jurakuji	Niigata	gas		rhyolite		Sakata *et al.* 1994
Minami Nagaoka- Katakai	Niigata	gas	10^6 m^3/d	altered rhyolite, dacite	horst/buried volcano	Magara 1968; Katahira & Ukai 1976; Komatsu *et al.* 1984; Sato *et al.* 1992; Uchida 1992
Mitsuke	Niigata	oil		dacite, tuff	horst/buried volcano	Powers 1932; Kujiraoka 1967; Magara 1968; Katahira & Ukai 1976; Komatsu *et al.* 1984
Myohoji	Niigata	gas		andesite		Katahira & Ukai 1976
Nakadori	Niigata	gas		rhyolite		Sakata *et al.* 1994
Nishi-Nagaoka	Niigata	oil, gas		andesite	anticline	Kujiraoka 1967; Magara 1968
Sarukawa	Akita	gas		volcanics		Sakata *et al.* 1994
Sekihara	Niigata	gas		pyroclastics	anticline	Magara 1968
Shiunji	Niigata	gas		tuff breccia		Sakata *et al.* 1994

Table 1. *Continued*

Field or show	Basin	Type	Size	Reservoir rock	Trap	Reference
Teradomari Oki 1A-1	Niigata	oil			drape over volcanic mound	Suzuki 1983
Yoshii-Higashi Kashiwazaki	Niigata	gas	529 BCF EUR; best wells 17.5 MMCFGPD	rhyolite	horst/buried volcano	Katahira & Ukai 1976; Komatsu *et al.* 1984
Yukihara	Akita	oil, gas		altered basalt		Hoshi & Liou 1988; Mitsuhata *et al.* 1999
Kazakhstan						
Oymasha		oil		granite	contraction void	Popkov *et al.* 1986, *cited in* Dimitriyevskiy *et al.* 1993
(unspecified localities)	Turgay Depression (W Siberian)	bitumen		limestone, tuff, flows		Meyerhoff & Meyer 1987
Mongolia						
(unspecified localities)	Choybalsan	oil, asphalt		basalt		Meyerhoff & Meyer 1987
Tsagaan-Els	East Govi	oil	7 to 15 MM bbl EUR; best wells up to 250 BOPD	sandstone near basalt intrusions	anticline	Meyerhoff & Meyer 1987; Penttila 1992; Patton 1995
Zuunbayan	East Govi	oil	45.6 MM bbl OIIP; 12.7 MM bbl EUR	sandstone near basalt intrusions	anticline	Meyerhoff & Meyer 1987; Penttila 1992; Patton 1995
Syria						
Khaldieh Volcano	southern	bitumen	seep	carbonatite lava	NA	Mahfoud & Beck 1991
Thailand						
Wichian Buri	Phetchabun	oil, gas	30 MM bbl	dolerite, sandstone	laccolith, anticline over laccolith	Remus *et al.* 1993; Williams *et al.* 1995; Anon. 2002a, b
Tibet						
(unspecified localities)	Lunpola	oil		tuff, sandstone		Meyerhoff & Meyer 1987
Turkey						
Bin Gol Mountains	30 mi from Katranly on Mediterranean	oil	seep	sediments near basaltic dyke	NA	Powers 1932
Chirali	40 mi south of Anatalia	gas	seep	serpentine-sediment contact	NA	Powers 1932
Karacaoglan	Thrace	oil, gas	typical wells 3 MMCFGPD and 40 BCPD	fractured rhyolitic tuff	transfer zones	Ozkanli & Kumsal 1993; Coskun 1997
Katranly	25 mi south of Erzurum	oil	seep	basalt-sediment contact	NA	Powers 1932

Vietnam						
Bach Ho (White Tiger)	Cuu Long (Mekong)	oil		altered granite	fault block	Dmitriyevskiy *et al.* 1993; Tran *et al.* 1994
15-2-RD 1X	Cuu Long (Mekong)	oil	10,000 BOPD from igneous reservoir	altered granite	fault block	Koen 1995
Australia						
Northern Territory						
Friendship 1	McArthur	oil, bitumen	show	dolerite		George & Jardine 1994
Moroak	McArthur	bitumen	seep	basalt?	NA	Wade 1926 *cited in* Jackson *et al.* 1988
Waggon Lagoon	McArthur	bitumen	seep	basalt	NA	Powers 1932
(unnamed locality)	Ord	bitumen	seep	basalt	NA	Bradshaw *et al.* 1999
(unnamed locality)	Victoria River, Birrindudu	bitumen	seep	basalt	NA	Jones 1976, *in* Bradshaw *et al.* 1994, p.101
Queensland						
Scotia	Bowen-Surat	gas	Scotia 1A = 6.3 MMCFGPD	fractured andesite	anticline	O'Sullivan 1992
Taylor 1	Bowen-Surat	oil, gas	minor gas, 49° API oil	fractured volcanics		Lindner 1987
South Australia						
Caroline 1	Otway	oil, CO_2		sandstone near maar volcano	fault trap	Mulready 1977; Chivas *et al.* 1987; McKirdy & Chivas 1992
Victoria						
Bream	Gippsland	oil		sandstone	anticlines over dolerite laccoliths	McKerron *et al.* 1998
Western Australia						
Ashmore Reef 1	Bonaparte	oil	shows	Upper Jurassic volcanics		Martinson *et al.* 1973
Ord River basin, 150 mi south of Wyndham	Ord	oil	seep	basalt	NA	Wade 1926; Powers 1932; Meyer 1987
Sulphur Springs	Pilbara	oil, bitumen	shows	massive sulphides in felsic to intermediate volcanics	NA	Rasmussen & Buick 2000
Indonesia						
Java						
Jatibarang	NW Java	oil, gas	1.2 MMM bbl, 2.7 TCF cumulative; discovery IP = 2200 BOPD; peak field production of 33,000 BOPD; 30° API	fractured basalt, andesitic tuff, tuff breccia, agglomerate	faulted anticline	Sutan Assin & Tarunadjaja 1972; Sembodo 1973; Nutt & Sirait 1985; Kartanegara *et al.* 1996
Bantam	NW Java	oil		tuff		Powers 1932

Table 1. *Hydrocarbons associated with igneous rocks or igneous activity*

Field or show	Basin	Type	Size	Reservoir rock	Trap	Reference
Indonesia						
Sumatra						
Palembang	southern Sumatra	oil		tuff		Powers 1932
New Caledonia						
Near Koumac, NW end		oil	seep	sediments near peridotite	NA	Powers 1932
Near Koumac		oil	shows	fractured peridotite		Kaufmann 1955
New Zealand						
Bay of Plenty	Rotorua-Taupo geothermal region	gas, oil	seeps	hydrothermal sulphides	NA	Stoffers *et al.* 2000
Kora	Taranaki	oil	1170 BOPD, 32° API; GOR = 430 SCF/bbl	andesite tuffs, volcaniclastics	buried volcano	Bergman *et al.* 1992; Batchelor 2000; Hart 2001
Moturoa (Taranaki)	Taranaki	oil	216,000 bbl ultimately recovered	sandstone	flanks of andesite intrusion	Clapp 1929; McBeath 1977; Pilaar & Wakefield 1984; Abbott 1990
Waiotapu	Rotorua-Taupo geothermal region	oil	seep	tuff	NA	Czochanska *et al.* 1986
Philippines						
Zambales	western Luzon	gas	seep	ophiolite	NA	Abrajano *et al.* 1988
Antarctica						
(unnamed locality)	Bransfield	oil	show	turbidites	NA	Whiticar *et al.* 1985; Brault & Simoniet 1990

Table 2. *Hydrocarbons beneath igneous rocks*

Field	Basin	Type	Size	Overlying Igneous Rocks	Exploration methods	Reference
United States						
Arizona						
Alpine	Holbrook	oil	show	basalt	geothermal test	Heylmun 1997, p. 131; Rauzi 2001
Colorado						
Kirby 1 Jynnifer	San Luis Valley	oil, gas	25 BOPD	volcanics		Ligrani *et al.* 1985
Kirby 1 LMG	San Luis Valley	oil	show	volcanics		Ligrani *et al.* 1985
Milestone 1 AMF	San Luis Valley	oil	show	volcanics		Ligrani *et al.* 1985
Idaho						
(unnamed localities)	Snake River Downwarp	gas	shows	Snake River basalts		Felts 1954
Oregon						
(unnamed well near Clarno)	Columbia	gas	4 MMCFGPD	flood basalt		L. H. Fisk (pers. comm.) *cited in* Schmoker *et al.* 1996
Texas						
Marcelina Creek		oil	1.6 MM bbl	altered basaltic volcanics	seismic	Hutchinson 1994
Utah						
Iron Springs	SW Utah	oil	show	granitic laccolith	seismic	Van Kooten 1987
Washington						
Shell BISSA 1-29	Columbia	gas	show	flood basalt	surface mapping, MT	Withers *et al.* 1994
Shell BN 1-9	Columbia	gas	2 MMCFGPD	flood basalt	surface mapping, MT	Haug & Bilodeau 1985; Lingley & Walsh 1986; Withers *et al.* 1994
Shell Yakima Minerals 1-33	Columbia	gas	show	flood basalt	surface mapping, MT	Dignes & Woltz 1982; Withers *et al.* 1994
Wyoming						
Aspen Creek	Bighorn	oil		andesitic volcanics		Schmoker *et al.* 1996
Baird Peak	Bighorn	oil		andesitic volcanics		Schmoker *et al.* 1996
Dickie	Bighorn	oil		andesitic volcanics		Schmoker *et al.* 1996
Prospect Creek	Bighorn	oil		andesitic volcanics		Schmoker *et al.* 1996
Prospect Creek South	Bighorn	oil		andesitic volcanics		Schmoker *et al.* 1996
Skelton Dome	Bighorn	oil		andesitic volcanics		Schmoker *et al.* 1996
Brazil						
Autas Mirim	Amazonas	oil	noncommercial	dolerite sill		Clark 1960
Barra Bonita	Parana	gas	7.06 MMCFGPD	flood basalt; dolerite sill		Anon. 1998; Figueiredo & Milani 2000
Chapeu do Sol	Parana	gas	subcommercial	flood basalt		Milani *et al.* 1990
Cuiaba Paulista	Parana	gas	3 MMCFGPD	flood basalt; dolerite laccolith	surface mapping, seismic	Yoshida & Gama 1982

Table 2. *Continued*

Field	Basin	Type	Size	Overlying Igneous Rocks	Exploration methods	Reference
Brazil						
Herval Velho	Parana	gas	subcommercial	flood basalt		Milani *et al.* 1990
Igarape Cuia	Amazonas	oil	25,000 bbl 42° API; 500 BOPD	dolerite sill		Neves 1990
Jandiatuba (1-JD-1-AM)	Solimoes	oil, gas	180,000 m^3/d and 47° API condensate	dolerite sill		Brazil 1990
Jandiatuba area (6 fields)	Solimoes	gas	20×10^9 m^3 gas in place	dolerite sill		Brazil 1990
Jurua	Solimoes	gas	282 BCF	dolerite sill		Anon. 1990; Kingston & Matzko 1995
Jurua area (10 fields)	Solimoes	gas	up to 20×10^9 m^3 gas in place	dolerite sill		Brazil 1990
Matos Costa	Parana	oil	subcommercial	flood basalt		Milani *et al.* 1990
Nova Olinda	Amazonas	oil	600 BOPD; 40°–50° API	dolerite sill		Anon. 1990; Kingston & Matzko 1995
Sao Mateus	Solimoes	oil, gas		dolerite sill	3D seismic	Marchi *et al.* 1999
Tres Pinheiros	Parana	oil	subcommercial	flood basalt		Milani *et al.* 1990
Urucu area	Solimoes	oil, gas	123 MM bbl in place; 44°–60° API; 33×10^9 gas in place	dolerite sill		Brazil 1990; Castro & da Silva 1990
Paraguay						
Mallorquin T-1	Parana	oil, gas	shows	flood basalt		Petzet 1997
Italy						
Ragusa	Ibleo	oil	160 MM bbl EUR	gabbro sill	surface mapping; photo geology; magnetic; gravity; seismic	Hedberg 1964; Vercellino & Rigo 1970
Russia						
Aran	Tunguska	oil, gas		dolerite sills		Benelmouloud & Zhuravlev 1989
Omorin	Tunguska	oil, gas		dolerite sills		Benelmouloud & Zhuravlev 1989
Prebrazhenko	Markovo-Angara Arch	gas				Bazanov 1973
Sobin	Tunguska	oil, gas		dolerite sills		Benelmouloud & Zhuravlev 1989
Yaraktin	Markovo-Angara Arch	oil	210 MM bbl recoverable	1500–1750 m basalt; dolerite	projected trend mapping	Bazanov 1973; Meyerhoff 1980; Benelmouloud & Zhuravlev 1989
Yurubchen-Tokhomo	Tunguska	oil, gas		dolerite sills		Kontorovich *et al.* 1990

Algeria						
Ben Khalala	Triassic/Oued Mya	oil	>250 MM bbl EUR	basalt		Benelmouloud & Zhuravlev 1989; Boote *et al.* 1998
Haoud Berkaoui	Triassic/Oued Mya	oil	>250 MM bbl EUR; 9.5 × 19 km	basalt		Benelmouloud & Zhuravlev 1989; Boote *et al.* 1998
Oulougga	Triassic/Oued Mya	oil		basalt		
(22 unnamed)	Triassic/Oued Mya	oil, gas		basalt		
Ghana						
Saltpond = (Bonsu)	Tanu	oil	5.9 MM bbl (through 1984) (not all horizons below sill); up to 2500 BOPD from subsill sands	dolerite sill		Meyerhoff & Meyerhoff 1974, p. 112; Clifford 1986; Kesse 1986; Patton 1995
Namibia						
Kudu	Orange	gas	3 TCF EUR	basalt	seismic	Abreu *et al.* 1997; Gladczenko *et al.* 1997; Bray *et al.* 1998; Stanistreet & Stollhofen 1999; Stollhofen *et al.* 2000
India						
Bantumilli	Krishna-Godavari	gas		flood basalt		Rao 2001
Chintalapalli	Krishna-Godavari	gas		flood basalt		Rao 2001
Kaikaluru	Krishna-Godavari	oil		flood basalt		Rao 2001
Mandapeta	Krishna-Godavari	gas		flood basalt		Rao 2001
Narsapur	Krishna-Godavari	gas		flood basalt		Mukherjee 1983
Razole	Krishna-Godavari	gas		flood basalt		Mukherjee 1983; Samanta & Shukla 1987
Syria						
Golan Heights		oil	show	basalt; (Harrat Ash-Shamah) volcanic field		May & Shulman 1989
Turkey						
Beykan	SE Turkey	oil	45 MM bbl (through 7/1/78)	flood basalt	trend?	Ala & Moss 1979
Kurkan	SE Turkey	oil	37 MM bbl (through 7/1/78)	flood basalt	trend?	Ala & Moss 1979

Table 2. *Continued*

Field	Basin	Type	Size	Overlying Igneous Rocks	Exploration methods	Reference
Australia						
Northern Australia						
Jamison 1	McArthur	oil, gas	shows	Cambrian flood basalts		Clementson 1994
Victoria						
Kipper	Gippsland	oil, gas	30 MM bbl, 750 BCF	altered basalt flows	seismic	Sloan *et al.* 1992
Western Australia						
Scott Reef	Browse	gas	13.7 TCF and 131 MM bbl condensate	U Jurassic flood basalt		Kamen-Kaye & Meyerhoff 1979; Forrest & Horstman 1986; Symonds *et al.* 1998
New Zealand						
Kapuni	Taranaki	gas	34 MM bbl; 630 BCF	andesitic volcanics (Mt. Egmont)	seismic	McBeath,1977

I would like to thank Claire Ivison who created the map.

References

ABBOTT, W. O. 1990. Maui Field. *In*: BEAUMONT, E. A. & FOSTER, N. H. (compilers) *Structural traps I—Tectonic fold traps*. American Association of Petroleum Geologists, Treatise of Petroleum Geology, Atlas of Oil and Gas Fields, 1–25.

ABRAJANO, T. A., STURCHIO, N. C., BOHLKE, J. K., LYON, G. L., POREDA, R. J. & STEVENS, C. M. 1988. Methane-hydrogen gas seeps, Zambales ophiolite, Philippines: deep or shallow origin? *Chemical Geology*, **71**, 211–222.

ABREU, V., VAIL, P. R., BALLY, A. & WILSON, E. 1997. Geologic evolution of conjugate volcanic passive margins: influence on the petroleum systems of the South Atlantic. *Houston Geological Society, Bulletin*, **40**, December (n. 4), 10–11.

AGÊNCIA NACIONAL DO PETROLEO 2002. Brazil Round 4, licensing round pamphlet.

ALA, M. A. & MOSS, B. J. 1979. Comparative petroleum geology of southeast Turkey and northeast Syria. *Journal of Petroleum Geology*, **1**, n. 4, 3–27.

ALI, O. 1973. Stratigraphy of Lower Triassic sandstone of northeast Algerian Sahara, Algeria. *American Association of Petroleum Geologists, Bulletin*, **57**, 528–540.

ANON. 1959. World wide oil report—Cuba. *Oil & Gas Journal*, **57**, 28 December (n. 53), 96–97.

ANON. 1990. Petrobras draws bead on Amazon region for more oil and gas. *Oil & Gas Journal*, **88**, 12 February (n. 7), 17–20.

ANON. 1993. International news is mainly good. *AAPG Explorer*, **14**, October (n. 10), 22–27.

ANON. 1998. Brazil. *AAPG Explorer*, **19**, January (n. 1), 10.

ANON. 2002a. Pacific Tiger roars. *Upstream*, **7**, 15 February (n. 7), 28.

ANON. 2002b. Wichian's riches look set to treble. *Upstream*, **7**, 28 March (n. 13), 23.

AOYOGI, K. 1985. Origin of the Miocene carbonate reservoir rocks, Fukubezawa oil field, Akita Province, northeast Honshu, Japan. *In*: ROEHL, P. O. & CHOQUETTE, P. W. (eds) *Carbonate petroleum reservoirs*. Springer-Verlag, New York, 561–569.

ARMENTROUT, J. M. & SUCK, D. H. 1985. Hydrocarbon exploration in western Oregon and Washington.

BAILEY, E. H. 1959. Froth veins, formed by immiscible hydrothermal fluids, in mercury deposits, California. *Geological Society of America, Bulletin*, **70**, 661–664.

BAILEY, M. H. & SUNDELL, K. A. 1986a. Preliminary results of wildcat drilling in Absaroka volcanic rocks, Hot Springs County, Wyoming. *American Association of Petroleum Geologists, Bulletin*, **70**, 1030–1031.

BAILEY, M. H. & SUNDELL, K. A. 1986b. Preliminary results of wildcat drilling in Absaroka volcanic rocks, Hot Springs County, Wyoming. *The Outcrop*, **35**, December (n. 12), 3–4.

BAKER, C. L. 1928. Panuco oil field, Mexico. *American Association of Petroleum Geologists, Bulletin*, **12**, 395–443.

BARRABÉ, L. 1932. Oil in Limagne area, France. *American Association of Petroleum Geologists, Bulletin*, **16**, 825–832.

BATCHELOR, A. S. (compiler) 2000. *Hydrocarbon production from fractured basement formations*. Geoscience Ltd., Falmouth, Cornwall, online publication, Version 6, November, 2000, unpaginated.

BAZANOV, E. A. 1973. [Geology of the Yaraktin oil field of the Irkutsk region]. *Geologiya Nefti i Gaza*, **11**, n. 7, 15–18 (English translation, *Petroleum Geology*, **11**, 282–283).

BEEBE, B. W. 1968. Natural gas in post-Paleozoic rocks of Mississippi. *In*: BEEBE, B. W. & CURTIS, B. F. (eds) *Natural gases of North America*. American Association of Petroleum Geologists, Memoirs, **9**, **1**, 1176–1226.

BELOTTI, H. J., REBORI, L. O. & IBAÑEZ, G. H. 1995. Igneous reservoirs in triangle zones, Neuquén Basin, Argentina. *American Association of Petroleum Geologists, Annual Convention, Meeting Abstracts*, **4**, 8A.

BENELMOULOUD, M. & ZHURAVLEV, E. 1989. Problems of petroleum exploration under plateau basalts. *In*: PRICE, R. A. (ed.) *Origin and evolution of sedimentary basins and their energy and mineral resources*. American Geophysical Union, Monographs, **4**, 197–202.

BERGMAN, S. C., TALBOT, J. P. & THOMPSON, P. R. 1992. The Kora Miocene submarine andesite stratovolcano hydrocarbon reservoir, northern Taranaki Basin, New Zealand. *New Zealand Oil Exploration Conference, Proceedings*. Ministry of Commerce, Wellington, New Zealand, 178–206.

BESKROVNI, N. S. 1958. [Indications of petroleum in volcanic pipes of the Siberian Platform]. *Akademi Nauk SSSR, Doklady*, **122**, n. 1, 119–122 (English translation, 1958, 733–736).

BESKROVNYY, N. S. & LEBEDEV, B. A. 1971. [Oil shows in the caldera of Uzon Volcano, Kamchatka]. *Akademi Nauk SSSR, Doklady*, **201**, n. 4, 953–956 (English translation, 1971, 266–268).

BIEDERMAN, E. W., JR. 1986. *Atlas of selected oil and gas reservoir rocks from North America*. John Wiley and Sons, New York, 1–399.

BOGDANOVA, L. A., KASPERKEVICH, E. P. & LETUSHOVA, I. A. 1977. A highly metamorphosed bitumen in the Cambrian deposits of the Siberian Platform. *Lithology and Mineral Resources*, **1977**, 334–340.

BOOTE, D. R. D., CLARK-LOWES, D. D. & TRAUT, M. W. 1988. Palaeozoic petroleum systems of North Africa. *In*: MACGREGOR, D. S., MOODY, R. J. T. & CLARK-LOWES, D. D. (eds) *Petroleum geology of North Africa*. Geological Society, London, Special Publications, **132**, 7–68.

BORTZ, L. C. 1983. Hydrocarbons in the northern Basin and Range province, Nevada. *In*: *The role of heat in development of energy and mineral resources in the northern Basin and Range province*. Geothermal Resources Council, Special Reports, **13**, 179–197.

BORTZ, L. C. 1987. Heavy-oil deposit, Great Salt Lake, Utah. *In*: MEYER, R. F. (ed.) *Exploration for heavy crude and natural bitumen*. American Association of Petroleum Geologists, Studies in Geology, **25**, 555–563.

BORTZ, L. C. 1994. Petroleum geology of the Eagle Springs oil field, Nye County, Nevada. *In*: SCHALLA, R. A. & JOHNSON, E. H. (eds) *Oil fields of the Great Basin*. Nevada Petroleum Society, Reno, 285–294.

BRADSHAW, M. T., BRADSHAW, J., *ET AL*. 1994. Petroleum systems in West Australian basins. *In*: PURCELL, P. G. & PURCELL, R. R. (eds) *The sedimentary basins of Western Australia, Symposium, Proceedings*. Petroleum Society of Australia, Perth, 93–118.

BRADSHAW, M. T., FOSTER, C. B., FELLOWS, M. E. & ROWLAND, D. C. 1999. Australia's search for petroleum: patterns of discovery. *Oil & Gas Journal*, **97**, 7 June (n. 23), 37–44, 14 June (n. 24), 111–116.

BRAULT, M. & SIMONEIT, B. R. T. 1990. Mild hydrothermal alteration of immature organic matter in sediments from the Bransfield Strait, Antarctica. *Applied Geochemistry*, **5**, 149–158.

BRAUNSTEIN, J. & MCMICHAEL, C. E. 1976. Door Point: a buried volcano in southeast Louisiana. *Gulf Coast Association of Geological Societies, Transactions*, **26**, 79–80.

BRAY, R., LAWRENCE, S. & SWART, R. 1998. Source rock, maturity data indicate potential off Namibia. *Oil & Gas Journal*, **96**, 10 August (n. 32), 84–89.

BRAZIL, I. R. 1990. O potentcial petrolifero e as perspectivas exploratórias do bacia do Solimões—norte do Brasil. *Boletim Geociências Petrobrás*, **4**, n. 1, 85–93.

BURLIN, J. K., ARKHIPOV, A. J., VASSOEVICH, N. B., TROFIMUK, A. A. & OVCHARENKO, A. V. 1975. Sedimentary basins of the Far East and the northeast of the USSR and their possible oil and gas content. *World Petroleum Congress, 9th, Proceedings*, **3**, 39–50.

BURYAKOVSKY, L. A. 1993. Outline of general and petroleum geology in Azerbaijan and the South Caspian Basin. *Houston Geological Society, Bulletin*, **35**, February (n. 5), 16–17, 19–23, 43–47.

CASTRO, L. O. C. & DA SILVA, O. B. 1990. Ocurrência de gás sulfidrico na bacia do Solimões—modelo geológico e implicações. *Boletim Geociências da Petrobrás*, **4**, 285–298.

CHEN, Z., YAN, H., LI, J., ZHANG, G., ZHANG, Z. & LIU, B. 1999. Relationship between Tertiary volcanic rocks and hydrocarbons in the Liaohe Basin, People's Republic of China. *American Association of Petroleum Geologists, Bulletin*, **83**, 1004–1014.

CHIVAS, A. R., BARNES, I., EVANS, W. C., LUPTON, J. E. & STONE, J. D. 1987. Liquid carbon dioxide of magmatic origin and its role in volcanic eruptions. *Nature*, **326**, 587–589.

CHRISTIANSEN, F. G., BOJESEN-KOEFOED, J., DAM, G., NYTOFT, H.-P., LARSEN, L. M., PETERSEN, A. K. & PULVERTAFT, T. C. R. 1996. The Marraat oil discovery on Nuussuaq, West Greenland; evidence for a latest Cretaceous—early Tertiary oil prone source rock in the Labrador Sea—Melville Bay region. *Bulletins of Canadian Petroleum Geology*, **44**, 39–54.

CHRISTIANSEN, F. G., DAM, G. & PEDERSEN, A. K. 1994. Discovery of live oil at Marraat, Nuussuaq: field work, drilling and logging. *Grønlands Geologiske Undersøgelse, Rapport*, **160**, 57–63.

CLAPP, F. G. 1929. Role of geologic structure in the accumulation of petroleum. *In*: *Structure of typical American oil fields*. American Association of Petroleum Geologists, **2**, 667–716.

CLARK, E. W. 1960. Oil and gas developments in South America and Caribbean area in 1959. *American Association of Petroleum Geologists, Bulletin*, **44**, 1014–1057.

CLEMENTSON, I. 1994. Precambrian frontier opportunity in Australia's Beetaloo Sub-basin. *Oil & Gas Journal*, **92**, 27 June (n. 26), 60–62.

CLIFFORD, A. J. 1986. African oil—past, present and future *In*: HALBOUTY, M. J. (ed.) *Future petroleum provinces of the world*. American Association of Petroleum Geologists, Memoirs, **40**, 339–372.

CLIFTON, C. G., WALTERS, C. C. & SIMONEIT, B. R. T. 1990. Hydrothermal petroleums from Yellowstone National Park, Wyoming, U.S.A. *Applied Geochemistry*, **5**, 169–191.

COSKUN, B. 1997. Oil and gas fields—transfer zone relationships, Thrace Basin, NW Turkey. *Marine and Petroleum Geology*, **14**, 401–416.

CREELY, R. S. & SATERDAL, A. O. 1956. Ojo Anticline, Huerfano County, Colorado. *In*: *Guidebook to the geology of the Raton Basin*. Rocky Mountain Association of Geologists, Denver, 68–70.

CZOCHANSKA, Z., SHEPPARD, C. M., WESTON, R. J., WOODHOUSE, A. D. & COOK, R. A. 1986. Organic geochemistry of sediments in New Zealand, Pt. 1. A biomarker study of the petroleum seepage in the geothermal region of Waitapu. *Geochimica et Cosmochimica Acta*, **50**, 507–515.

DANIEL, G. A. & HVALA, O. 1982. Analisis de rocas tobaceas y volcanicas a partir de los perfiles sonicos y de resistividad. *Primer Congreso Nacional de Hidrocarburos, Petroleo y Gas*. Trabajos Tecnicos, Explotacion, **4**, 99–126.

DE GOYLER, E. 1915. The Furbero oil field, Mexico. *American Institute of Mining Engineering, Transactions*, **52**, 268–280.

DE GOYLER, E. 1932. Oil associated with igneous rocks in Mexico. *American Association of Petroleum Geologists, Bulletin*, **16**, 799–808.

DES MARAIS, D. J., DONCHIN, J. H., NEHRING, N. L. & TRUESDELL, A. H. 1981. Molecular carbon isotopic evidence of the origin of geothermal hydrocarbons. *Nature*, **292**, 826–828.

DIDYK, B. & SIMONEIT, B. R. T. 1989. Hydrothermal oil of Guaymas Basin and implications for petroleum formation mechanisms. *Nature*, **342**, 65–69.

DIGNES, T. W. & WOLTZ, D. 1982. Oil and gas developments in West Coast area. *American Association of Petroleum Geologists, Bulletin*, **66**, 1784–1805.

DMITRIYEVSKIY, A. N., KIREYEV, F. A., BOCHKO, R. A. & FEDOROVA, T. A. 1993. Hydrothermal origin of oil and gas reservoirs in basement rock of the

South Vietnam continental shelf. *International Geology Review*, **35**, 621–630.

DONS, J. A. 1956. Coal blend and uraniferous hydrocarbon in Norway. *Norsk Geologisk Tidsskrift*, **36**, 249–266.

DOTT, R. H., SR. & REYNOLDS, M. J. 1969. *Sourcebook for Petroleum Geology*, American Association of Petroleum Geologists, Memoirs, **5**.

DOU, L. 1997. The Lower Cretaceous petroleum system in NE China. *Journal of Petroleum Geology*, **20**, 475–488.

DU, Y., SHENG, Z., CHEN, Y. & XIONG, X. 1984. [Type and characters of reservoir rocks in the Erlian Basin]. *Petroleum Exploration and Development*, **11**, 1–7.

DUEY, H. D. 1983. Oil generation and entrapment in Railroad Valley, Nye County, Nevada. *In*: *The role of heat in development of energy and mineral resources in the northern Basin and Range province*. Geothermal Resources Council, Special Reports, **13**, 199–205.

EASTON, H. D. 1935. Volcanoes of northeastern Louisiana account for accumulation of oil. *Oil & Gas Journal*, **33**, 28 March (n, 45), 28, 30, 43.

ECHEVARRIA-RODRIGUEZ, G., HERNANDEZ-PEREZ, G., ET AL. 1991. Oil and gas exploration in Cuba. *Journal of Petroleum Geology*, **14**, 259–274.

ENGLISH, R. M. & GROGAN, R. M. 1948. Omaha pool and mica-peridotite intrusives, Gallatin County, Illinois. *In*: HOWELL, J. V. (ed.) *Structure of Typical American Oil Fields*. American Association of Petroleum Geologists, Tulsa, **3**, 189–212.

EVANS, W. D., MORTON, R. D. & COOPER, B. S. 1964. Primary investigations on the oleiferous dolerite of Dypvika. *In*: COLOMBO, U. & HOBSON, G. D. (eds) *Advances in organic geochemistry*. Proceedings of the International Meeting, Milan, 1962 (Earth Science Series, Monograph No. 15). Pergamon Press, Oxford, 141–144.

FALLIN, J. A. 1989. Hydrogeologic evolution and hydrocarbon potential, southern Rio Grande Rift, Texas. *American Association of Petroleum Geologists, Bulletin*, **73**, 1154–1155.

FELTS, W. M. 1954. Occurrence of oil and gas and its relation to possible source beds in continental Tertiary of Intermountain region. *American Association of Petroleum Geologists, Bulletin*, **38**, 1661–1670.

FIGUEIREDO, A. M. F. & MILANI, E. J. 2000. Petroleum systems of the South American basins. *In*: CORDANI, U. G., MILANI, E. J., THOMAZ FILHO, A. & CAMPOS, D. A. (eds) *Tectonic evolution of South America*. International Geological Congress, 31st, Rio de Janiero, 689–718.

FLANIGAN, T. 1989. Borehole evaluation and completion: carbonate, volcanic, clastic reservoirs of Nevada. *Oil & Gas Journal*, **87**, 23 October (n. 43), 92, 94–96.

FORREST, J. T. & HORSTMAN, E. L. 1986. The Northwest Shelf of Australia—geologic review of a potential major petroleum province of the future. *In*: HALBOUTY, M. L. (ed.) *Future petroleum provinces of the world*. American Association of Petroleum Geologists, Memoirs, **40**, 457–486.

FOX, C. S. 1922. The occurrence of bitumen in Bombay Island. *Geological Survey of India, Records*, **54**, part 1, 117–128.

FRENCH, D. E. 1994. Discovery of Duckwater Creek oil field, Railroad Valley, Nye County, Nevada. *In*: SCHALLA, R. A. & JOHNSON, E. H. (eds) *Oil fields of the Great Basin*. Nevada Petroleum Society, Reno, 271–277.

FRENCH, D. E. & FREEMAN, K. J. 1979a. Tertiary volcanic stratigraphy and reservoir characteristics of Trap Spring Field, Nye County, Nevada. *In*: NEWMAN, G. W. & GOODE, H. D. (eds) *Basin and Range Symposium*. Rocky Mountain Association of Geologists—Utah Geological Association, 487–502.

FRENCH, D. E. & FREEMAN, K. J. 1979b. Volcanics yield another oil field. *World Oil*, **188**, May (n. 6), 58–63.

GARFIAS, V. R. 1923. *Petroleum resources of the world*. John Wiley and Sons, New York, 1–243.

GEORGE, S. C. & JARDINE, D. R. 1994. Ketones in a Proterozoic dolerite sill. *Organic Geochemistry*, **21**, 829–839.

GERLACH, T. M. 1980. Chemical characteristics of the volcanic gases from Nyiragongo lava lake and the generation of CH_4-rich fluid inclusions in alkaline rocks. *Journal of Volcanology and Geothermal Research*, **8**, 177–189.

GIZE, A. P. 1986. The development of a thermal mesophase in bitumens from high temperature ore deposits. *In*: DEAN, W. E. (ed.) *Organics and Ore Deposits*. Denver Region Exploration Geologists, Symposium, Proceedings, 137–150.

GIZE, A. P. & MACDONALD, R. 1993. Generation of compositionally atypical hydrocarbons in CO_2-rich geologic environments. *Geology*, **21**, 129–132.

GLADCZENKO, T. P., HINZ, K., ELDHOLM, O., MEYER, H., NEBEN, S. & SKOGSIED, J. 1997. South Atlantic volcanic margins. *Journal of the Geological Society of London*, **154**, 465–470.

GOODARZI, F., ECKSTRAND, O. R., SNOWDON, L., WILLIAMSON, B. & STASIUK, L. D. 1992. Thermal metamorphism of bitumen in Archean rocks by ultramafic flows. *International Journal of Coal Geology*, **20**, 165–178.

GORDON, D. 1931. Richland gas field, Richland Parish, Louisiana. *American Association of Petroleum Geologists, Bulletin*, **15**, 939–952.

GRABB, R. 1994a. Sand Spring oil field, Railroad Valley, Nye County, Nevada. *In*: SCHALLA, R. A. & JOHNSON, E. H. (eds) *Oil fields of the Great Basin*. Nevada Petroleum Society, Reno, 241–252.

GRABB, R. 1994b. Three Bar oil field, Eureka County, Nevada. *In*: SCHALLA, R. A. & JOHNSON, E. H. (eds) *Oil fields of the Great Basin*. Nevada Petroleum Society, Reno, 333–338.

GRIES, R. R. 1985. San Juan Sag: Cretaceous rocks in a volcanic-covered basin, south-central Colorado. *Mountain Geologist*, **22**, n. 4, 167–179.

GRYNBERG, M. E., PAPAVA, D., SHENGELIA, M., TAKAISHVILI, A., NANADZE, A. & PATTON, D. K. 1993. Petrophysical characteristics of the Middle Eocene laumontite tuff reservoir, Samgori Field, Republic of Georgia. *Journal of Petroleum Geology*, **16**, 312–322.

GUAIGUATA, P., R. C. & RICHARDSON, J. A. 1960. Produccion petrolifera del basamento en el oeste del lago de Maracaibo. *Congreso Geologico Venezolano, III, Memoria*, **3**, 985–1008.

HAMILTON, T. S. & CAMERON, B. E. B. 1989. Hydrocarbon occurrences on the western margin of the Queen Charlotte Basin. *Bulletins of Canadian Petroleum Geology*, **37**, 443–466.

HAMMER, A. A. 1934. Rattlesnake Hills gas field, Benton County, Washington. *American Association of Petroleum Geologists, Bulletin*, **18**, 847–859.

HARRELSON, D. W. & BICKER, A. R., JR. 1979. Petrography of some subsurface igneous rocks of Mississippi. *Gulf Coast Association of Geological Societies, Transactions*, **29**, 244–251.

HART, A. 2001. New Zealand potential. *Oil & Gas Journal*, **99**, 16 July (n. 29), 38–44; 23 July (n. 30), 40–43.

HAUG, G. A. & BILODEAU, B. J. 1985. Oil and gas developments in West Coast in 1984. *American Association of Petroleum Geologists, Bulletin*, **69**, 1567–1575.

HAUGHTON, S. H., BLIGHNAUT, J. J. G., ROSSOUW, P. J., SPIES, J. J. & ZAGT, S. 1953. *Results of an investigation into the possible presence of oil in parts of the Union of South Africa.* Union of South Africa, Department of Mines, Geological Survey, Memoirs, **45**, 1–114.

HAWLEY, A. S. 1968. Wild Goose gas field, Butte County, California. *In*: BEEBE, B. W. & CURTIS, B. F. (eds) *Natural gases of North America.* American Association of Petroleum Geologists, Memoirs, **9**, **1**, 642–645.

HAZLETT, D. P. & HERSCH, J. B. 1993. Petroleum potential of east central Nevada. *Houston Geological Society, Bulletin*, **35**, March (n. 7), 16.

HAZZARD, J. C. 1957. Petroleum seepage from volcanic rocks, Carolina District, Department of San Miguel, El Salvador, Central America. *In*: *Geologia del Petroleo.* International Geological Congress, 20th, Section 3, 59–64.

HEATON, R. L. 1929. Relation of accumulation to structure in northwestern Colorado. *In*: *Structure of typical American oil fields.* American Association of Petroleum Geologists, **2**, 93–114.

HEDBERG, H. D. 1964. Geologic aspects of origin of petroleum. *American Association of Petroleum Geologists, Bulletin*, **48**, 1755–1803.

HEYLMUN, E. B. 1997. Arizona's Holbrook salt basin holds oil, gas opportunities. *Oil & Gas Journal*, **95**, 5 May (n. 18), 127–132.

HODGE, E. T. 1927. Native asphalts in Oregon. *American Association of Petroleum Geologists, Bulletin*, **11**, 395–406.

HOLLISTER, V. F. 1980. Origin of graphite in the Duluth Complex. *Economic Geology*, **75**, 764–766.

HOPKINS, A. A., LAPERLE, G. R., MATHEWS, J. W. & SCHWADE, I. T. 1956. Marysville Buttes: geological and geophysical analysis. *American Association of Petroleum Geologists, Bulletin*, **40**, 431–432.

HOSHI, K. & LIOU, J. G. 1988. Miocene ocean-floor metamorphism during back-arc spreading in the Japan Sea. *EOS*, **69**, n. 44, 1492.

HU, J., DOU, L. & LI, Q. 1999. The geology of natural gas in the People's Republic of China. *Journal of Petroleum Geology*, **22**, 191–213.

HULEN, J. B., BERESKIN, R. S. & BORTZ, L. C. 1990. High-temperature origin of fractured carbonate reservoirs in the Blackburn oil field, Nevada. *American Association of Petroleum Geologists, Bulletin*, **74**, 1262–1272.

HULEN, J. B., COLLISTER, J. A., DAHDAH, N. F., WANNAMAKER, P. E. & QUICK, J. C. 1998. Hydrocarbons in Miocene lamproite dikes of the San Rafael Desert, Utah—implications for fault-controlled oil migration and accumulation in the western Colorado Plateau. *American Association of Petroleum Geologists, Annual Meeting, online extended abstract*, 1–7.

HUNTER, G. W. 1955a. Marysville Buttes gas field. *California Oil Fields*, 41st Annual Report of the State Oil and Gas Supervisor, Department of Natural Resources, Division of Oil and Gas, **41**, n. 1, 11–15.

HUNTER, G. W. 1955b. Wild Goose gas field. *California Oil Fields*, 41st Annual Report of the State Oil and Gas Supervisor, Department of Natural Resources, Division of Oil and Gas, **41**, n. 1, 5–9.

HUTCHINSON, P. J. 1994. Upper Cretaceous (Austin Group) volcanic deposits as a hydrocarbon trap. *Gulf Coast Association of Geological Societies, Transactions*, **44**, 293–303.

IHS ENERGY GROUP 2001. Global Exploration and Production Service (on-line news service), 6 December.

IHS ENERGY GROUP 2002. Weekly Highlights (on-line news service), 7 January.

IRIZARRY, O. B. 1954. Cuban picture hazy. *Oil & Gas Journal*, **53**, 28 June, 95.

JACKSON, M. J., SWEET, I. P. & POWELL, T. C. 1988. Studies on petroleum geology and geochemistry, Middle Proterozoic, McArthur Basin, Northern Australia I. Petroleum potential. *Australian Petroleum Exploration Association, Journal*, **28**, pt. 1, 283–302.

JAKOBBSSON, S. P. & FRIDLEIFSSON, G. D. 1990. Asphaltic petroleum in amygdules in Skyndidalur, Lon, SE Iceland. *Natturufraedingurinn*, **59**, 169–188.

JONES, P. J. 1976. Ord Basin. *In*: LESLIE, R. B., EVANS, H. J. & KNIGHT, C. L. (eds) *Economic Geology of Australia and Papua New Guinea, 3. Petroleum.* The Australian Institute of Mining and Metallurgy, Monograph Series **7**, 242–246.

KAMEN-KAYE, M. & MEYERHOFF, A. A. 1979. Flood basalts: implications for global-tectonic hypotheses and petroleum exploration. *Journal of Petroleum Geology*, **1**, 29–37.

KARTANEGARA, A. L., BAIK, R. N. & IBRAHIM, M. A. 1996. Volcanics oil bearing in Indonesia. *American Association of Petroleum Geologists, 1996 Annual Convention, Meeting Abstracts*, **5**, A73.

KATAHIRA, T. & UKAI, M. 1976. Petroleum fields of Japan with volcanic-rock reservoirs—summary *In*: HALBOUTY, M. T., MAHER, J. C. & LIAN, H. M. (eds) *Circum-Pacific Energy and Mineral Resources.* American Association of Petroleum Geologists, Memoirs, **25**, 276–279.

KAUFMANN, G. F. 1955. Petroleum developments in the Far East in 1954. *American Association of Petroleum Geologists, Bulletin*, **39**, 1385–1402.

KAWKA, O. E. & SIMONIET, B. R. T. 1990. Polycyclic aromatic hydrocarbons in hydrothermal petroleums form the Guaymas Basin spreading center. *Applied Geochemistry*, **5**, 17–27.

KEENER, W. 1987. Basin hides oil beneath volcanics. *Western Oil World*, **44**, n. 3, 13–15.

KENDELL, C. F. 1994a. Rozel Point oil field, Box Elder County, Utah. *In*: SCHALLA, R. A. & JOHNSON, E. H. (eds) *Oil fields of the Great Basin*. Nevada Petroleum Society, Reno, 361–369.

KENDELL, C. F. 1994b. West Rozel oil field, Box Elder County, Utah. *In*: SCHALLA, R. A. & JOHNSON, E. H. (eds) *Oil fields of the Great Basin*. Nevada Petroleum Society, Reno, 370–372.

KENT, P. E. 1954. Oil occurrences in coal measures of England. *American Association of Petroleum Geologists, Bulletin*, **38**, 1699–1713.

KESSE, G. D. 1986. Oil and gas possibilities on- and offshore Ghana. *In*: HALBOUTY, M. T. (ed.) *Future petroleum provinces of the world*. American Association of Petroleum Geologists, Memoirs, **40**, 427–444.

KHAVARI-KHOROSANI, G. & MURCHISON, D. G. 1978. Thermally metamorphosed bitumen from Windy Knoll, Derbyshire, England. *Chemical Geology*, **22**, 91–105.

KING, R. J. & FORD, T. D. 1968. Mineralization. *In*: SYLVESTER-BRADLEY, P. C. & FORD, T. D. (eds) *The geology of the East Midlands*. Leicester University Press, 112–137.

KINGSTON, J. & MATZKO, J. R. 1995. Undiscovered petroleum of the Brazilian interior sag basins. *International Geology Review*, **37**, 959–980.

KIRKHAM, V. R. D. 1935. Natural gas in Washington, Idaho, eastern Oregon, and northern Utah. *In*: LEY, H. A. (ed.) *Geology of natural gas*. American Association of Petroleum Geologists, Tulsa, 221–243.

KOEN, A. D. 1995. Discoveries, production add luster to offshore Vietnam's outlook. *Oil & Gas Journal*, **93**, 17 July (n. 29), 17–21.

KOMATSU, N., FUJITA, Y. & SATO, O. 1984. Cenozoic volcanic rocks as potential hydrocarbon reservoirs. *World Petroleum Congress, 11th, Proceedings*, **2**, 411–422.

KONNERUP-MADSEN, J., LARSEN, E. & ROSE-HANSEN, J. 1979. Hydrocarbon-rich fluid inclusions in minerals from the alkaline Ilimaussaq intrusion, South Greenland. *Bulletin de Minérologie*, **102**, 642–653.

KONTOROVICH, A. E., MANDEL'BAUM, M. M., SURKOV, V. S., TROFIMUK, A. A. & ZOLOTOV, A. N. 1990. Lena-Tunguska Upper Proterozoic–Palaeozoic petroleum super province. *In*: BROOKS, J. (ed.) *Classic Petroleum Provinces*, Geological Society, London, Special Publications, **50**, 473–489.

KORNFELD, J. A. & TRAVIS, M. M. 1967. Arizona's spectacular oil strike tops Rocky Mountain field interest. *World Oil*, **164**, May (n. 6), 180–190.

KRIBEK, B., HOLUBÁR, PARNELL, J., POUBA, Z. & HLADIKOVA, J. 1993. Interpretation of thermal mesophase in vanadiferous bitumens from Upper Proterozoic lava flows (Mitov, Czechoslovakia). *In*: PARNELL, J., KUCHA, H. & LANDAIS, P. (eds) *Bitumen in ore deposits*. Springer-Verlag, Berlin, 61–77.

KUJIRAOKA, A. 1967. Volcanic activity and its influence on the migration and accumulation of oil and gas in the Nagaoka Plain, Japan. *In*: *3rd Symposium on the Development of Petroleum Resources of Asia and the Far East, Proceedings*. United Nations Economic Commission For Asia and the Far East (ECAFE), Mineral Resources Development Series, **26**, 1, 262–274.

KUUSKRAA, V. A. & SCHMOKER, J. W. 1998. Diverse gas plays lurk in gas resource pyramid. *Oil & Gas Journal*, **96**, 8 June (n. 23), 123–130.

KVENVOLDEN, K. A., RAPP, J. B., HOTSTETTLER, F. D., MORTON, J. L., KING, J. D. & CLAYPOOL, G. E. 1986. Petroleum associated with polymetallic sulfide in sediment from Gorda Ridge. *Science*, **234**, 1231–1234.

LAIER, T., NYTOFT, H. P., JORGENSEN, O. & ISAKSEN, G. H. 1997. Hydrocarbon traces in the Tertiary basalts of the Faeroe Islands. *Marine and Petroleum Geology*, **14**, 257–266.

LEE, K. Y. 1989. *Geology of petroleum and coal deposits in the North China Basin, eastern China*. United States Geological Survey, Bulletin, **1871**, 1–36.

LEWIS, J. W. 1932. Occurrence of oil in igneous rocks of Cuba. *American Association of Petroleum Geologists, Bulletin*, **16**, 809–818.

LIGRANI, L. D., ADAMS, C. W. & NIBBELINK, K. A. 1985. Oil and gas developments in eastern and northwestern Colorado in 1984. *American Association of Petroleum Geologists, Bulletin*, **69**, 1576–1579.

LINDNER, A. W. 1987. Oil and gas developments in Australia in 1986. *American Association of Petroleum Geologists, Bulletin*, **71**, n. 10B, 176–189.

LINGLEY, W. S., JR. & WALSH, T. J. 1986. Issues relating to petroleum drilling near the proposed high-level nuclear waste repository at Hanford. *Washington Geologic Newsletter*, **14**, n. 3, 10–19.

LIU, Z., XIN, Q., WANG, Y., XU, P. & ZHANG, X. 1989. Formation conditions and distribution regularities of oil-gas pools in Tertiary volcanic rocks in western Huimin Depression, Shandong Province. *Chinese Journal of Geochemistry*, **8**, n. 2, 146–154.

LOPEZ, D. A. 1995. *Geology of the Sweet Grass Hills, north-central Montana*. Montana Bureau of Mines and Geology, Memoirs, **68**, 1–35.

LOVE, J. D. & GOOD, J. M. 1970. *Hydrocarbons in thermal areas, northwestern Yellowstone National Park, Wyoming*. United States Geological Survey, Professional Paper **644-B**, 1–23.

MACGREGOR, D. S. 1998. Giant fields, petroleum systems and exploration maturity of Algeria. *In*: MACGREGOR, D. S., MOODY, R. T. J. & CLARK-LOWERS, D. D. (eds) *Petroleum geology of North Africa*. Geological Society, London, Special Publications, **132**, 79–96.

MACKEVETT, N. H. 1998. Gas reserves growth boosts Sacramento Basin. *Oil & Gas Journal*, **96**, 26 January (n. 4), 102–105.

MAGARA, K. 1968. Compaction and migration of fluids in Miocene mudstone, Nagaoka Plain, Japan. *American Association of Petroleum Geologists, Bulletin*, **52**, 2466–2501.

MAHFOUD, R. F. & BECK, J. N. 1991. Inorganic origin in upper mantle seen likely for solid hydrocarbon in Syria plateau basalt. *Oil & Gas Journal*, **89**, 28 October (n. 43), 88–92.

MALLORY, W. W. 1977. *Oil and gas from fractured shale reservoirs in Colorado and northwest New Mexico*. Rocky Mountain Association of Geologists, Special Publications, **1**, 1–38.

MARCHI, L. C., LEFAILLER, S. & POSTEL, J. J. 1999. 3-D seismic survey in the Amazonian rain forest: the challenges of the São Mateus prospect. *Geophysics—the Leading Edge of Exploration*, **18**, 823–825.

MARTINSON, N. W., MCDONALD, D. R. & KAYE, P. 1973. Exploration on continental shelf off Northwest Australia. *American Association of Petroleum Geologists, Bulletin*, **57**, 972–989.

MASTERS, J. A. 2000. The slow discovery of Dineh-bi-Keyah. *Mountain Geologist*, **37**, 91–100.

MATJASIC, W. L. 1954. Case history of Wild Goose gas field, Butte County, California. *Geophysics*, **19**, 509–516.

MATTHEWS, T. F. 1986. The petroleum potential of "serpentine plugs" and associated rocks, Central and South Texas. *Baylor Geological Studies, Bulletin*, **44**, 1–43.

MATTICK, R. E., TELEKI, P. G., *ET AL*. 1996. Structure, stratigraphy, and petroleum geology of the Little Plain Basin, northwestern Hungary. *American Association of Petroleum Geologists, Bulletin*, **80**, 1780–1800.

MAY, P. R. & SHULMAN, H. 1989. Oil exploration in the Golan Heights. *Israel Geological Society, Annual Meeting Abstracts*. Israel Geological Survey, 103.

MCBEATH, D. M. 1977. Gas-condensate fields of the Taranaki Basin, New Zealand. *New Zealand Journal of Geology and Geophysics*, **20**, 99–127.

MCCASLIN, J. C. 1971. Union may start new Queen Charlotte play. *Oil & Gas Journal*, **69**, 19 July (n. 29), 91.

MCFARLAND, C. 1979. History of oil and gas exploration in the state of Washington. *Washington Geologic Newsletter*, **7**, n. 3, 1–6.

MCKENNY, J. W. 1969. Oil and gas on Toadlena Anticline, Arizona and New Mexico. *American Association of Petroleum Geologists, Bulletin*: **53**, 217.

MCKENNY, J. W. & MASTERS, J. A. 1968. Dineh-bi-Keyah field, Apache County, Arizona. *American Association of Petroleum Geologists, Bulletin*, **52**, 2045–2057.

MCKERROW, A. J., DUNN, V. L., FISH, R. M., MILLS, C. R. & VANDER LINDEN-DHONT, S. K. 1998. Bass Strait's Bream B reservoir development: success through a multifunctional team approach. *APPEA (Australian Petroleum Production and Exploration Association), Journal*, **38**, Pt. 1, 13–35.

MCKIRDY, D. M. & CHIVAS, A. R. 1992. Nonbiodegraded aromatic condensate associated with volcanic supercritical carbon dioxide, Otway Basin: implications for primary migration from terrestrial organic matter. *Organic Geochemistry*, **18**, 611–627.

MENCHER, E., FICHTER, H. J., RENZ, H. H., WALLIS, W. E., PATTERSON, J. M. & ROBIE, R. H. 1953. Geology of Venezuela and its oil fields. *American Association of Petroleum Geologists, Bulletin*, **37**, 690–777.

MEYER, R. F. 1987. Heavy crude fields and bitumen deposits of the Middle and Far East and Oceania. *In*: MEYER, R. F. (ed.) *Exploration for heavy crude oil and natural bitumen*. American Association of Petroleum Geologists, Studies in Geology, **25**, 143–154.

MEYERHOFF, A. A. 1980. Geology and petroleum fields in Proterozoic and Lower Cambrian strata, Lena-Tunguska petroleum province, eastern Siberia, USSR. *In*: HALBOUTY, M. T. (ed.) *Giant oil and gas fields of the decade 1968–1978*. American Association of Petroleum Geologists, Memoirs, **30**, 225–252.

MEYERHOFF, A. A. & MEYER, R. F. 1987. Geology of heavy crude oil and natural bitumen in the USSR, Mongolia, and China. *In*: MEYER, R. F. (ed.) *Exploration for heavy crude oil and natural bitumen*. American Association of Petroleum Geologists, Studies in Geology, **25**, 31–101.

MEYERHOFF, A. A. & MEYERHOFF, H. A. 1974. Tests of plate tectonics. *In*: KAHLE, C. F. (ed.) *Plate tectonics—assessments and reassessments*. American Association of Petroleum Geologists, Memoirs, **23**, 43–145.

MICHAELIS, W., JENISCH, A. & RICHNOW, H. H. 1990. Hydrothermal petroleum generation in Red Sea sediments from the Kebrit and Shaban deeps. *Applied Geochemistry*, **5**, 103–114.

MILANI, E. J., KINOSHITA, E. M., DE ARAÚJO, L. M. & CUNHA, P. R. DA C. 1990. Bacia do Paraná: possibilidades petrolifera da calha central. *Boletim Geociências Petrobrás*, **4**, n. 1, 21–34.

MITSUHATA, Y., MATSUO, K. & MINEGISHI, M. 1999. Magnetotelluric survey for exploration of a volcanic-rock reservoir in the Yurihara oil and gas field, Japan. *Geophysical Prospecting*, **47**, 195–218.

MOISEYEV, A. N. 1968. The Wilbur Springs quicksilver district (California): example of a study of hydrothermal processes by combining field geology and theoretical geochemistry. *Economic Geology*, **63**, 169–181.

MORRIS, A. E. L., MEYERHOFF, A. A., TANER, I., BUENO-SALAZAR, R. & YOUNG, G. A. 1990. Energy resources of the Caribbean region. *In*: DENGO, G. & CASE, J. E. (eds) *The geology of North America*, **H**—*The Caribbean region*. Geological Society of America, Boulder, Colorado, 483–510.

MUELLER, G. 1964. Discussion of evidences indicating the existence of petroleums of distinct genetical histories. *International Geological Congress, 22nd, Report*, Part 1, 38–63.

MUKHERJEE, M. K. 1983. Petroleum prospects of Cretaceous sediments of the Cambay Basin, Gujarat, India. *Journal of Petroleum Geology*, **5**, 275–286.

MULREADY, J. 1977. The Caroline carbon dioxide field and associated carbon dioxide occurrences, Gambier Embayment, South Australia. *APEA Journal*, **17**, pt. 1, 121–127.

NAGLE, H. E. & PARKER, E. S. 1971. Future oil and gas potential of the onshore Ventura Basin, California. *In*: CRAM, I. A. (ed.) *Future petroleum provinces of the United States—their geology and potential*. American Association of Petroleum Geologists, Memoirs, **15**, **1**, 254–297.

NEVES, C. A. DE O. 1990. Prospectos potencias e áreas prioritárias para exploração na bacia do Amazonas. *Boletim Geociências Petrobrás*, **4**, n. 1, 95–103.

NOGUERA, M. A. 1946. Cuba's Jarahueca field. *Oil Weekly*, 2 December, International Section, 32.

NURMI, R., IMRAN, H., AKBAR, M., CHEUNG, P., ETCHECOPAR, A. & HELIOT, D. 1991. Grabens. *In*: CHARNOCK, G. & GAMBLE, S. (eds) *Middle East Well Evaluation Review*. Special Supplement–Structural Geology, Schlumberger Technical Services, Dubai, 20–31.

NUTT, W. L. & SIRIAT, J. 1985. Application of offset seismic profiles in the Jatibarang volcanic reservoir. *Indonesian Petroleum Association, 14th Annual Convention, Proceedings*, **2**, 385–406.

ORDOÑEZ, E. 1915. The oil fields of Mexico. *American Institute of Mining Engineering, Transactions*, **61**, 532–543.

OREMLAND, R. S., MILLER, L. G. & WHITICAR, M. J. 1987. Sources and flux of natural gases from Mono Lake, California. *Geochimica Cosmochimica Acta*, **51**, 2915–2929.

O'SULLIVAN, T. 1992. Scotia—gas resource in a fractured andesite reservoir. *American Association of Petroleum Geologists, Bulletin*, **76**, 1120–1121.

OZKANLI, M. & KUMSAL, K. 1993. Petrophysical evaluation of volcanic gas reservoirs of Upper Eocene formations in Thrace Basin, Turkey. *American Association of Petroleum Geologists, 1993 Annual Convention, Program*, 162.

P'AN, C.-H. 1982. Petroleum in basement rocks. *American Association of Petroleum Geologists, Bulletin*, **66**, 1597–1643.

PARNELL, J. 1983. The distribution of hydrocarbon minerals in the Welsh borderlands and adjacent areas. *Journal of Geology*, **18**, 129–139.

PARNELL, J. 1984.

PARNELL, J. 1988. Migration of biogenic hydrocarbons into granites: a review of hydrocarbons in British plutons. *Marine and Petroleum Geology*, **5**, 385–396.

PARNELL, J. 1990. Mineralogy of rare-earth-bearing "thucholite", Parry Sound, Ontario. *Canadian Mineralogist*, **28**, 357–362.

PARNELL, J. 1992. Burial histories and hydrocarbon source rocks on the North West Seaboard. *In*: PARNELL, J. (ed.) *Basins of the Atlantic Seaboard: petroleum geology, sedimentology and basin evolution*. Geological Society, London, Special Publications, **62**, 449–464.

PARNELL, J., MONSON, B. & BUCKMAN, J. 1992. Excursion guide: basins and petroleum geology in the north of Ireland. *In*: PARNELL, J. (ed.) *Basins of the Atlantic Seaboard. petroleum geology, sedimentology and basin evolution*. Geological Society, London, Special Publications, **62**, 3–16.

PATTON, D. K. 1993. Samgori Field, Republic of Georgia: critical review of island-arc oil and gas. *Journal of Petroleum Geology*, **16**, 153–168.

PATTON, D. K. 1995. *Petroleum exploration opportunities in Africa and countries beyond*. Pennwell, Tulsa, 1–293.

PEABODY, C. E. 1993. The association of cinnabar and bitumen in mercury deposits of the California Coast Ranges. *In*: PARNELL, J., KUCHA, H. & LANDAIS, P. (eds) *Bitumen in ore deposits*. Springer-Verlag, Berlin, 178–209.

PEDERSEN, A. K. 1986. Indication of migrated hydrocarbons in Tertiary volcanic rocks from western Nûgssauq, central West Greenland. *Grønlands Geologiske Undersøgelse, Rapport*, **130**, 32–35.

PENTTILA, W. C. 1992. Newly democratic Mongolia offering exploration contracts. *Oil & Gas Journal*, **90**, 7 December (n. 49), 39–43.

PERCIVAL, J. C. 1878. Indurated bitumen; geology of Connecticut. *American Journal of Science*, series 3, **16**, 130.

PEREA, D. A. & GIORDANO, J. D. 1988. Igneous rocks as hydrocarbon reservoirs in Malargue, Argentina. *Society of Professional Well Log Analysts, 29th Annual Logging Symposium, Transactions*, Z1–Z24.

PEREA, D. A., SCHLUMA, M. & VILOSO, J. L. 1984. Las rocas igneas como reservorio de hidrocarburos. *9° Congreso Geologico Argentino, Actas*, **7**, 20–38.

PETERSILIE, I. A. & SÖRENSEN, H. 1970. Hydrocarbon gases and bituminous substances in rocks from the Ilimaussaq alkaline intrusion, South Greenland. *Lithos*, **3**, 59–76.

PETERSILIE, V. I. 1973. [Physical and reservoir parameters of tuffs]. *Geologiya Nefti i Gaza*, **1973**, n. 6, 46–50 (English translation, *Petroleum Geology*, **11**, 249–253).

PETROLEUM AGENCY SA 2000. South Africa—petroleum exploration opportunities. 24 pp.

PETZET, G. A. 1997. Nonproducing Paraguay to get rare wildcats. *Oil & Gas Journal*, **95**, 21 April (n. 16), 76–77.

PILAAR, W. F. H. & WAKEFIELD, L. L. 1984. Hydrocarbon generation in the Taranaki Basin, New Zealand. *In*: DEMAISON, G. & MORRIS, R. J. (eds) *Petroleum geochemistry and basin evolution*. American Association of Petroleum Geologists, Memoirs, **35**, 405–423.

PONNAMPERUMA, C. & PERING, K. 1966. Possible abiogenic origin of some naturally occurring hydrocarbons. *Nature*, **209**, 979–982.

POPKOV, V. I., RABINOVICH, A. A. & TUROV, N. I. 1988. [Reservoir model of an oil pool in a granite pluton]. *Geologiya Nefti I Gaza*, **1988**, n. 8, 27–30.

POREDA, R. J., JENDEN, P. D., KAPLAN, I. R. & CRAIG, H. 1986. Mantle helium in Sacramento Basin natural gas wells. *Geochimica Cosmochimica Acta*, **50**, 2847–2853.

POWERS, S. 1932. Notes on minor occurrences of oil, gas, and bitumen with igneous and metamorphic rocks. *American Association of Petroleum Geologists, Bulletin*, **16**, 837–858.

PRATT, L. M. & BURRUSS, R. C. 1968. Evidence for petroleum generation and migration in the Hartford and Newark basins. *In*: FROELICH, A. J. & ROBINSON, G. R. (eds) *Studies of the Early Mesozoic basins of the eastern United States*. United States Geological Survey, Bulletin, **1776**, 74–79.

PUCCI, J. C. 1996. Relinquishments in Argentina will boost exploration interest. *Oil & Gas Journal*, **94**, 30 September (n. 40), 103–107.

PYE, W. D. 1967. How unique Arizona syenite oil reservoir formed. *World Oil*, **165**, July (n. 1), 81–83.

QIANG, J. & MCCABE, P. 1998. Genetic features of petroleum systems in rift basins of eastern China. *Marine and Petroleum Geology*, **15**, 343–358.

RAO, G. N. 2001. Sedimentation, stratigraphy and petroleum potential of Krishna-Godavari basin, east coast of India. *American Association of Petroleum Geologists, Bulletin*, **85**, 1623–1643.

RASMUSSEN, B. & BUICK, R. 2000. Oily old ores: evidence for hydrothermal petroleum generation in an Archean volcanogenic massive sulfide deposit. *Geology*, **28**, 731–734.

RAUZI, S. L. 2001. *Arizona has oil and gas potential!* Arizona Geological Survey Circular, **29**, 1–40.

RAY, R. R. 1989. Seismic expressions of igneous sills in San Juan Sag, south-central Colorado. *American Association of Petroleum Geologists, Bulletin*, **73**, 1159.

REEVES, F. 1953. Italian oil and gas resources. *American Association of Petroleum Geologists, Bulletin*, **37**, 601–653.

REMUS, D., WEBSTER, M. & KEAWKAN, K. 1993. Rift architecture and sedimentology of the Phetchabun intermontane basin, central Thailand. *Journal of Southeast Asian Earth Sciences*, **8**, 421–432.

RICHARDS, M. E. 1985. Structural configuration of the Santa Clara Avenue oil field, Ventura Basin, California. *Geological Society of America, Abstracts with Programs*, **17**, n. 6, 404.

RILETT, M. P. H. 1956. An occurrence of oil in the Middle Ecca beds near Dannhauser, Natal. *In*: GUZMAN, E. J. (ed.) *Simposio sobre petroleo y gaz*. International Geological Congress, 20th, 13–21.

RIVES, J. S. 1968. Structural and stratigraphic traps related to extrusive rocks in south central Texas. *South Texas Geological Society, Bulletin*, **9**, September (n. 1), 4–14.

RODRIQUEZ, N. 1996. Analysis of well logging methods in volcanic and volcano sedimentary rocks from Piña petroleum field. *Gulf Coast Association of Geological Societies, Transactions*, **46**, 478.

ROOS, M. 1986. Tamarugal Basin exploration under way in northern Chile. *Oil & Gas Journal*, **94**, 30 December (n. 53), 89–91.

RUSSELL, I. C. 1878. On the occurrence of solid hydrocarbon in the eruptive rocks of New Jersey. *American Journal of Science*, series 3, **16**, 112–114.

RUSTAMOV, R. I. 1982. [Formation of the Muradkhanly oil field]. *Geofizika*, **1982**, n. 10, 15–17 (English translation in *Petroleum Geology*, 1984, **20**, 171–173).

SAKATA, S., MATSUSHISA, Y. & IGARI, S.-I. 1994. Organic geochemistry of petroleum from the Green Tuff basin, Japan. *Organic Geochemistry*, **21**, 27–34.

SALAS, G. P. 1968. Petroleum evaluation of north-central Mexico. *American Association of Petroleum Geologists, Bulletin*, **52**, 665–674.

SAMANTA, U. & SHUKLA, R. K. 1987. Trends of organic matter maturation in Krishna-Godavari Basin. *In*: KUMAR, R. K., DWIVEDI, P., BANERJEE, V. & GUPTA, V. (eds) *Petroleum geochemistry and exploration in the Afro-Asian region*. Balkema, Rotterdam, 297–300.

SATERDAL, A. 1955. Tow Creek oil field. *In*: RITZMA, H. R. & ORIEL, S. S. (eds) *Guidebook to the geology of northwest Colorado*. 6th Annual Field Conference, Intermountain Association of Petroleum Geologists/Rocky Mountain Association of Geologists, 110–112.

SATO, O., SEKIGUCHI, K. & WATANABE, T. 1992. Reservoir quality of rhyolite in the Minami Nagaoka gas field, Japan. *International Geological Congress, 29th, Abstracts*, **3**, 810.

SATO, O., TAKASAKI, M., KAWAMOTO, T., HASEGAWA, K. & WATANABE, T. 1992. Lithofacies and reservoir characteristics of rhyolite core samples from the Minami Nagaoka gas field, Japan. *International Geological Congress, 29th, Abstracts*, **3**, 810.

SAWYER, J. H. 1975. Latin America after 1920. *In*: OWEN, E. L. (ed.) *Trek of the oil finders: a history of exploration for petroleum*. American Association of Petroleum Geologists, Memoirs, **6**, 960–1251.

SCHMOKER, J. W., DYMAN, T. S., DOLTON, G. L., FOX, L. E., LAW, B. E., PATERSON, J. A. & TENNYSON, M. E. 1996. *Gas plays of the U. S. Geological Survey 1995 National Assessment that underlie regionally extensive surface volcanics*. United States Geological Survey, Open-File Report **90-099**, 1–13.

SEMBODO 1973. Notes on formation evaluation in the Jatibarang volcanic reservoir. *Indonesian Petroleum Association, 2nd Annual Convention, Proceedings*, 131–147.

SEYLER, B. & CLUFF, R. M. 1990. Petroleum traps in the Illinois Basin. *In*: LEIGHTON, M. W., KOLATA, D. R., OLTZ, D. F. & EIDEL, J. J. (eds) *Interior Cratonic Basins*. American Association of Petroleum Geologists, Memoirs, **51**, 361–401.

SHREVEPORT GEOLOGICAL SOCIETY 1946. Cary Field. *In*: *1945 reference report on certain oil and gas fields of North Louisiana, South Arkansas, Mississippi and Alabama*, **1**, 247–248.

SHREVEPORT GEOLOGICAL SOCIETY 1968. Stratigraphy and selected gas-field studies of North Louisiana. *In*: BEEBE, B. W. & CURTIS, B. F. (eds) *Natural gases of North America*. American Association of Petroleum Geologists, Memoirs, **9**, **1**, 1099–1175.

SIKANDER, A. H. 1975. Occurrence of oil in the Devonian rocks of eastern Gaspé, Quebec. *Bulletins of Canadian Petroleum Geology*, **23**, 278–294.

SIMONIET, B. R. T. 1985. Hydrothermal petroleum: genesis, migration and deposition in the Guaymas Basin, Gulf of California. *Canadian Journal of Earth Sciences*, **22**, 1919–1929.

SIMONIET, B. R. T. & DIDYK, B. M. 1978. Organic geochemistry of a Chilean paraffin dirt. *Chemical Geology*, **23**, 21–40.

SLOAN, M. W., MOORE, P. S. & MCCUTCHEON, A. 1992. Kipper—a unique oil and gas discovery, Gippsland Basin, Australia. *APEA Journal*, **32**, pt. 1, 1–8.

SMITH, J. E. 1956. Basement reservoir of La Paz-Mara oil fields, western Venezuela. *American Association of Petroleum Geologists, Bulletin*, **40**, 380–385.

SPARLIN, M. A. & LEWIS, R. D. 1994. Interpretation of the magnetic anomaly over the Omaha oil field, Gallatin County, Illinois. *Geophysics*, **59**, 1092–1099.

SPENCE, H. S. 1930. A remarkable occurrence of thucholite and oil in a pegmatite dyke, Parry Sound District, Ontario. *American Mineralogist*, **15**, 499–520.

SPOONER, W. C. 1928. Production in East Carroll Parish, Louisiana. *American Association of Petroleum Geologists, Bulletin*, **12**, 765.

STANISTREET, I. G. & STOLLHOFEN, H. 1999. Onshore equivalents of the main Kudu gas reservoir in Namibia. *In*: CAMERON, N. R., BATE, R. H. & CLURE, V. S. (eds) *The oil and gas habitats of the South Atlantic*. Geological Society, London, Special Publications, **153**, 345–365.

STOFFERS, P., HANNINGTON, M., WRIGHT, I., HERZIG, P. & DE RONDE, C. 1999. Elemental mercury at submarine hydrothermal vents in the Bay of Plenty, Taupo volcanic zone, New Zealand. *Geology*, **27**, 931–934.

STOLLHOFEN, H., STANISTREET, I. G. & GERSCHÜTZ, S. 2000. Early Jurassic rift-valley-related alkaline lake deposits interbedded with Karoo flood basalts, southern Namibia. *In*: GIERLOWSKI-KORDESCH, E. H. & KELTS, K. R. (eds) *Lake basins through space and time*. American Association of Petroleum Geologists, Studies in Geology, **46**, 167–180.

SUMMER, N. S. & VEROSUB, K. L. 1992. Diagenesis and organic maturation of sedimentary rocks under volcanic strata, Oregon. *American Association of Petroleum Geologists, Bulletin*, **76**, 1190–1199.

SUNDELL, K. A. & LOVE, J. D. 1986. Oil and gas seeps within Absaroka volcanics of northwestern Wyoming. *American Association of Petroleum Geologists, Bulletin*, **70**, 1058–1059.

SUTAN ASSIN, N. A. D. & TARUNADJAJA, A. N. S. 1972. Djatibarang, the discovery and development of a new oil field. *Indonesian Petroleum Association, 1st Annual Convention, Proceedings*, 125–137.

SUZUKI, U. 1983. The volcanic mound. *In*: BALLY, A. W. (ed.) *Seismic expression of structural styles*. American Association of Petroleum Geologists, Studies in Geology, **15**, 1, 3.1–15–3.1–18.

SYLVESTER-BRADLEY, P. C. & KING, R. J. 1963. Evidence for abiogenic hydrocarbons. *Nature*, **198**, 728–731.

SYMONDS, P. A., PLANKE, S., FREY, D. & SKOGSEID, J. 1998. Volcanic evolution of the Western Australian continental margin and its implications for basin development. *In*: PURCELL, P. G. & PURCELL, R. R. (eds) *The sedimentary basins of Western Australia 2, Symposium, Proceedings*. Petroleum Exploration Society of Australia, Perth, 33–54.

TALIAFERRO, N. L., HUDSON, F. S. & CRADDOCK, W. N. 1924. The oil fields of Ventura County, California. *American Association of Petroleum Geologists, Bulletin*, **8**, 789–829.

TIEMAN, D. J. 1992. A glimpse of the suspected oldest commercial reservoir in the Gulf Coast province. *Gulf Coast Association of Geological Societies, Transactions*, **42**, 861.

TIERCELIN, J.-J., BOULÉGUE, J. & SIMONIET, B. R. T. 1993. Hydrocarbons, sulphides, and carbonate deposits related to sublacustrine hydrothermal seeps in the North Tanganyika Trough, East Africa Rift. *In*: PARNELL, J., KUCHA, H. & LANDAIS, P. (eds) *Bitumen in ore deposits*. Springer-Verlag, Berlin, 96–113.

TIERCELIN, J.-J., THOUIN, C., KALALA, T. & MONDEGUER, A. 1989. Discovery of sublacustrine hydrothermal activity and associated massive sulfides and hydrocarbons in the North Tanganyika Trough, East African Rift. *Geology*, **17**, 1053–1056.

TRAN, C., DO, V. H., CARSTENS, H. & BERSTAD, S. 1994. Viet Nam—attractive plays in a new geological province. *Oil & Gas Journal*, **93**, 14 March (n. 11), 78–83.

TRAYNOR, J. J. & SLADEN, C. 1995. Tectonic and stratigraphic evolution of the Mongolian People's Republic and its influence on hydrocarbon geology and potential. *Marine and Petroleum Geology*, **12**, 35–52.

TURIC, M., FREYTES, E. & YRIGOYEN, M. R. 1985. Argentina. *In*: TAPPMEYER, D. M. (compiler) Oil and gas developments in South America, Central America, Caribbean area and Mexico in 1984. *American Association of Petroleum Geologists, Bulletin*, **69**, 1632.

TURIC, M., FREYTES, E. & YRIGOYEN, M. R. 1986. Argentina. *In*: WIMAN, W. D. (compiler) Oil and gas developments in South America, Central America, Caribbean area and Mexico in 1985. *American Association of Petroleum Geologists, Bulletin*, **70**, 1371.

UCHIDA, T. 1992. Mode of occurrences of secondary porosities in volcanic rock reservoirs distributed along the west coast of the Northeastern Honshu Arc. *International Geological Congress, 29th, Abstracts*, **3**, 810.

VAN KOOTEN, G. K. 1987. Structure and hydrocarbon potential beneath the Iron Springs laccolith, southwestern Utah. *Geological Society of America, Abstracts with Programs*, **19**, n. 7, 876.

VERCELLINO, J. & RIGO, F. 1970. Geology and exploration of Sicily and adjacent areas. *In*: HALBOUTY, M. T. (ed.) *Geology of giant petroleum fields*. American Association of Petroleum Geologists, Memoirs, **14**, 388–398.

VERNIK, L. 1990. A new type of reservoir rock in volcaniclastic sequences. *American Association of Petroleum Geologists, Bulletin*, **74**, 830–836.

WAAGSTEIN, R. 1988. Structure, composition and age of the Faeroe basalt plateau. *In*: MORTON, A. C. & PARSON, L. M. (eds) *Tertiary volcanism and the opening of the NE Atlantic*. Geological Society, London, Special Publications, **39**, 225–238.

WADE, A. 1926. The search for oil in Australia. *Institute of Petroleum Technology Journal*, **12**, 145–172.

WASSALL, H. 1957. The relationship of oil and serpentine in Cuba. *In*: *Geologia del Petroleo*, International Geological Congress, 20th, Section 3, 65–77.

WATT, W. S. & WRANG, P. 1984. Migrated hydrocarbons in basalt on the south side of Scoresby Sund. *Grønlands Geologiske Undersøgelse, Rapport*, **120**, 84–85.

WEISMANN, T. J. 1971. Stable carbon isotope investigation of natural gases from Sacramento and Delaware-Val Verde basins—possible igneous origins. *American Association of Petroleum Geologists, Bulletin*, **55**, 369.

WHITE, D. E. & WARING, G. A. 1963. Volcanic emanations. *In*: *Data on Geochemistry*, 6th edn. United States Geological Survey, Professional Paper **440K**, K1–K27.

WHITICAR, M. J., SUESS, E. & WEHNER, H. 1985. Thermogenic hydrocarbons in surface sediments of the Bransfield Strait, Antarctic Peninsula. *Nature*, **314**, 87–90.

WILLIAMS, H. H., FOWLER, M. & EUBANK, R. T. 1995. Characteristics of selected Palaeogene and Cretaceous lacustrine source basins of Southeast Asia. *In*: LAMBAISE, J. J. (ed.) *Hydrocarbon habitat in rift basins*. Geological Society, London, Special Publications, **80**, 241–282.

WIMAN, W. D. 1987. Oil and gas developments in South America, Central America, Caribbean area and Mexico in 1986. *American Association of Petroleum Geologists, Bulletin*, **71**, n. 10B, 337–363.

WITHERS, R., EGGERS, D., FOX, T. & CREBS, T. 1994. A case study of integrated hydrocarbon exploration through basalt. *Geophysics*, **59**, 1666–1679.

YATES, R. G. & THOMPSON, G. A. 1959. *Geology and quicksilver deposits of the Terlingua District*. United States Geological Survey, Professional Paper **312**, 1–114.

YOSHIDA, R. & GAMA, E., JR. 1982. Potencialidade em hidrocarbonetos da Bacia do Paraná: uma reavaliação. *Ciências da Terra*, n. 4, 28–32.

YU, J. 1989. [Water saturation in the pays of the Menggulin Breccia of Erlian Basin]. *Exploration and Petroleum Geology*, **1**, 273–276.

ZEZIN, R. B. & SOKOLOVA, M. N. 1967. [Macroscopic occurrences of carbonaceous matter in hydrothermal deposits of the Khibiny pluton]. *Akademi Nauk SSSR, Doklady*, **177**, n. 4, 921–924 (English translation, 1968, 217–221).

ZHANG, G. 1992. The formation of hydrocarbon pools in Mesozoic andesite reservoir in Fenghuadian structure, Huanghua Basin, eastern China. *International Geological Congress, 29th, Abstracts*, **3**, 809.

ZHANG, Y., WEI, Z., XU, W., TAO, R. & CHEN, R. 1989. The North Jiangsu-South Yellow Sea Basin. *In*: ZHU, X. (ed.) *Chinese sedimentary basins. Sedimentary basins of the world*, **1**. Elsevier, Amsterdam, 122–123.

ZHOU, Z. 1987. [Study on the reservoir porosity in Mesozoic lamprophyric intrusive body, Yibei oil field]. *Oil and Gas Geology*, **8**, n. 2, 145–152 [English abstract *in Petroleum Abstracts*, **28**, 2647, Abstract No. 449408 (1988)].

ZHOU, Z. 1988. [Formation conditions of a Lower Tertiary basalt reservoir, Binnan oil field, Shandong]. *Oil and Gas Geology*, **9**, n. 2, 133–139 [English abstract *in Petroleum Abstracts*, **29**, 1428 (1989)].

ZHOU, Z., *ET AL*. 1984.

ZIEGLER, D. G. 1983. Hydrocarbon potential of the Newark Rift System, eastern North America. *Northeastern Geology*, **5**, 200–208.

ZIMMERMAN, R. K. & SASSEN, R. 1993. Hydrocarbon transfer pathways from Smackover source rocks to younger reservoir traps in the Monroe Gas Field, northeast Louisiana. *Gulf Coast Association of Geological Societies*, **43**, 473–480.

Volcanic reservoir rocks of northwestern Honshu island, Japan

KINJI MAGARA

University of Shimane, Hamada, Shimane, Japan

Abstract: The main Japanese oil producing region lies on the Japan Sea side of northern Honshu island. Although the total reserve is small and production supplies only three-tenths of a percent of total Japanese oil consumption, it has two distinguishing features: (1) the main reservoir rocks are volcanic, pyroclastic, or tuffaceous, and (2) primary oil and gas migration seems to have taken place downward from the overlying source rocks. Marine volcanic activity since 15 Ma formed the main reservoir sections together with significant secondary porosity development. Thick and continuous deposition of organic-rich shales and mudstones followed and lower parts of these fine-grained rocks became the main source rocks. The principal direction of primary hydrocarbon migration occurred vertically downward from them. These fine-grained rocks seem to have acted as pressure seals as well as capillary seals over the oil/gas saturated zones below.

The Japanese Archipelagos consist of four main islands: (1) Honshu or main land, (2) Hokkaido or northern land, (3) Kyushu or nine provinces, and (4) Shikoku or four districts (Fig. 1). These four islands form an arc approximately 2,000 km long, and are surrounded by both the Pacific Ocean and the Sea of Japan. Because mountains and hills cover approximately six-sevenths of Japan, only about 15% of the land remains suitable for farming and building cities and towns. Japan has little mineral wealth, including energy resources such as oil, gas and coal. Among Japan's primary energy supplies, oil ranks first at approximately 50%; virtually its entirety is imported. Domestic oil production supplies only about three-tenths of a percent of the nation's total oil consumption.

Although both the reserve and annual production are rather small, the history of petroleum exploration in Japan is relatively old, dating back to 1888 when the first oil was discovered by a modern drilling method at Amaze in the Niigata Prefecture on the Japan Sea side, approximately 200 km NW of Tokyo (Fig. 2). Oil exploration has continued primarily along the Japan Sea coast, where more than sixty oil and gas fields have been found so far. The only exception is the Joban Gas Field in the Pacific Ocean off northeastern Honshu (Fig. 2). The largest oil reserve has been discovered at the Yabase Oil Field in the Akita Prefecture in the north (Fig. 2), with a recoverable oil reserve of some 50 million bbl. Figure 3 shows east–west cross-sections through the central Akita region. Principal reservoirs in this region are lavas (liparite or rhyolite, andesite and basalt), pyroclastics, tuffs and tuffaceous sandstones. Most oil fields in the northern region are found in north–south elongated, narrow anticlines.

All of the prime reservoirs in the Japanese oil and gas fields in the southern Niigata Prefecture are of volcanic origin and share a number of common features. They are predominantly acidic (rhyolite, dacite and some andesite) reservoirs, characterized by domal shapes which form buried hills. The prime source rocks are fine-grained, organic-rich shales and mudstones which overlie the volcanic reservoirs, and also act as both pressure and capillary seals. Finally, geothermal gradients are high, ranging on average from 3.8 °C to 4.5 °C/100 m, but reaching 5.5 °C/100 m at some locations. In the following sections, examples of these igneous reservoirs are given, along with a discussion of the processes responsible for primary and secondary hydrocarbon migration.

Volcanic reservoirs (Niigata), with domal shapes and downward fluid migration from the overlying marine, organic-rich shales

The volcanic reservoirs of the Niigata Prefracture comprise Neogene marine lavas and intrusives which range in composition from basic to acidic (basalt, andesite, dacite and rhyolite, see Fig. 4). Associated pyroclastic and tuffaceous rocks were also deposited. In the Nagaoka Plain district of the Niigata Prefecture, the more viscous dacite and rhyolite lavas formed domal shapes. These domes were later covered by thick, marine, organic-rich shales and mudstones. Primary porosity in the reservoir rocks is considered to have formed during cooling, spalling and autobrecciation as the magma crystallized in water on the sea floor. A regional NE–SW geological section through oil and gas fields of the Niigata Prefecture is shown in

From: PETFORD, N. & MCCAFFREY, K. J. W. (eds) 2003. *Hydrocarbons in Crystalline Rocks*. Geological Society, London, Special Publications, **214**, 69–81. 0305-8719/03/$15 © The Geological Society of London.

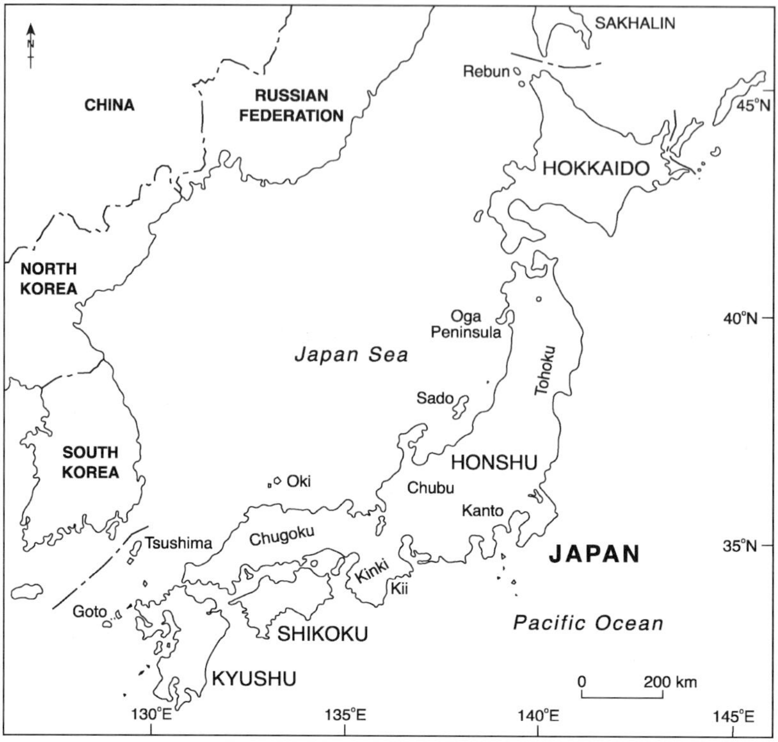

Fig. 1. Map showing the geographical regions of Japan.

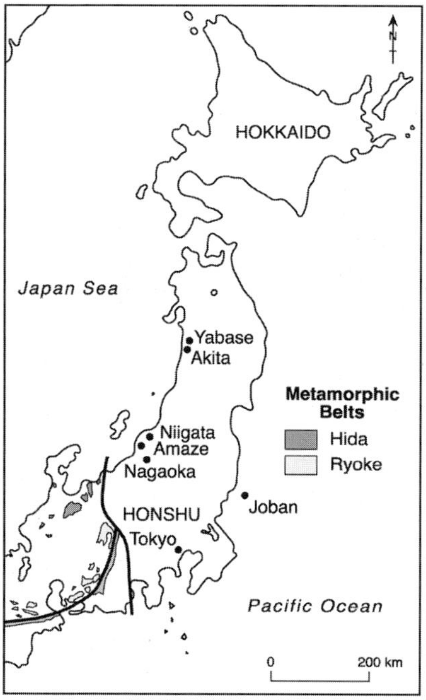

Fig. 2. Map of the oil-producing regions of Japan.

Figure 5, highlighting several of the volcanic domes that form the reservoirs. Figure 6 is a more detailed example of a cross-section through an oil field in this southern region, showing the draping of sediments over a volcanic dome.

Secondary pore spaces were created by hydrothermal activity accompanying later volcanism, with mega fractures and vugs developed in lava and pillow breccia facies. These structures are particularly common in the Mitsuke Oil Field, while microfractures and vugs predominate in the Minami Nagaoka Gas Field (Fig. 5). Figure 7 shows the porosity–permeability relationships in rhyolites and basalts from the nearby Katakai gas field. Permeability values in excess of 80 mD at porosity values of a few percent in the rhyolites suggests that fracture permeability is important in these rocks. In general, the rhyolites make better reservoirs than the basalts in this region.

Discussion

A model for the stages leading to hydrocarbon accumulation in the Niigata district is shown in

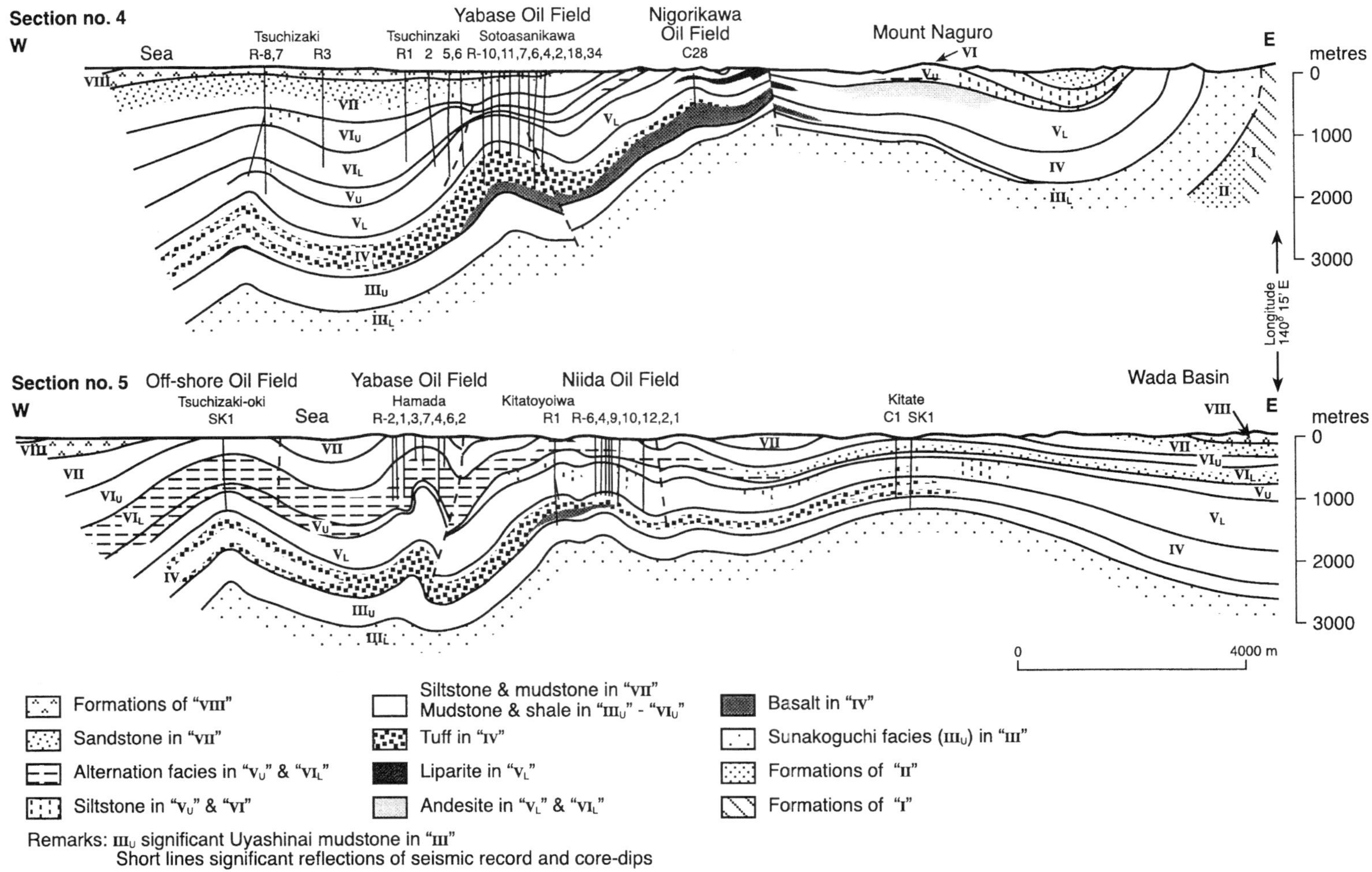

Fig. 3. Geological cross-sections through northern Akita district, Japan (from Ikebe 1963).

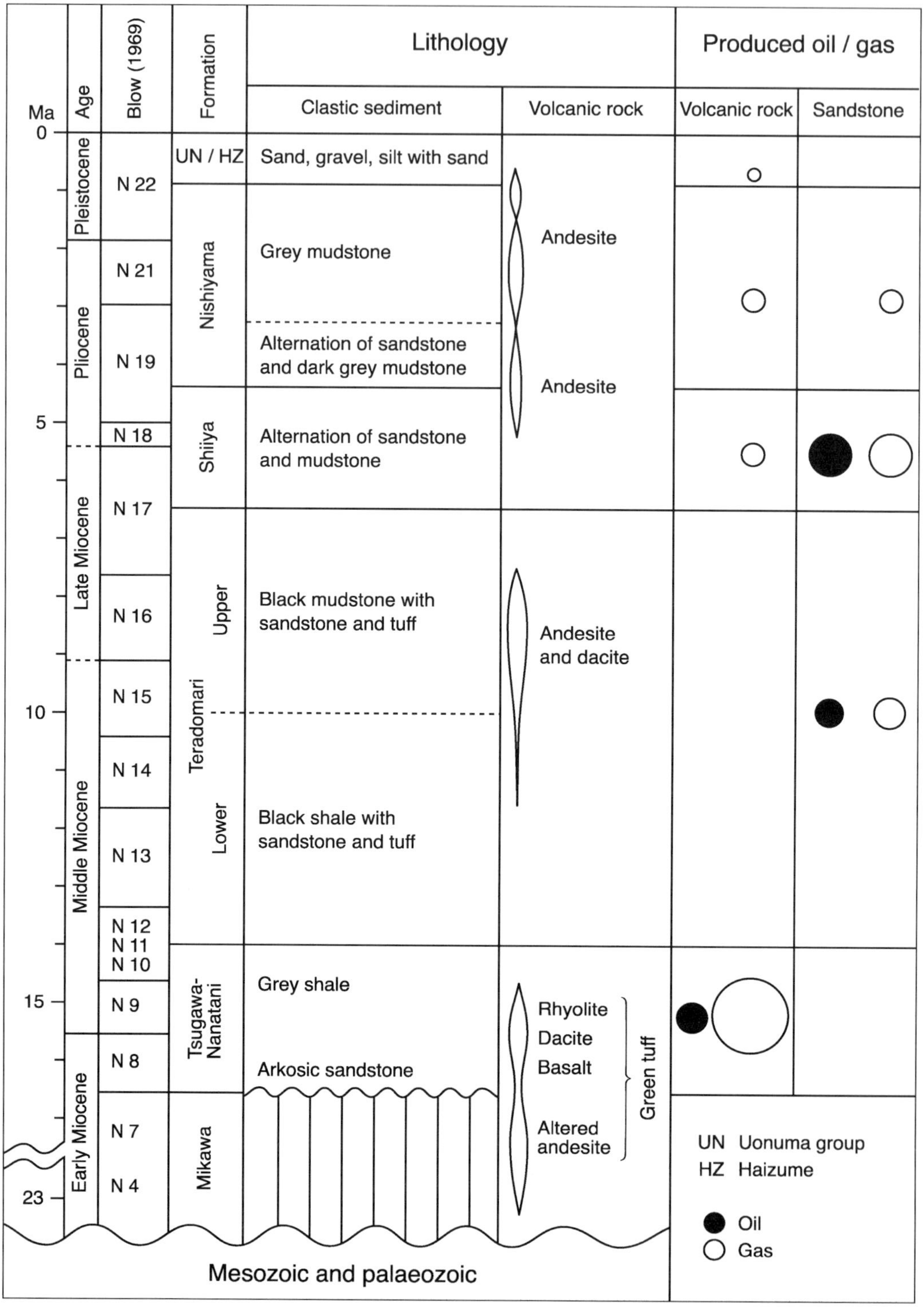

Fig. 4. Generalized stratigraphical column with oil and gas producing horizons in Niigata district, Japan (from Komatsu *et al.* 1983).

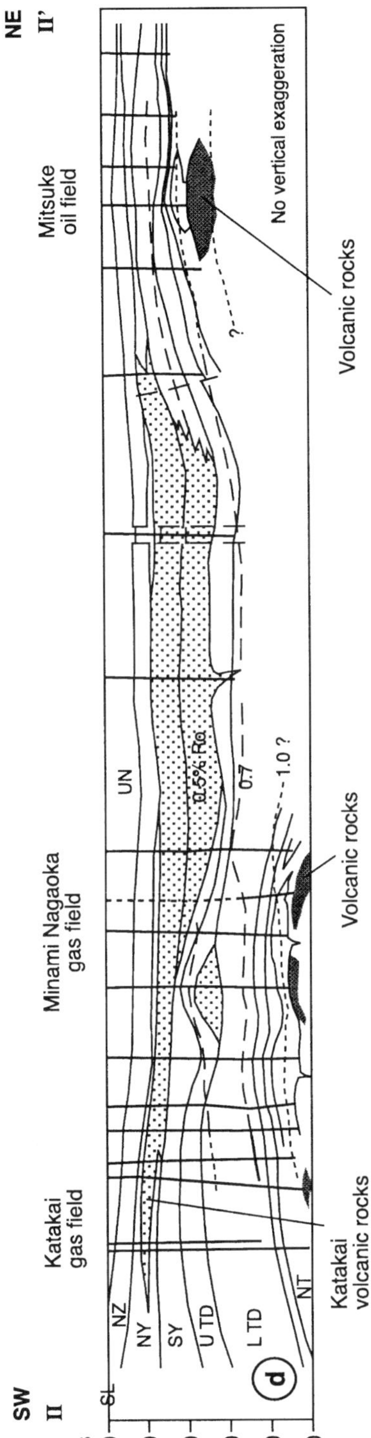

Fig. 5. NE–SW geological section through Mitsuke and Minami Nagaoka fields, Niigata district (from Komatsu *et al.* 1983).

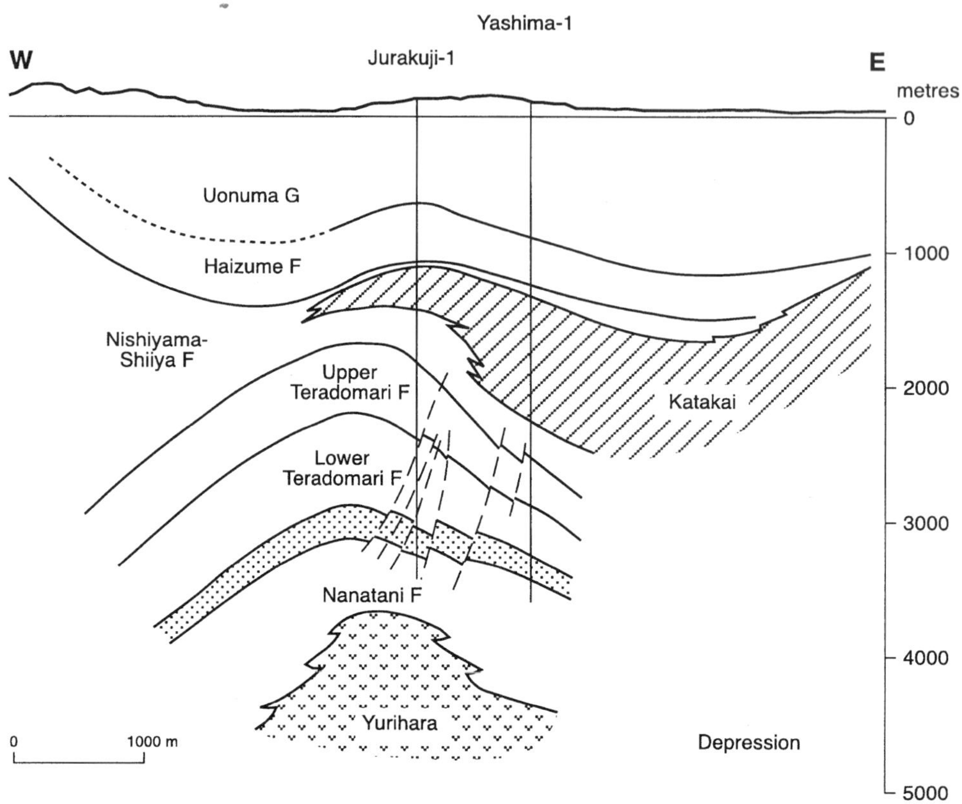

Fig. 6. Geological cross-section through an oil field in the Niigata district showing the dome structure of the volcanic reservoir at depth.

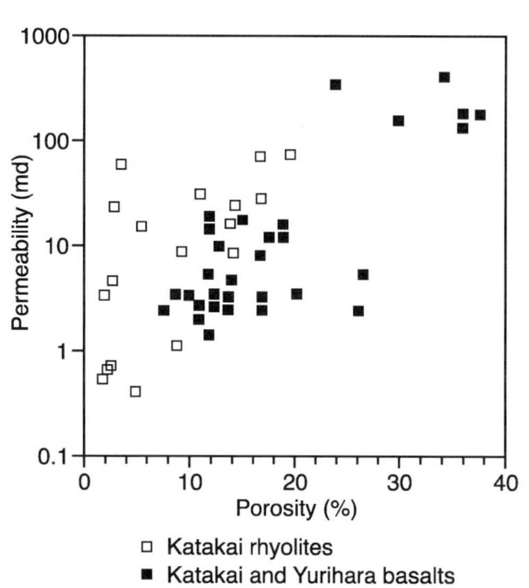

Fig. 7. Porosity–permeability relationship of volcanic reservoirs in the Niigata district (from Yamada & Uchida 1997).

Figure 8. It is assumed that any potential source rocks which may have existed prior to the volcanic eruption would have been destroyed, and their source-rock potential lost. After volcanism, marine, organic-rich shale deposition continued (Stage 1). Because the underlying volcanic rocks have significant primary and secondary porosity and permeability, some compaction fluid could have migrated downward (black arrows) as well as upward. During Stage 2, fine-grained clastic sediments (A–B) accumulated above the original shale layer, driving more compaction fluid downward. At the final stage (Stage 3), the organic-rich source rocks below level A reached their maturity level, with continued downward fluid migration, the prime cause of oil accumulation in the volcanic dome reservoir.

Pressure-seal and capillary-seal development

Shales immediately above the volcanic reservoirs in the Niigata region are commonly slightly

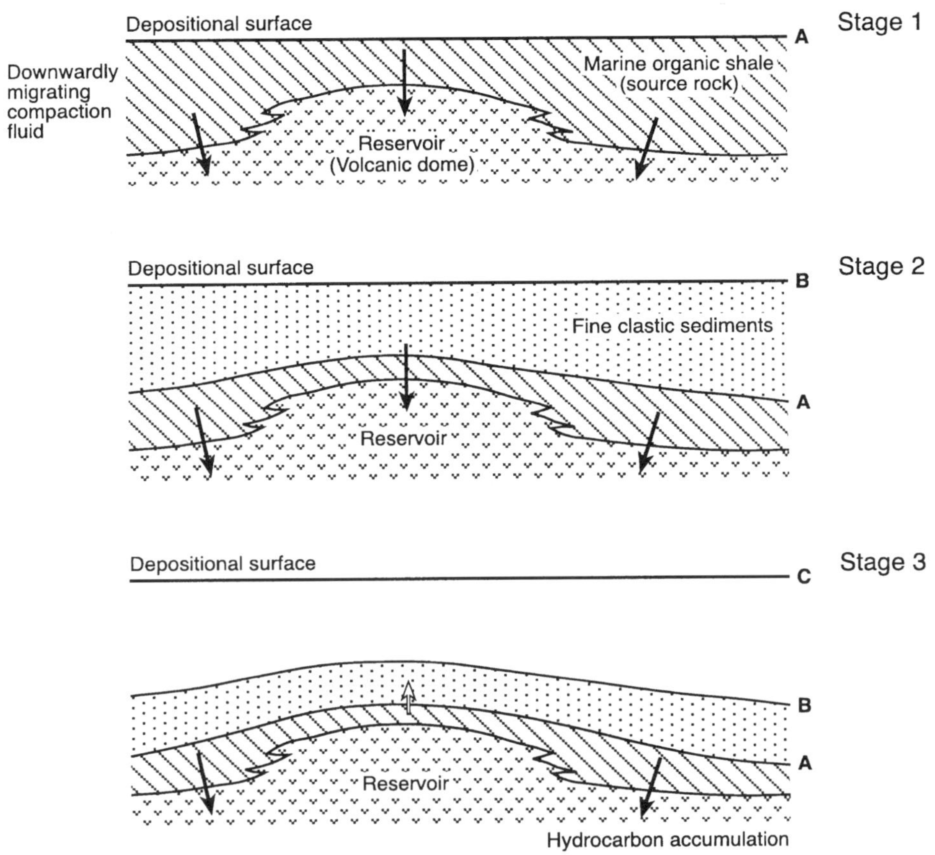

Fig. 8. Schematic model showing the history at three stages during the deposition and burial of volcanic reservoir and source rocks of Niigata district.

undercompacted and classified by Magara (1978) as pressure seals. Examples of shale porosity and fluid pressures as a function of depth from the Shiunji gas field in this region are shown in Figure 9. More regional pressure-seal developments are shown in the fluid-pressure profiles of Figures 10 and 11. The shales immediately above the volcanic reservoir section are usually well compacted, suggesting that the shales have lost most of their free-moving water, leaving behind semi-solid or structured water. They would thus have a high capillary sealing capacity against hydrocarbons in the underlying reservoirs. The shales at higher stratigraphic levels (further from the volcanic reservoirs) are, however, softer, undercompacted and more ductile. Fluid pressure in such shales is abnormally high and the potential direction of fluid movement is downward, as shown by black arrows in Figures 10 and 11. Note that this fluid flow direction is opposite to any upward hydrocarbon loss due to buoyancy forces.

As shown in Figure 12, a high hydrocarbon column in the reservoir would result in excess hydrocarbon pressure, P_h. However, this pressure can in principle be contained by the combined effect of both capillary sealing by overlying compacted shale and the pressure seal (P_{sh}). Should the value of P_h exceed the combined capillary and pressure sealing effect, some hydrocarbons will be lost through any open fractures in the cap rock. Due to the undercompacted and ductile nature of the pressure-sealing shales, however, the fractures would quickly close up, and upward hydrocarbon loss would be minimized.

High geothermal gradients and effective oil maturation and migration

Source rocks in the Niigata region commonly have up to 2 wt% organic carbon. Organic matter is either type I or II, which is considered

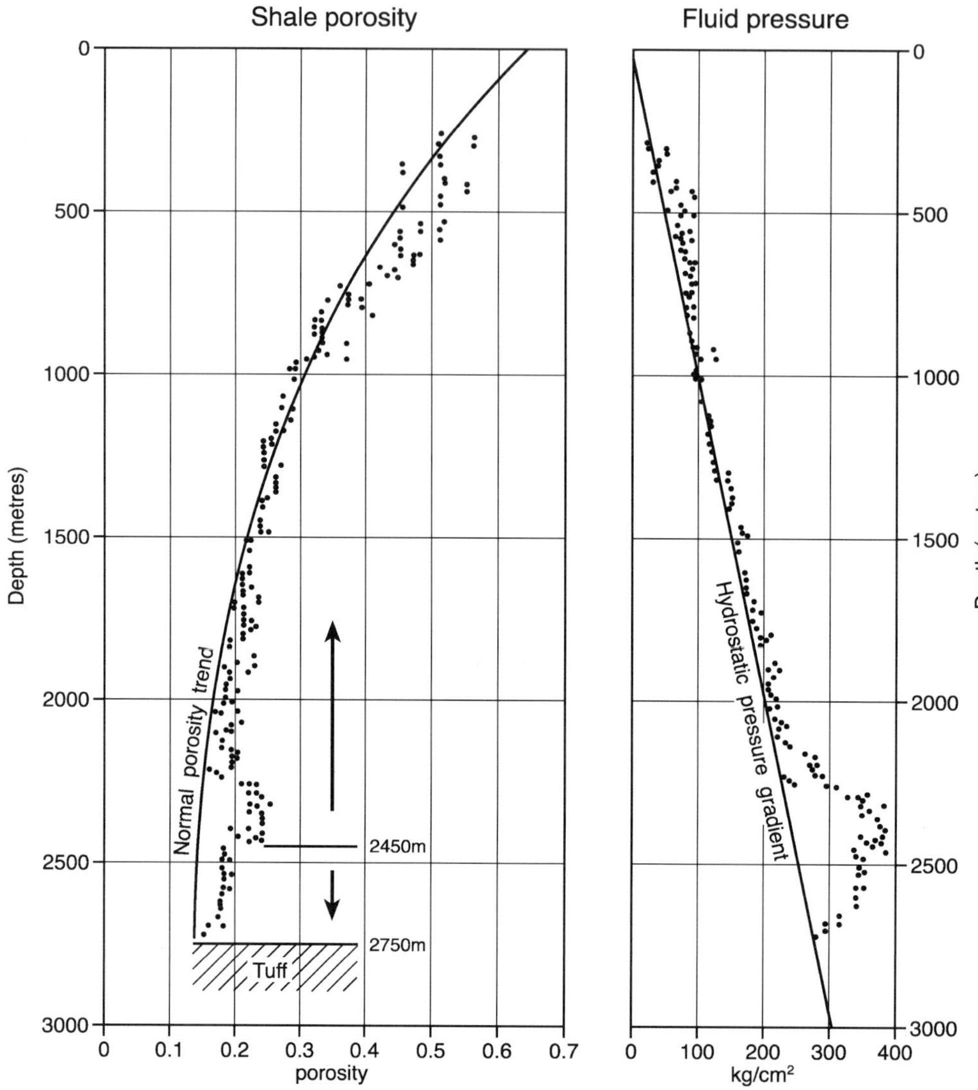

Fig. 9. Porosity–depth and calculated pressure–depth plots of Shiunji gas field, Niigata district (from Magara 1978).

to be an effective source for oil. As shown in the geological column (Fig. 4), the age of most reservoir and source rocks ranges from Middle Miocene (15 Ma) to Pliocene (3 Ma). For source rocks to achieve thermal maturation over such a short time interval (*c.* 12 Ma), requires temperatures in excess of 180 °C (see Fig. 13). Although uncommon, examples from other parts of the world where rapid heating on short time scales has resulted in hydrocarbon generation include the Los Angeles and Ventura basins, California. The measured high geothermal gradients in the study area are caused by both volcanism and associated hydrothermal effects, and are compatible with thermal maturation of the source rocks driven by heat related to volcanic activity, with the same heat source also providing at a later stage the reservoir rock.

Such high geothermal gradients may also have caused relatively fast compaction and effective fluid (water, oil and gas) expulsion from the shales. Figure 14 compares porosity–depth relationships of various sedimentary basins around the world; the rate of compaction of curve 7 (Nagaoka Plain, Japan, Magara, 1978) below 5,000 ft (the oil and gas generation zone), is

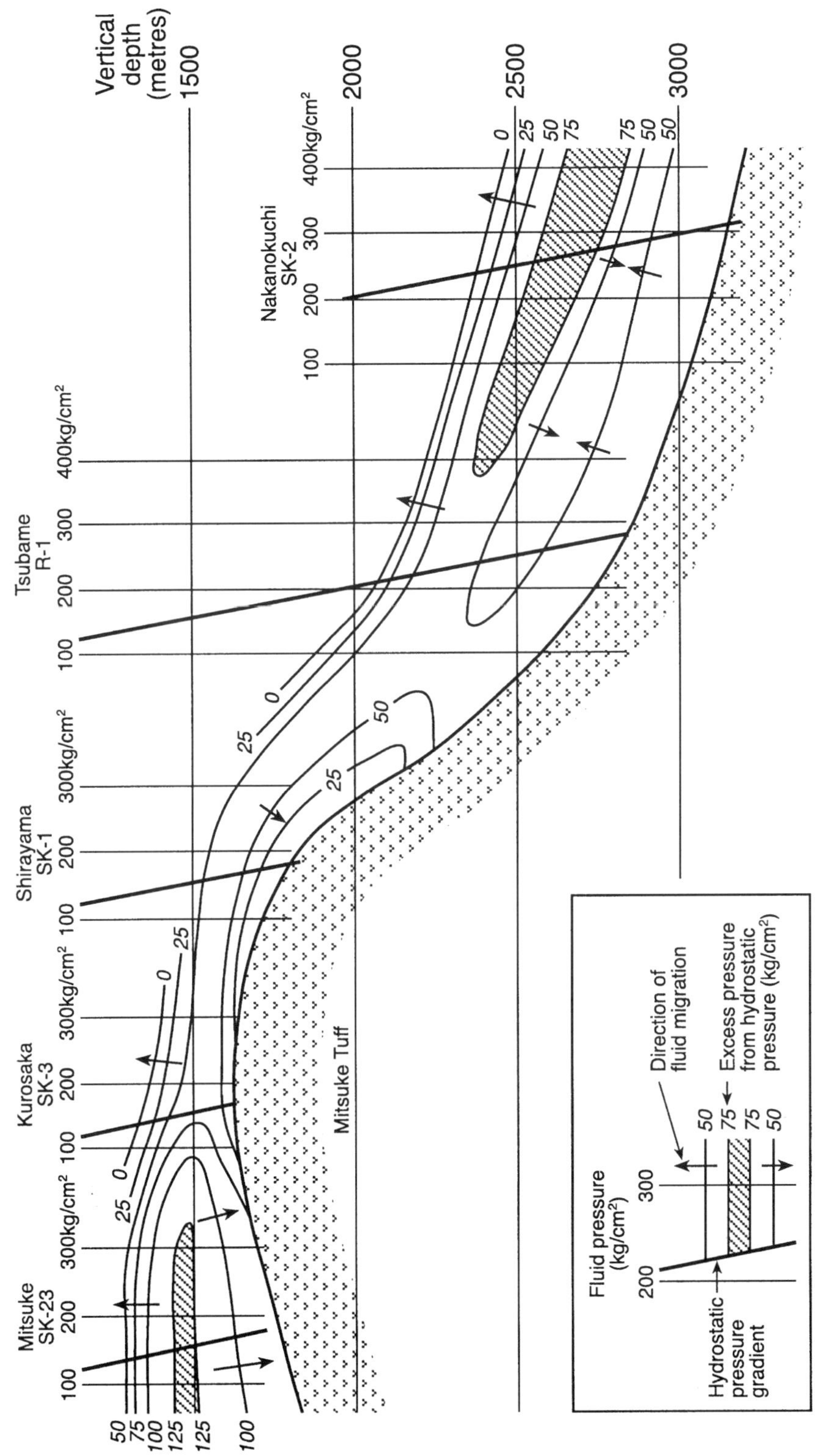

Fig. 10. Calculated fluid-pressure profile in Mitsuke Field, Niigata (from Magara 1978).

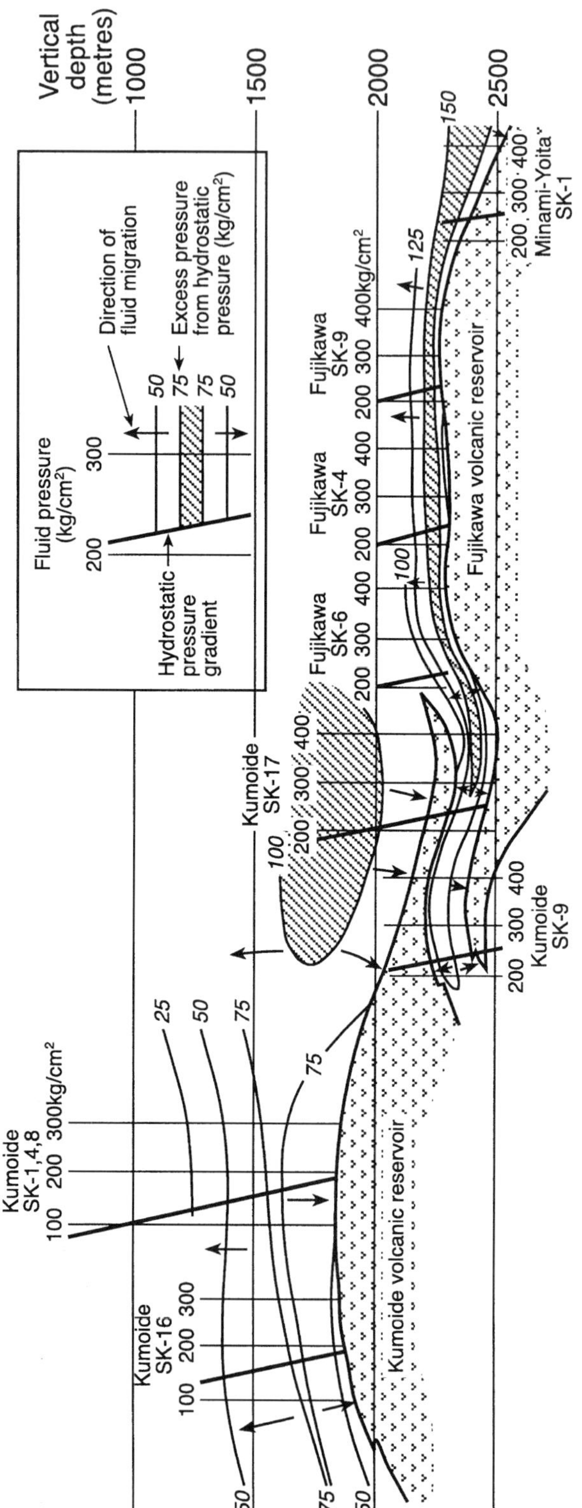

Fig. 11. Calculated fluid-pressure profile in Fujikawa–Kumoide Fields, Niigata (from Magara 1978).

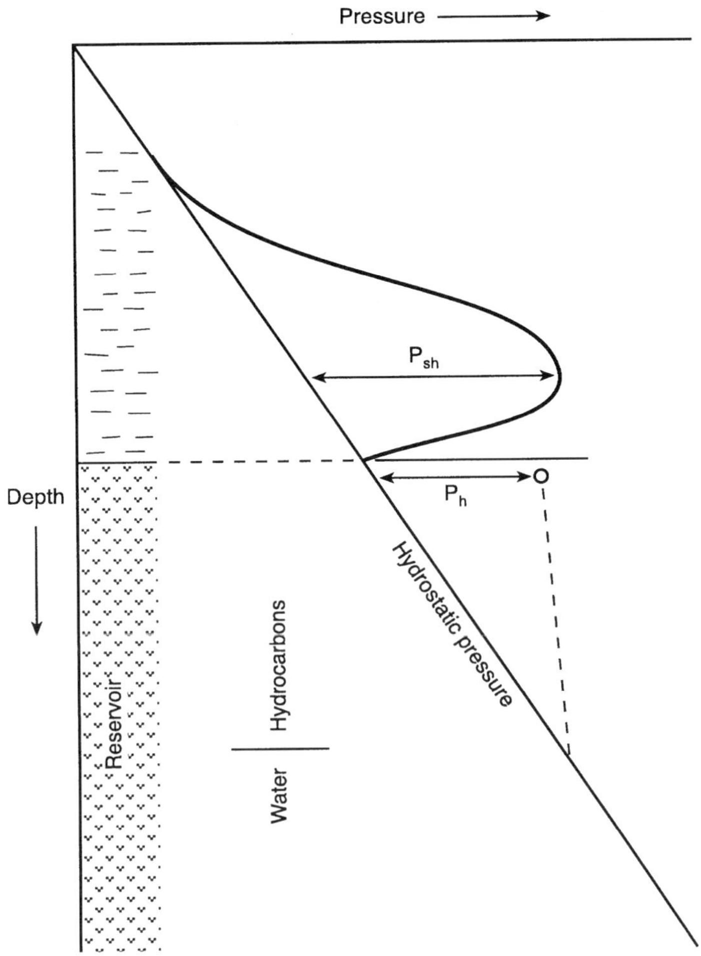

Fig. 12 Schematic diagram showing maximum sealing pressure P_{sh} and excess Hydrocarbon pressure P_h (from Magara 1978).

much faster than curve 10, for sediments of similar geological ages in the Lousiana Gulf Coast (e.g. Foster & Whalen 1966).

Summary and conclusions

Although reserves are relatively small, the volcanic reservoirs of northwestern Honshu island, Japan, share a number of common factors that together have resulted in oil and gas accumulations:

(1) A high geothermal gradient up to 5.5 °C/100 m promoted rapid thermal maturation of source shales and fast and efficient hydrocarbon migration.
(2) Marine volcanism, in particular the extrusion of viscous rhyolitic magmas which cooled to form buried hills or synchronous highs, was followed by thick deposition of organic-rich marine shales from which effective downward fluid migration took place. Differential compaction and radial fluid migration to the centre of the synchronous highs from the surrounding area was an important process.
(3) The combined effects of the capillary seal and an undercompacted pressure seal prevented significant vertical loss of accumulated hydrocarbons.
(4) Rapid chilling, spalling, autobrecciation and hydrothermal processes that typically accompany marine volcanism helped to form both primary and secondary porosity in volcanic rocks that rendered them suitable as reservoir rocks for hydrocarbon accumulations.

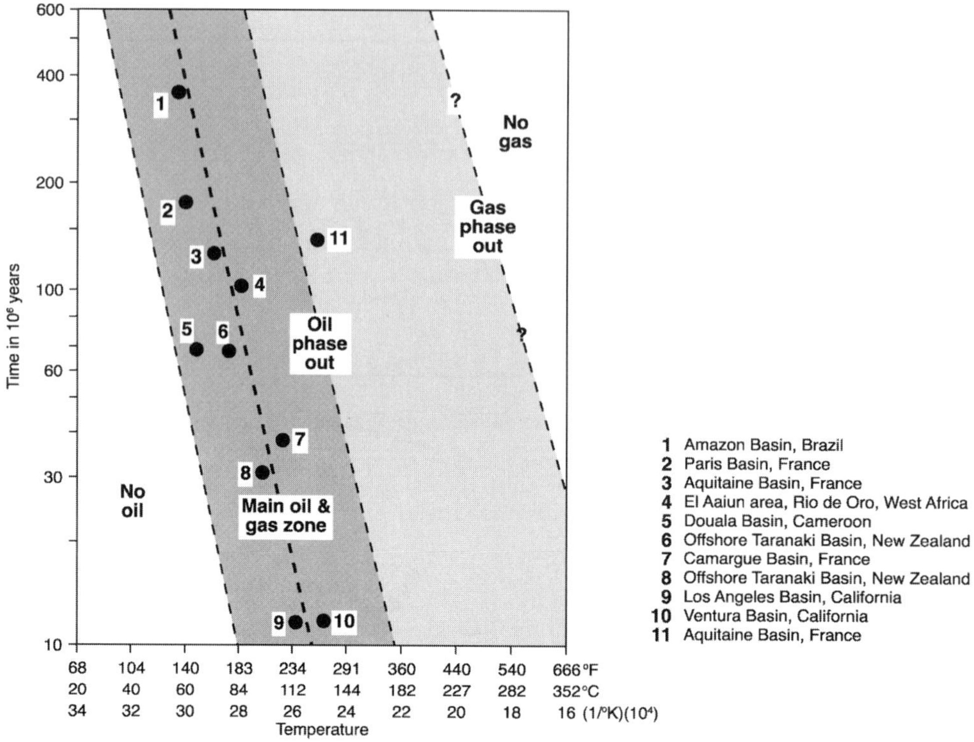

Fig. 13. Time–temperature relationship of petroleum genesis (from Hunt 1974).

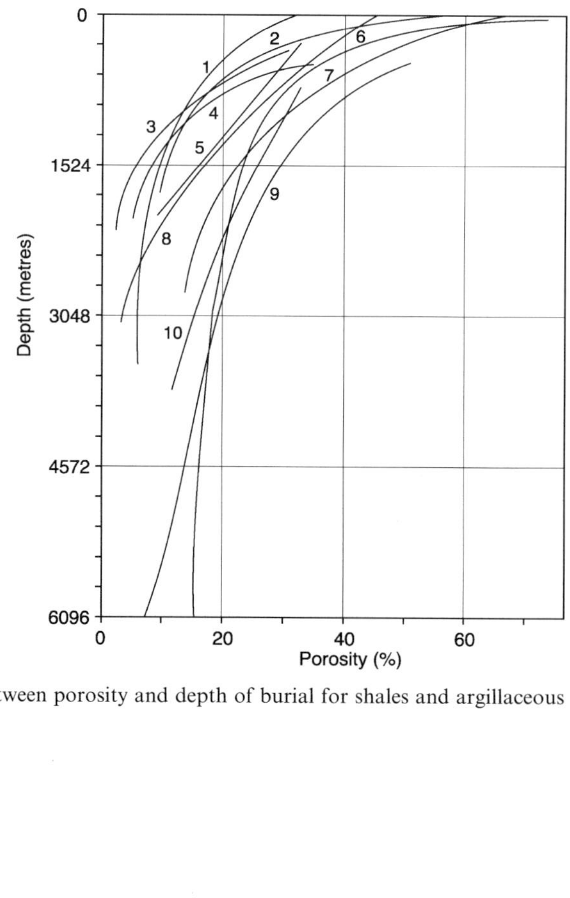

Fig. 14 Relationship between porosity and depth of burial for shales and argillaceous sediments (from Rieke & Chilingarian 1974).

References

FOSTER, J. B. & WHALEN, H. E. 1966. Estimation of formation pressures from electrical surveys—offshore Louisiana, *Journal Petrol. Technology* **18**, 165–171.

HUNT, J. M. 1974. How deep can we find economic oil and gas accumulations? *SPE5177, 1974 Deep Drilling and Production Symposium, Preprint*, 103–110.

IKEBE, Y. 1963, Geologic sections of Akita Region. *In*: *Oil Mining Manual (Sekiyuougyou Bunran)*, Japanese Association for Petroleum Technology, 498.

KOMATSU, N., FUJITA, Y. & SATO, O. 1983. Cenozoic volcanic rocks as potential hydrocarbon reservoirs. *Proceedings of the 11th World Petroleum Congress*, **2**, 411–420.

MAGARA, K. 1978. *Compaction and Fluid Migration, Practical Petroleum Geology*. Elsevier Scientific Publishing Co., Amsterdam.

RIEKE, III, H. H. & CHILINGARIAN, G. V. 1974. *Compaction of Argillaceous Sediments*. Elsevier Scientific Publishing Co., Amsterdam.

YAMADA, Y. & UCHIDA, T. 1997. Characteristics of hydrothermal alteration and secondary porosities in volcanic rock reservoirs, the Katakai gas field. *Journal of the Japanese Association for Petroleum Technology*, **62**, 311–320.

Oil and gas production from basement reservoirs: examples from Indonesia, USA and Venezuela

TAKO KONING

Texaco Angola Inc., Luanda, Angola

Abstract: Basement rocks are important oil and gas reservoirs in various areas of the world. Such reservoirs include fractured or weathered granites, quartzites, or metamorphics. In South America, basement reservoirs occur in Venezuela and Brazil. In North Africa, basement oil and gas production occurs in Morocco, Libya, Algeria and Egypt. Significant basement reservoirs occur in the West Siberia basin as well as in China. In the USA, basement-derived oil production occurs in a number of areas, including California (Wilmington and Edison fields), Kansas (El Dorado and Orth fields) and Texas (Apco field). In Southeast Asia, basement reservoirs are the main contributor of oil production in Vietnam. In Indonesia, to date oil and gas production from basement rocks has been minimal. However, the recent large gas discovery in pre-Tertiary fractured granites in southern Sumatra has led to a focusing of exploration in Indonesia for basement reservoirs.

The term 'basement rocks' generates a variety of definitions by geologists depending on the specific sedimentary basin discussed as well as the individual's experience in that area. Most workers consider basement as any metamorphic or igneous rock (regardless of age) which is unconformably overlain by a sedimentary sequence. Oil or gas may have migrated into older porous metamorphic or igneous rocks, thereby forming a basement reservoir. However, in some basins such as the Central Sumatra basin, the basement rocks may be partially or completely unmetamorphosed. Therefore, the most appropriate definition of 'basement' is that of Landes *et al.* (1960) which stated: 'the only major difference between basement rock and the overlying sedimentary rock oil deposits is that in the former case the original oil-yielding formation (source rock) cannot underlie the reservoir'. A final comment on the definition of basement rocks is that further exploration, geological and geochemical studies in a specific area may result in revisions of commonly accepted definitions of basement rocks in that area. Further exploration may indeed prove the existence of hydrocarbon source rocks located stratigraphically within rocks previously regarded as basement. Accordingly, the explorationists' definition of basement rocks cannot be rigid but must be responsive to new geological ideas and data.

Indonesia

Central Sumatra

Oil and gas production from pre-Tertiary basement rocks is rare within the Tertiary back arc (foreland) basins of western Indonesia. The Beruk Northeast field (Fig. 1) is the only field in the prolific Central Sumatra basin that produces from basement (Koning & Darmono 1984). The field was discovered in 1976 by the drilling of Beruk Northeast No. 1 which tested 1680 BOPD from fractured basement quartzites (Fig. 2). Approximately 2 million barrels of oil have been produced from quartzites, weathered argillites and weathered granite. The basement rocks have K–Ar radiometric age dates varying from Early Permian to Early Cretaceous, indicating a complex pre-Tertiary geological history. The Beruk Northeast field presents challenging production problems due to the great variability in reservoir rocks, the presence of at least four separate oil–water contacts, and a possible unrecognized water-bearing fracture system.

Southern Sumatra

Exploration targeted for basement hydrocarbons has met with recent success in southern Sumatra, where operator Gulf Indonesia has reported the significant Suban gas discovery (Fig. 1). Three wells drilled in 1999 in the Suban field have defined a gas pool located within fractured pre-Tertiary granites. Gas flow rates of 26 million cubic feet of gas per day were obtained from the Durian Mabok-2 well. Test data combined with seismic mapping indicates a gas pool with a minimum gas column of 500 m covering an area of at least 72 km^2 (Gulf Indonesia 1999; Koning 2000). Reserves are estimated at approximately 5 trillion cubic feet of gas. On an oil equivalency basis, using 1000 cubic feet of gas to 1 barrel of oil, this field has oil equivalent reserves of 500 million barrels which places it in

From: PETFORD, N. & MCCAFFREY, K. J. W. (eds) 2003. *Hydrocarbons in Crystalline Rocks*. Geological Society, London, Special Publications, **214**, 83–92. 0305-8719/03/$15

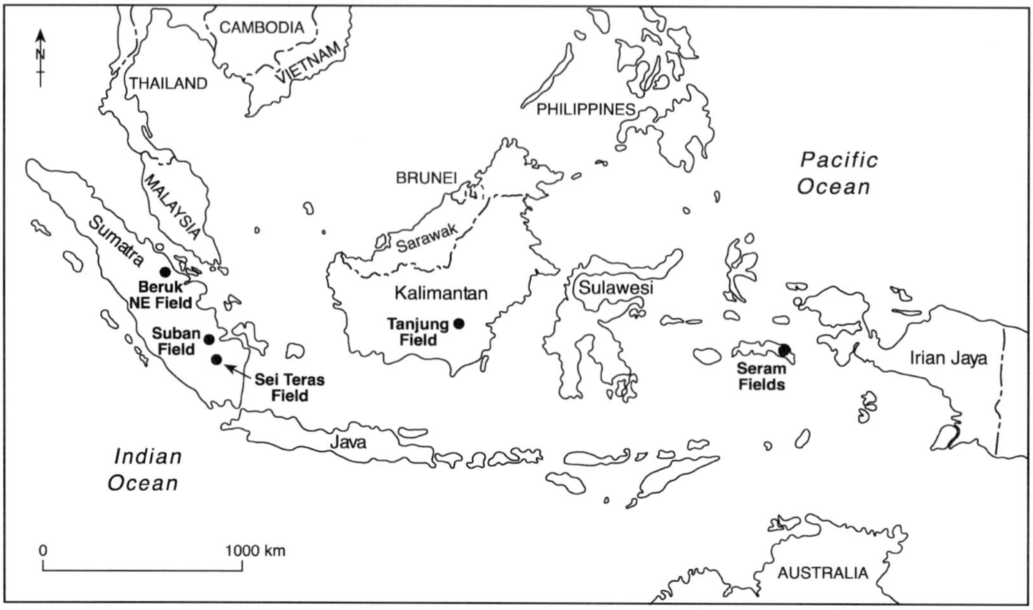

Fig. 1. Locations of Indonesian oil fields producing from pre-Tertiary basement rocks.

the ranks of a 'giant' oil field (AAPG definition of a giant oil field is one with reserves >0.5 billion barrels of oil). Likely markets for the gas include the Duri steam flood project in central Sumatra as well as power generation projects in Singapore.

The area where the Suban field was discovered was previously subjected to a number of exploration campaigns by various operators. The search was for oil in structural highs in the Tertiary Talang Akar and Batu Raja formations. A number of wells 'tagged' into basement in order to tie the top of basement into seismic data. None of these wells penetrated sufficiently deep into basement to discover the giant Suban gas field until Gulf discovered the field in 1999.

Kalimantan

The Tanjung field in the Barito basin, southern Kalimantan, was discovered in 1938 and has produced over 21 million barrels of oil from pre-Tertiary basement rocks (Figs 3, 4 and 5). Oil occurs in volcanics, pyroclastics and metamorphosed sandstones and claystones which are locally deeply weathered and fractured. A general structural section through the oil field is shown in Figure 3.

The Beruk Northeast and Tanjung fields share many similarities. For example, both fields occur within faulted anticlines. The overlying thickness of Tertiary sediments in both fields is less than 2000 m. The likely oil source rocks for these fields are the adjacent and deeper Tertiary shales. The Beruk Northeast, Tanjung and Suban fields indicate that pre-Tertiary basement is a valid oil exploration objective in the Tertiary basins in western Indonesia and that, whenever feasible, exploration wells in these basins should be drilled into basement.

USA

Kansas

In Kansas, oil is produced from Precambrian basement rocks in the central Kansas uplift (Fig. 6). The Precambrian rocks include quartzite, schist, gneiss and granite; however, fractured quartzite is the reservoir rock most often penetrated since it occurs on the summits of many buried Precambrian hills (Landes *et al.* 1960 and Fig. 7). Basement oil pools include the Orth, Ringwald, Kraft-Prusa, Beaver, Bloomer, Trapp, Eveleigh and Silica fields. The source rocks are flanking Cambro-Ordovician shales or overlying Pennsylvanian shales. Production from fields such as the Orth and Ringwald fields is relatively low at production rates varying between 120 and 190 BOPD.

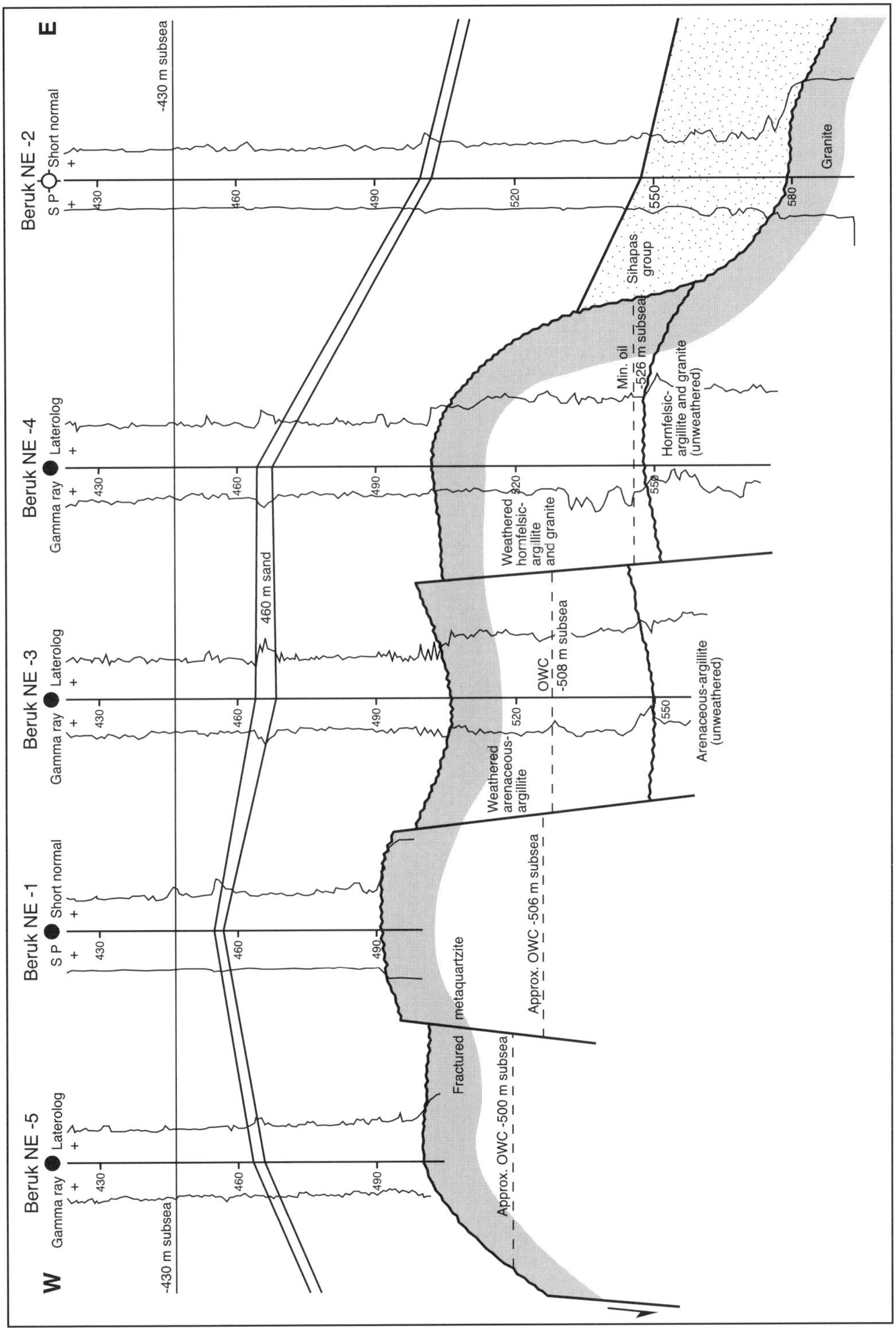

Fig. 2. Structural cross-section through the Beruk northeast field, Sumatra. (Koning and Darmono, 1984).

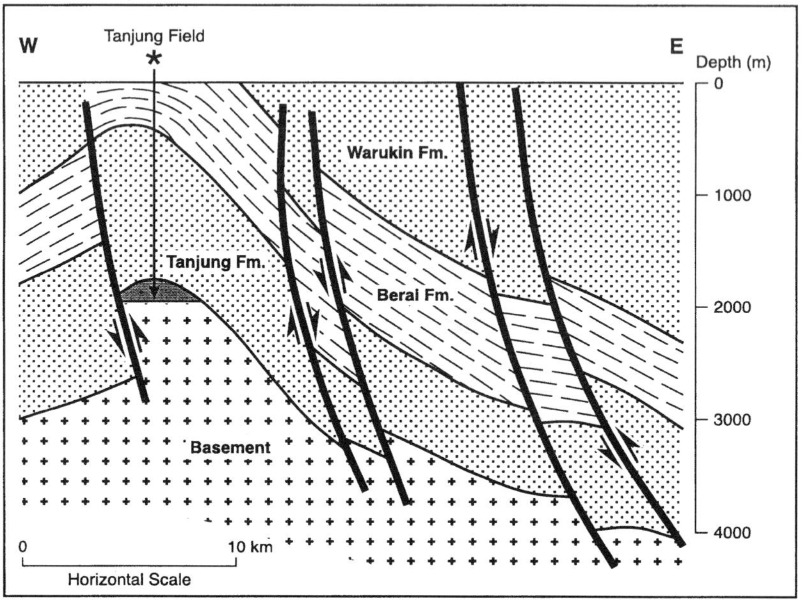

Fig. 3. Structural cross-section through the Barito basin (Tanjung area), Kalimantan (Koning, 2000).

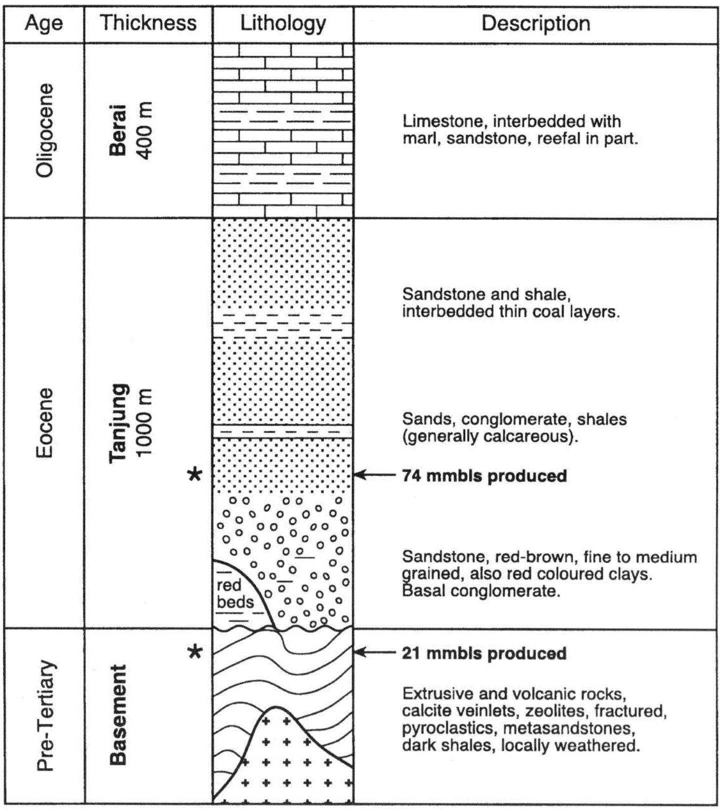

Fig. 4. The general stratigraphy of the Tanjung field, Kalamantan (Koning, 2000).

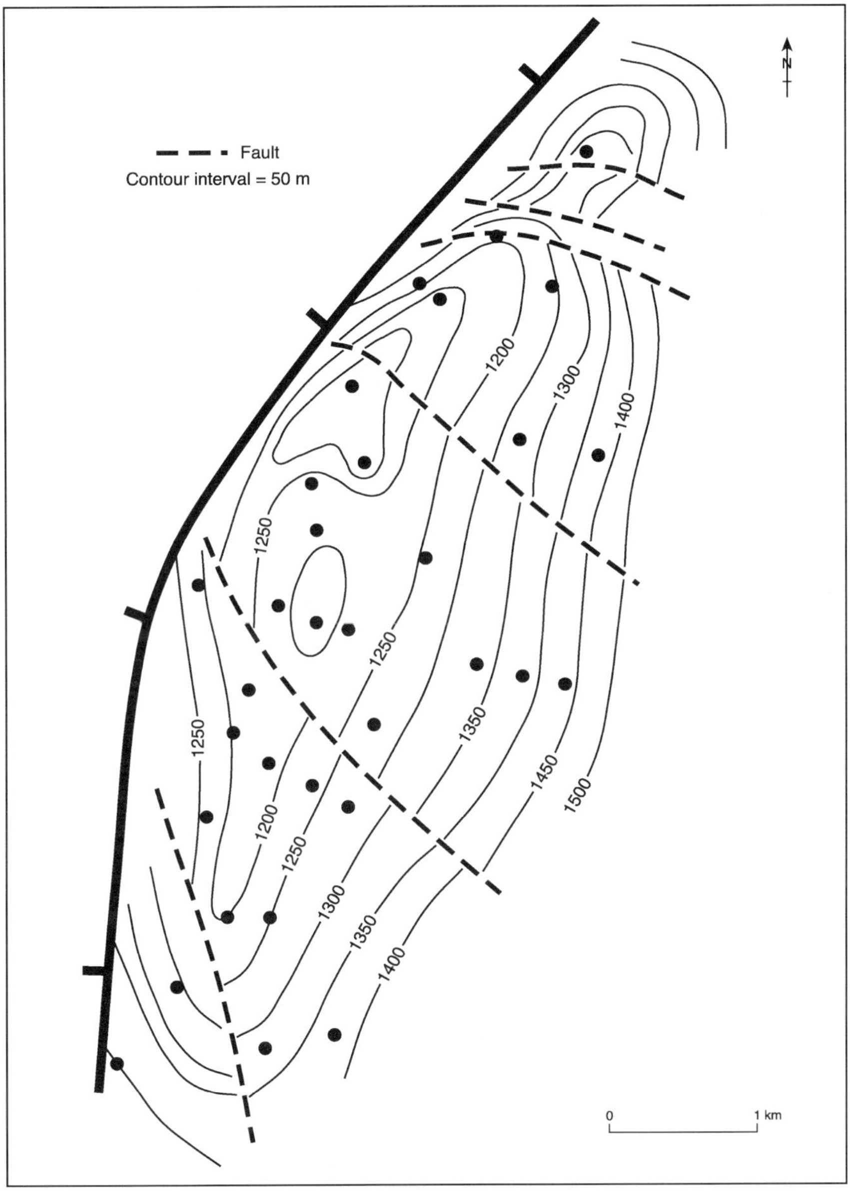

Fig. 5. Structure on top of basement, Tanjung field, Kalamantan (Koning, 2000).

California

In California, oil is produced from basement consisting of fractured Jurassic schists. Fields which contain basement reservoirs include the Playa del Rey, El Segundo, Santa Maria, Wilmington and Edison fields (Fig. 8). Relatively few wells produce from basement rocks alone; most are multiple completions in both the basement schist and overlying schist conglomerate and Tertiary sandstones. The majority of the oil-producing schists are in a relatively high position and have usually undergone weathering and erosion, thus increasing the porosity (Landes *et al.* 1960). The wells in the Edison field had an average production rate of about 1,000 BOPD and cumulative production in the field has exceeded 20 million barrels of oil. The Wilmington field has produced more than 22 million barrels of oil from basement with rates

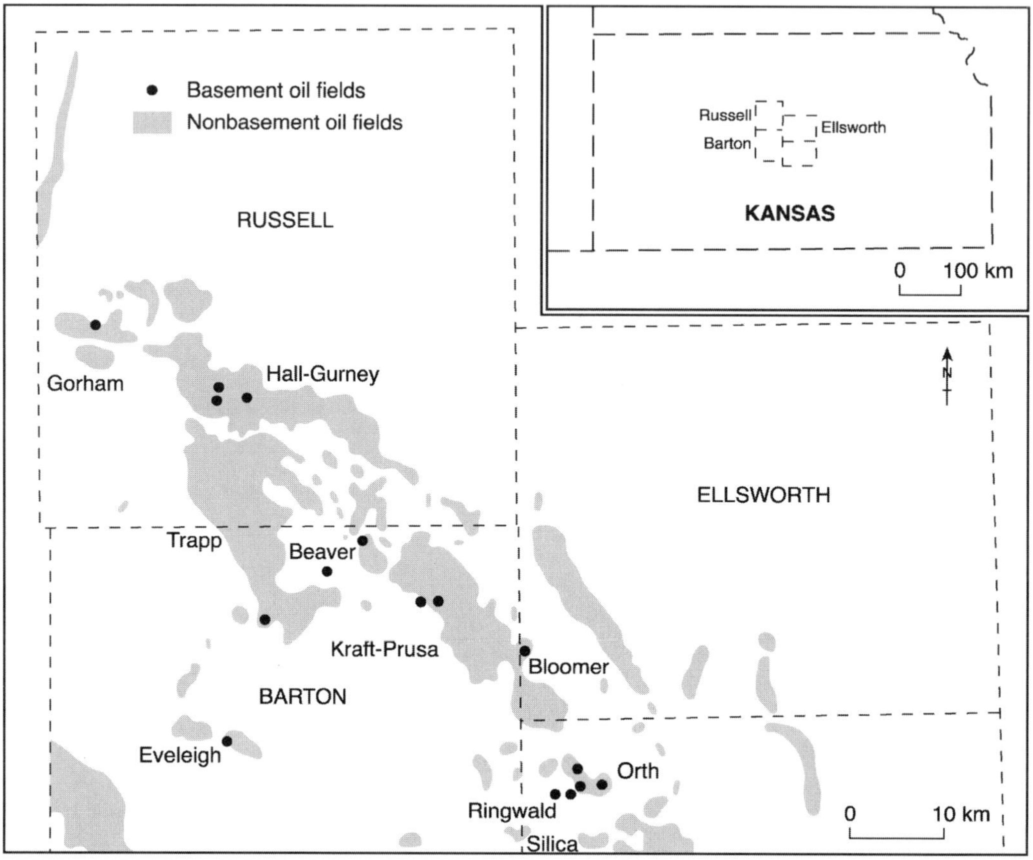

Fig. 6. Map showing the location of the Kansas basement oil fields. (Landes *et al.* 1960).

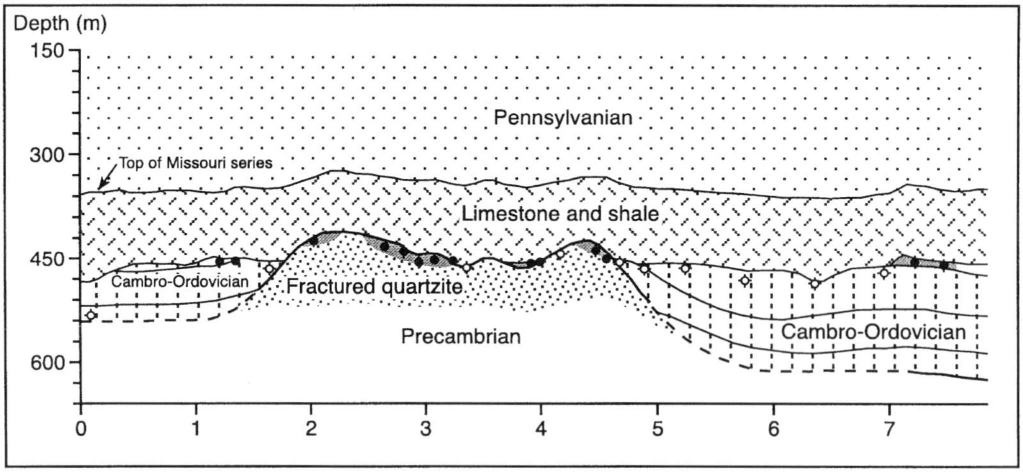

Fig. 7. Kansas basement oil production. Oil is produced from Precambrian basement (in section), most commonly fractured quartzites. Oil is sourced from flanking Cambro-Ordovician or overlying Pennsylvanian rocks. (Landes *et al*, 1960).

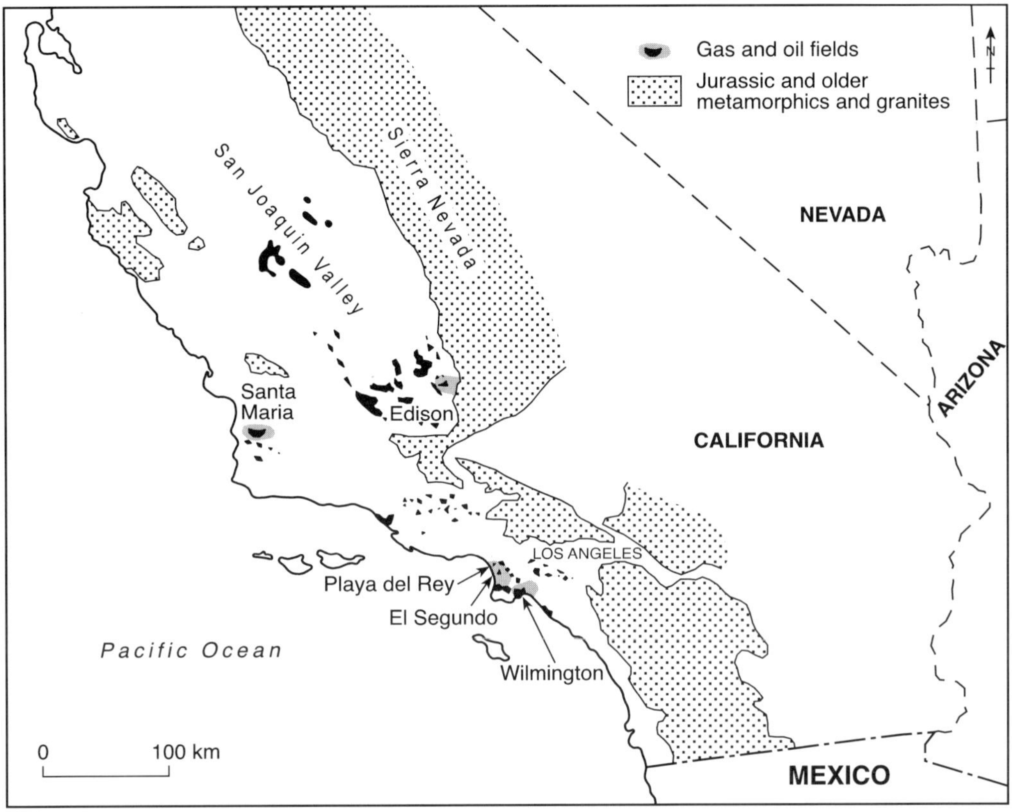

Fig. 8. Map showing the main Californian gas and oil wells of El Segundo, Santa Maria, Wilmington, Playa del Ray and Edison. (Landes *et al.* 1960).

of production varying from 1,200 to 2,000 BOPD. Figure 9 shows a cross-section through the El Segundo field, highlighting the hydrocarbon accumulations in the upper layer of porous schists.

Venezuela

Within Venezuela's Maracaibo basin, oil is produced from fractured granitic and metamorphic basement rocks in the La Paz and Mara fields, which are located 50 km NW of Lake Maracaibo (Fig. 10). The depth to basement in the two fields ranges from 2750 m to 3500 m. These two fields occur along the crest of a NE–SW trending, strongly folded and faulted anticline (Stevenson 1951).

La Paz field

The La Paz field was discovered in 1923 and has produced more than 830 million barrels of oil from low porosity Cretaceous limestones and underlying granitic basement (Nelson *et al.* 2000). The first basement reservoir wells were not drilled until 1953. Cumulative oil production from basement is approximately 245 million barrels and estimated remaining reserves of 80 million barrels occur within basement (Talukdar & Marcano 1994). During the initial development of the field, wells were drilled into basement with an average penetration of 500 m (Chung-Hsiang P'an 1982). Maximum initial production was 11,500 BOPD and the average initial production was 3,600 BOPD. A geological cross-section through the La Paz field is shown in Figure 11.

The Mara field, discovered in 1944, lies on the northeastern extension of the La Paz anticline and has produced 27 million barrels of oil from basement. Remaining reserves are estimated at 5 million barrels. Average penetration into basement was 360 m and initial production averaged 2,200 BOPD. The combined production from basement rocks in these fields exceeded 75,000 BOPD. Cores of basement rocks show intense fracturing, most commonly in vertical planes,

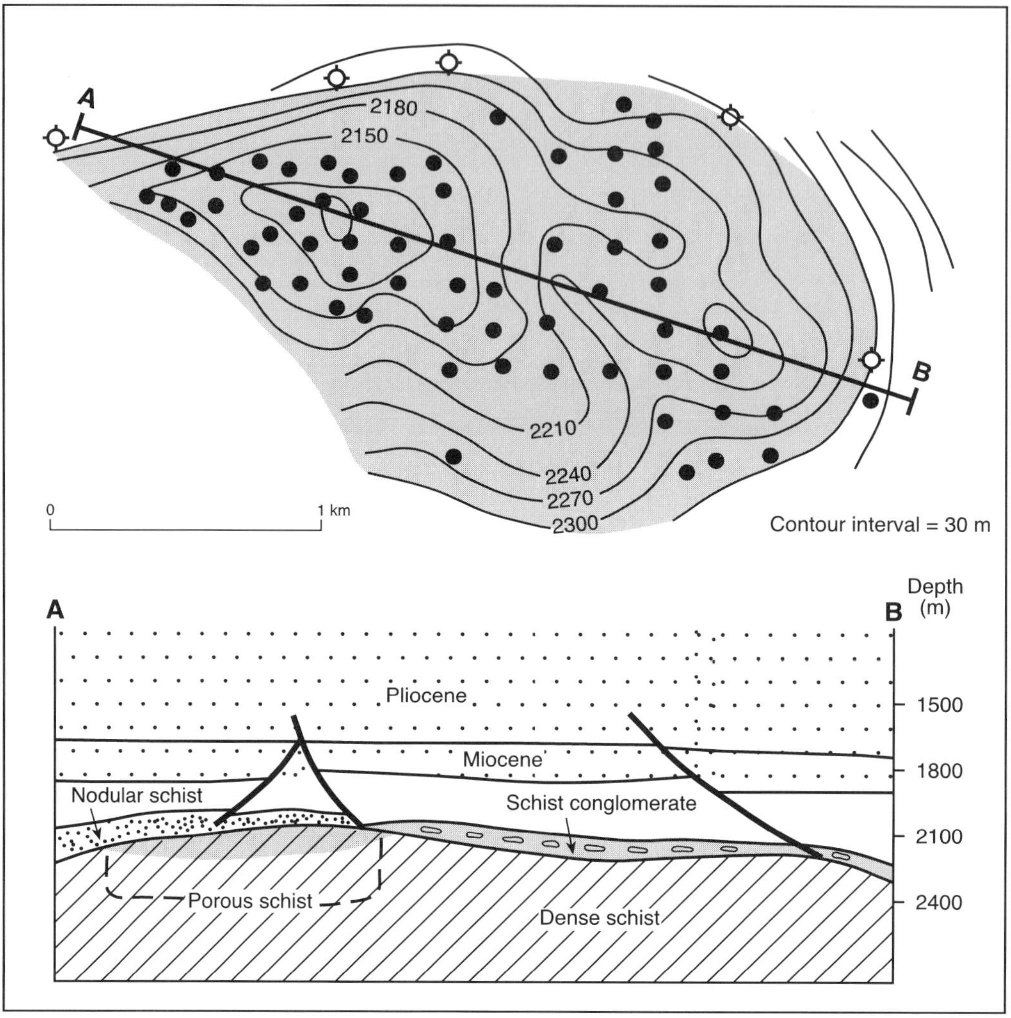

Fig. 9. Cross-section through the El Segundo field, California. The reservoir is in fractured Jurassic schists in the west and schist and conglomerate in the east. The average depth of oil basement production is *c.* 2300 m. (Landes *et al.* 1960).

and many core recoveries are poor (Smith 1955). The oil source rocks are overlying Upper Cretaceous La Luna marine shales which are immature at the fields, but mature to the south of the fields.

Discussion

Characteristics of oil and gas recovery in basement rocks

The following is a generalized summary of the oil field experiences of companies dealing with basement oil and gas fields:

(1) Oil and gas fields in basement rocks are usually discovered 'by accident'. Typically during drilling operations, the well will reach total depth (TD) in basement, encounters oil or gas shows in basement and the well is tested resulting in a basement oil or gas discovery. For example, the Beruk Northeast-1 well, central Sumatra, was targeted for oil in Tertiary sediments. Oil shows were noticed in fractured basement quartzites which resulted in a drill stem test on the top of the basement. This led to the Beruk Northeast oil discovery.

(2) Basement reservoirs can be very prolific if basement is highly faulted and fractured, as in the case of a quartzite reservoir. For

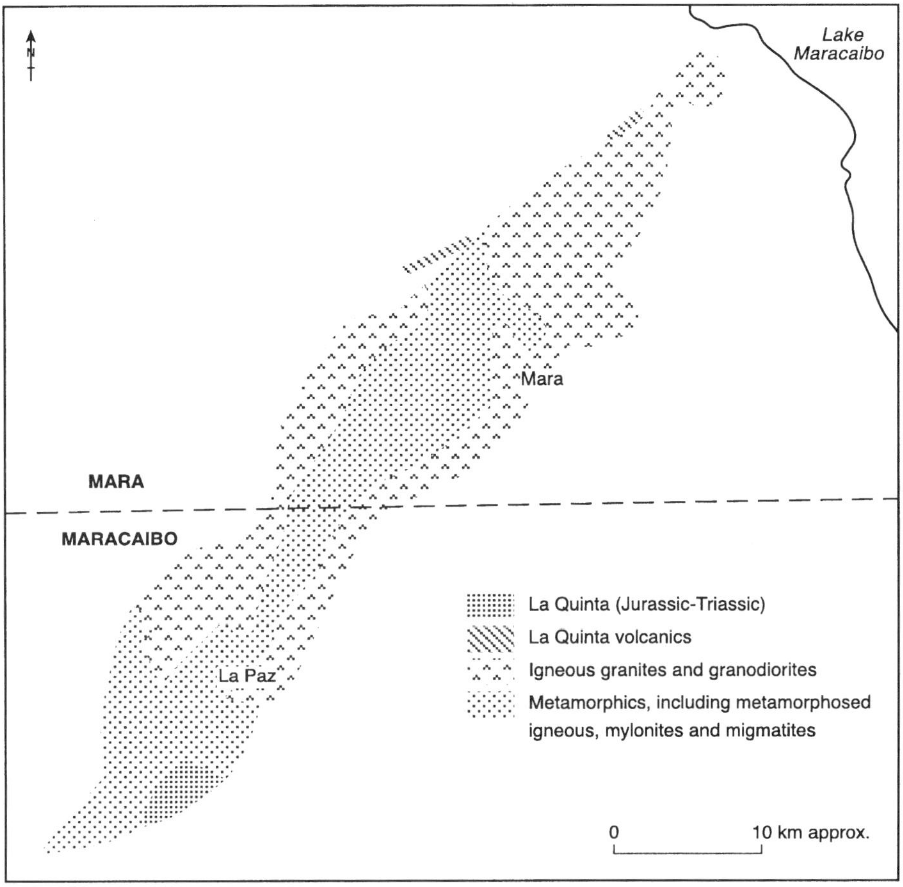

Fig. 10. Summary map showing the main lithologies comprising the Mara and Maracaibo oil fields, Venezuela. (Landes *et al.* 1960).

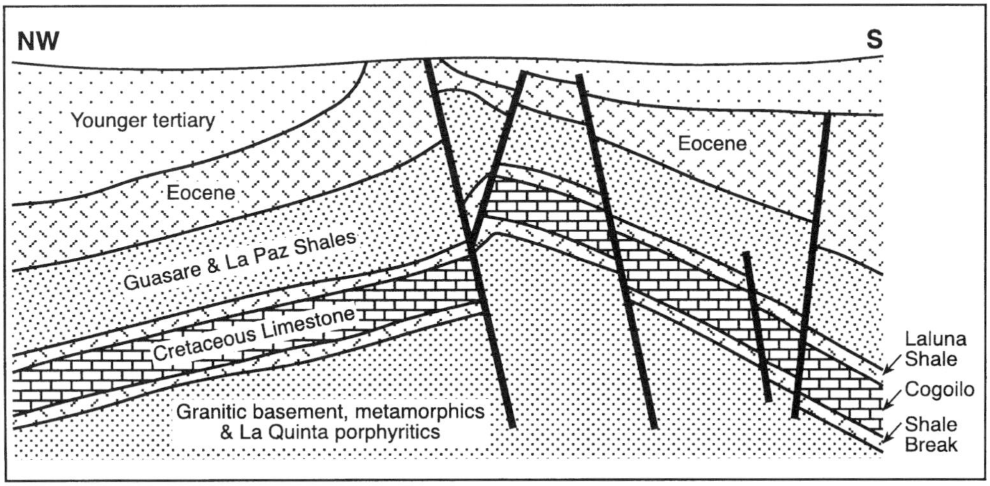

Fig. 11. Vertical section through the La Paz field, Venezuela. (Landes *et al.* 1960).

example, the Bach Ho (White Tiger) field in Vietnam produced in the range of 130,000 BOPD from fractured granite basement (*Offshore Magazine*, August 1994). Also weathered granite can be a highly effective reservoir. Weathering of granites, especially under humid tropical conditions, can result in very porous secondary porosity penetrating 100–200 m into the granite. These weathered granites can appear like coarse sandstones (granite wash sandstones) in hand specimen or core.

(3) Due to basement being often highly fractured and highly permeable, initial oil or gas flow rates are often very high and indeed deceptively high. This may lead to the operator overbuilding the production facilities. When the field is placed on production, rapid production declines may be experienced since the reservoir may have only fracture porosity and minimal matrix porosity. Also, early high volume water influx may be experienced.

(4) Basement oil fields are typically very complex reservoirs with multiple lithologies, possibly two or more fracture systems and multiple oil–water or gas–water contacts. Accordingly, these reservoirs need to be closely studied. Extensive core coverage is critically important, as are full log suites. Coring is typically difficult due to the fractured nature of the reservoir (Lamb 1997). Effective development of basement reservoirs can be a major challenge for reservoir engineers and geoscientists.

(5) Exploration for basement reservoirs should provide provision in the drilling programme to allow for adequate penetration of basement of at least 100 m into the top of basement. The top of basement may be tight but porosity may occur below the overlying tight zone.

(6) Prolific oil and gas fields in basement rocks in Libya, Vietnam, Indonesia, USA, Venezuela and elsewhere serve as a reminder to evaluate, if possible, the underlying basement, especially if the top of basement is structurally high and indications from seismic or regional geology are that the basement may be weathered or faulted.

Summary

The above-mentioned oil and gas accumulations in basement rocks are examples of such fields in three important oil-producing countries. These fields serve as a reminder of the Landes *et al.* (1960) classic paper on petroleum resources in basement rocks, in which the authors succinctly state the following: 'commercial oil deposits in basement rocks are not geological "accidents" but are oil accumulations which obey all the rules of oil sourcing, migration and entrapment; therefore in areas of not too deep basement, oil deposits within basement rocks should be explored with the same professional skill and zeal as accumulations in the overlying sediments.'

References

CHUNG-HSIANG P'AN 1982. Petroleum in basement rocks. *American Association of Petroleum Geologists Bulletin*, **66**, 1597–1643.

GULF INDONESIA RESOURCES LIMITED 1999. Press Release: Gulf Indonesia, Talisman and Pertamina announce substantial gas discovery in South Sumatra.

KONING, T. 2000. Oil production from basement reservoirs—examples from Indonesia, USA and Venezuela. *Proceedings of the 16th World Petroleum Congress, Calgary*.

KONING, T. & DARMONO, F. X. 1984. The geology of the Beruk Northeast Field, Central Sumatra—oil production from pre-Tertiary basement rocks. *Proceedings of the Thirteenth Annual Convention, May 29–30, 1984*. Indonesian Petroleum Association, Jakarta, Indonesia.

LAMB, C. F. 1997. Basement reservoirs—an overlooked opportunity. *Canadian Society of Petroleum Geologists and Society of Economic and Petroleum Mineralogists Joint Convention, Calgary, Proceedings*.

LANDES, K. K., AMORUSO, J. J., CHARLESWORTH, L. J., HEANY, F. & LESPERANCE, P. J. 1960. Petroleum resources in basement rocks. *American Association of Petroleum Geologists, Bulletin*, **44**, 1682–1691.

NELSON, R. A., MOLDOVANUI, E. P., MATCEK, C. C., AZPIRITXAGA, I. & BUENO, E. 2000. Production characteristics of the fractured reservoirs of the La Paz field, Maracaibo basin, Venezuela. *American Association of Petroleum Geologists*, **84**(11), 1791—1809.

SMITH, J. E. 1955. Basement reservoir of La Paz—Mara Oil Fields, Western Venezuela. *American Association of Petroleum Geologists, Bulletin*, **40**, 380–385.

STEVENSON, M. 1951. The Cretaceous limestone producing areas of the Mara and Maracaibo Districts, Venezuela—reservoir and production engineering. *Third World Petroleum Congress*, section 1, preprint 14.

TALUKDAR, S. C. & MARCANO, F. 1994. Petroleum systems of the Maracaibo Basin, Venezuela. *In*: MAGOON, L. B. & DOW, W. G. (eds) *The Petroleum System—From Source to Trap*. American Association of Petroleum Geologists, Memoirs, **60**, 463–481.

Controls on primary porosity and permeability development in igneous rocks

NICK PETFORD

Centre for Earth and Environmental Science Research, Kingston University, Kingston upon Thames, Surrey KT1 2EE, UK

Abstract: Some of the more important processes leading to the development of primary igneous porosity due to the cooling and crystallization of magma are reviewed. A distinction is made between volcanic and plutonic rocks, and crystalline and granular volcanic material. Porosity in each rock type is classified according to a proposed effective length scale and geometry into diffusive (Class D) and macroscopic flow (Class F) features. Estimated ranges in values of porosity and permeability are given for a wide selection of igneous rock types, and comparison is made with permeability variations (Δk) derived for both the continental and oceanic crust. While fracture porosity is dominant in most crystalline materials, primary porosity development may play an important role in the final (total) porosity in igneous basement. Some types of primary porosity and permeability in igneous rocks will be strongly time- and scale-dependent due to thermal effects associated with the emplacement and cooling of magmas and volcanic material. Tectonic reworking of the primary petrophysical properties of basement-forming igneous rocks may be significant in the development of regions of anisotropy and enhanced porosity.

Crystalline basement comprises a wide range of rock types formed under quite different geological conditions, of which igneous rocks, both plutonic and extrusive (volcanic), form an important component (e.g. Landes *et al.* 1960). Once *in-situ*, basement rocks collectively are exposed to tectonic stresses that will form discontinuities and other zones of elevated fluid transmissivity at shallow levels in the Earth's crust (Odling 1997). However, material properties alone will mean that different igneous lithologies will respond differently to applied stress, and that bulk petrophysical properties, including porosity and permeability, will differ accordingly (e.g. Zhang & Sanderson 1996). Importantly, igneous rocks can develop a primary (or intrinsic) porosity and permeability during their crystallization that, if preserved, would contribute to the transport and storage of fluids in the subsurface. A particularly important class of primary structures is joint sets that form in and around plutons due to internal stresses related to cooling and crystallization of magma (Gerla 1988; Bergbauer *et al.* 1998). Joints produced in this way, and implications for fluid flow and mineralization in and around plutons, have been the subject of numerous earlier studies, notably those by Norton & Knapp (1977) and Norton & Knight (1977), Segall & Pollard (1983), and more recently Bergbauer & Martel (1999).

Some primary features, notably cooling joints and fractures, may act as a template by localizing structures formed in the rock mass during subsequent tectonic deformation. This theme is developed more fully for the special case of a tabular-shaped intrusion in a companion paper (Koenders & Petford 2000). However, it is also true that not all primary porosity in igneous rocks is fracture porosity (Chen *et al.* 1999). Thus, in an attempt to provide a frame of reference for further investigations, it is proposed that primary igneous porosity and permeability be assigned to one of two general classes in accordance with the effective lengthscale of the relevant transport process.

This article sets out to identify and classify some of the more common examples of *primary* igneous porosity and permeability features relevant to fluid transmissivity with reference to hydrocarbon exploration and production. 'Primary' in this context is defined as those features formed entirely during the emplacement and cooling of magma or pyroclastic flows and preserved in the solid state. While distinct from *secondary* porosity that results from weathering, hydrothermal alteration, tectonic stresses and mineral dissolution by percolating groundwater (processes not addressed here in detail), the two are nevertheless complementary and closely associated. I begin with a brief review of the permeability of crystalline crustal material, followed by a generalized classification of primary porosity features in igneous materials based on an earlier model first proposed by Norton & Knapp (1977). The scheme is illustrated with field examples of primary porosity

From: PETFORD, N. & MCCAFFREY, K. J. W. (eds) 2003. *Hydrocarbons in Crystalline Rocks*. Geological Society, London, Special Publications, **214**, 93–107. 0305-8719/03/$15

and permeability in volcanic and plutonic rocks from a range of tectonic settings.

Crystalline basement and crustal permeability

Before considering in detail the types of porosity and permeability features that can develop locally in igneous rocks, it is helpful first to view the problem from a wider perspective. A number of investigations into the permeability of the Earth's crust, both continental and oceanic, have been made in recent years (for two comprehensive reviews, see Fisher 1998; Manning & Ingebritsen 1999), with a view to constraining permeability variation with depth. The data presented by Manning & Ingebritsen (1999), based on a comparison of *in-situ* geothermal measurements from deep drill holes, core analysis, heat flow modelling and metamorphic hydrology, show that in general, permeability decreases with

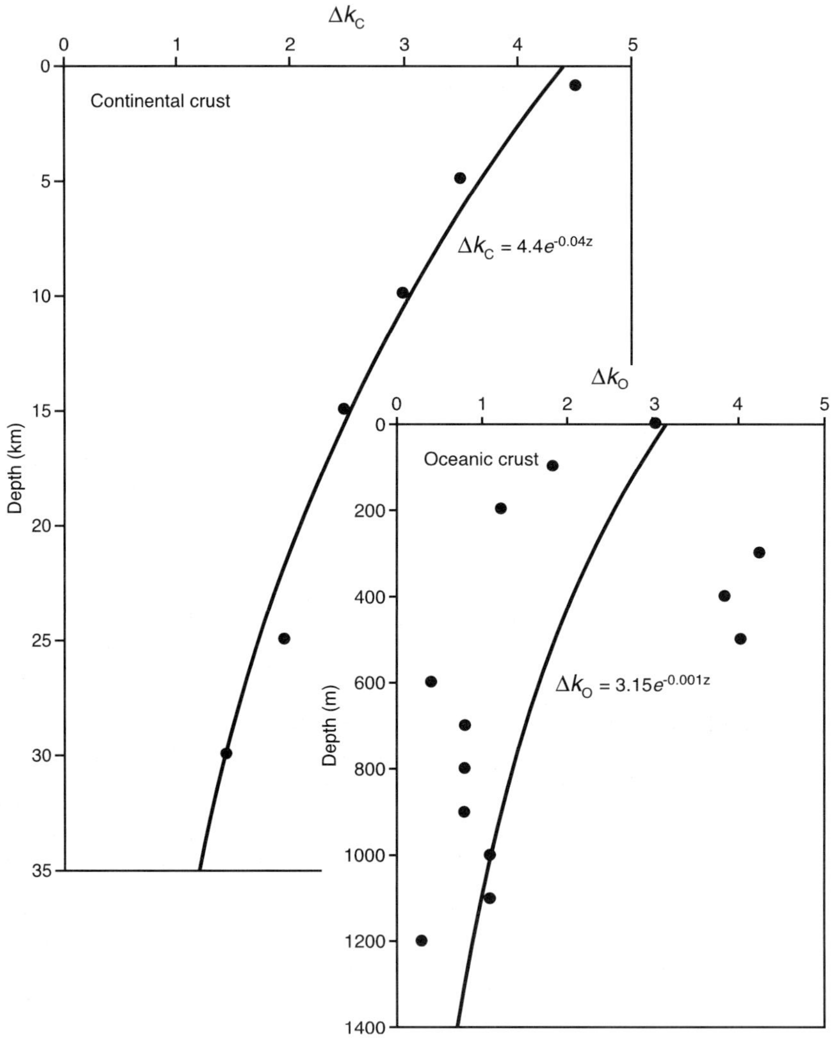

Fig. 1. Plot of the variation in crustal permeability expressed as Δk_C as a function of depth for two profiles through the continental and oceanic crust, constructed from data presented in Manning & Ingebritsen (1999) (continents) and Fisher (1998) (oceans). The variation in Δk is well described for both continental and ocean crust by the exponential relationship $y = a\mathrm{e}^{-bx}$, where the coefficients $a = 4.44$ and 3.15 ($r = 0.994$), and $b = 0.04$ and 1.04×10^{-3} ($r = 0.552$) for the continents and oceans respectively. Note the large difference in depth scale between plots.

depth, or increased confining pressure, following a quasi-exponential decay defined (in log form) as:

$$\log k = -14 - 3.2 \log z, \quad (1)$$

where k is permeability in metres squared and z is depth in kilometres, giving a mean value of permeability in the upper crust of *c.* $10^{-16}\,\mathrm{m}^2$ (but see also Brace 1984; Clauser 1992; Townend & Zobak 2000 for compilations that give different values). Additionally, Townend & Zobak (2000) have proposed that intraplate continental crust is in a state of failure equilibrium, with pore pressures that are close to hydrostatic and critically stressed faults that limit its strength. Their conclusion that permeability in the upper 10 km is *c.* 10^{-17} to $10^{-16}\,\mathrm{m}^2$ requires the brittle crust to be effectively permeable over timescales of 10 to 1000 years.

Figure 1 shows the observed variation in the range of log permeability, defined here as Δk, with increasing depth. The relationship between Δk and z in the continental crust (k_{C}) can be expressed according to the following exponential relationship:

$$\Delta k_{\mathrm{C}} = 4.4\,\mathrm{e}^{-0.04z}. \quad (2)$$

The parameter Δk_{C} ranges from 4.5 at $z = 1$ km to 1.5 at $z = 30$ km (Fig. 1). For comparison, in the oceans, the picture is less clear, with a wide range in permeability (k_{O}) close to the surface that decays to a relatively uniform value of *c.* $10^{-17}\,\mathrm{m}^2$ at depths $>$ *c.* 600 m (Fisher 1998). A rough estimate of this variation is given by:

$$\Delta k_{\mathrm{O}} = 3.15\,\mathrm{e}^{-0.001z}. \quad (3)$$

Although a relatively poor indicator ($r = 0.552$), equation (3) predicts that the permeability variation becomes asymptotic for $z > 3500$ m. The much wider scatter in values in oceanic crust compared to the continents is simply a reflection of the higher resolution and shallower depth interval (the former represents only *c.* 5% of the depth of the continental profile, Fig. 1). Indeed, studies of sedimentary basins in the upper continental crust ($z < 5$ km) show a similar permeability ($\Delta k_{\mathrm{C}} \approx 10^4$) variation in sedimentary basins (Ingebritsen & Manning 1999). Although a full understanding of the causes of variation in crustal permeability with depth is still incomplete, two interesting features relevant to this study can be extracted from these data. First, despite their widely differing depth profiles and sampling intervals, the maximum variation in both layers is similar ($4.2 < \Delta k < 4.5$). Secondly, as the oceanic crust is comprised almost entirely of young igneous material, it can be argued that much of the variation in Δk_{O} is due to an intrinsic, or igneous, porosity and permeability, albeit modified by sea-floor tectonic processes. The possible cause of this variation is reviewed later. However, it is tempting to conclude that similar variations are present in igneous continental crust on the hectometre scale, but are averaged out due to the larger scale of observation and its much greater mean age.

Porosity classification in igneous rocks

The majority of primary porosity and permeability development in igneous rocks, regardless of their emplacement level, develops as a result of physical and chemical processes that accompany the cooling of magma. However, in contrast to sedimentary rocks, little work has been done to classify the likely range of *primary* igneous porosity or to determine the relevant lengthscales. Most published work relating to the transmissivity of fluids in crystalline rocks is to be found in the hydrology literature (e.g. Evans & Nicholson 1987; Ingebritsen & Scholl 1993; Manga 1997; Burgdorff & Goldberg 2001), or in studies relating to fracture mineralization (e.g. Roberts *et al.* 1998; McCaffrey *et al.* 1999). Furthermore, as fracture permeability in general is dependent strongly on the stress conditions in the crust during uplift and burial (Zhang & Sanderson 1996), a distinction between fractures formed during cooling of magma, and those generated afterwards by regional tectonic stresses, may not at first seem necessary. However, some recent offshore exploration studies have concluded that primary fracture sets in igneous basement may contribute significantly to the porosity of hydrocarbon reservoirs (see Sanders *et al.* 2003).

From the analysis of Norton & Knapp (1977), and more recent work by Sanford (1997), two principal modes of fluid transport are envisaged to take place in igneous rocks, irrespective of composition: (1) flow along macro fractures, and (2) interfracture diffusion. Following Norton & Knapp (1977), the total porosity (ϕ_{total}) in fractured media can be conveniently expressed as the sum of the various porosity fractions, that is:

$$\phi_{\mathrm{total}} = \phi_{\mathrm{F}} + \phi_{\mathrm{D}} + \phi_{\mathrm{R}} \quad (4)$$

where the subscripts F, D and R are the flow, diffusion and residual (non-contributory) porosity, respectively. While this is a good description of the storage properties of crystalline rock, not all igneous materials capable of transmitting fluids are crystalline. Examples

Table 1. *Classification of primary porosity in igneous rocks according to class (see text for definitions)*

ϕ	Volcanic		Plutonic
	Granular	Crystalline	
Class D	Grain size variations $k \propto$ to particle diameter	Vesicles k independent of bubble diameter	Vesicles Miarolitic cavities Magmatic foliations
Class F	Bedding/layers Cross lamination	Flow/unit tops Columnar joints	Cooling joints Internal contacts Pluton-country rock contacts

include unconsolidated ash and tuff deposits, and layered and bedded granular flows. Thus, a more general classification is proposed to cover all igneous rocks, regardless of mode of emplacement. This is achieved most simply by expanding upon the petrophysical categories termed Class D ($=\phi_D$) and Class F ($=\phi_F$) corresponding with the aforementioned diffusion and flow porosities. Diffusion (Class D) porosity in this context refers to those regions of the rock mass where diffusional transport exceeds that of fluid flow. These conditions may occur in zones of discontinuous or small aperture fractures, and *discontinuous* pore space (Norton & Knapp 1977). Flow porosity (Class F) is characterized by *continuous* pore features including planar joints, lithological contacts, faults and bedding planes. In the nomenclature of flow regimes proposed by Bickle & McKenzie (1987), Class F would equate with advective transport while in Class D, advection is negligible. In practical terms, on production timescales, only Class F regions will flow.

In the following section, the primary igneous porosity and permeability in a range of common volcanic and plutonic igneous rock types are assessed by class (Table 1). Estimates of the likely range in petrophysical properties of each class, based on a literature survey, are summarized in Table 2.

Volcanic rocks: crystalline

Crystalline volcanic rocks comprise lavas and high-level dykes and sills. The magmas that cool to produce these rocks encompass a wide range in viscosity, from *c.* 10^1–10^2 Pa s for basalts to $>10^6$ Pa s for some silica-rich rhyolites, and initial gas (volatile) contents may comprise several weight percent of the magma (e.g. Burnham 1979). The interaction between these two magmatic properties results in a diverse range of lava flow morphologies and subsequent petrophysical properties. For example, volatile species can escape much more easily from low viscosity

Table 2. *Compilation showing the general range of values of primary porosity (ϕ) and permeability (k) in igneous rocks derived from various methods including pump, core, disk, drawdown, fracture and heat flow modelling.*

		Volcanic		Plutonic
		Granular	Crystalline	
Class D	ϕ	10–60%	0.1–1% *10–40%*	0.01–1%
	k	10^{-16}–10^{-12} m^2	10^{-16}–10^{-14} m^2	?
Class F	ϕ	15–25% *20–30%*	1–25%	10^{-6}–1% *0.5–17%*
	k	10^{-16}–10^{-12} m^2	10^{-14}–10^{-9} m^2	10^{-21}–10^{-8} m^2 *10^{-12}–10^{-8} m^2*

Note: All original values of k now expressed in metres squared for consistency and ease of comparison (1 darcy = ~1 μm^2). Data from: Norton & Knapp 1977; Freeze & Cherry 1979; Cas & Wright 1987; Dmitryeveskiy *et al.* 1993; Chen *et al.* 1999; Zhu & Wong 1999. Available secondary values (italics) due to hydrothermal alteration and weathering are also shown.

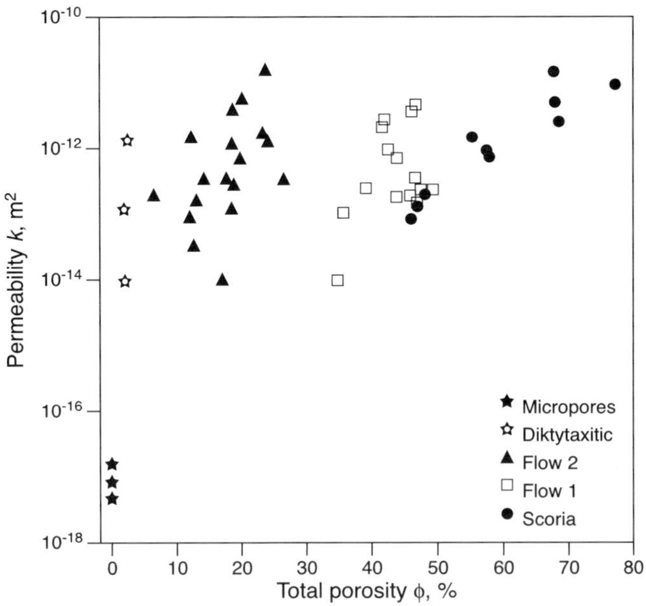

Fig. 2. Permeability–porosity variations in vesicular volcanic rocks (Holocene–Pleistocene basaltic andesite and scoria, Cascades, USA). Only the scoria shows a clear relationship between porosity and permeability (after Sarr & Manga 1999), implying that the empirical Kozney–Carman model may not hold for vesicular material.

magmas, resulting in higher concentrations of vesicles (unfilled gas cavities) contributing to primary diffusion porosity in basaltic rocks compared to rhyolites. Freeze & Cherry (1979) report values of permeability in the range 10^{-14} to 10^{-9} m^2 for vesicular basalts.

Diffusive porosity (Class D)

Primary porosity consists mainly of vesicles resulting from exsolution of magmatic volatiles (most commonly H_2O, CO_2 and SO_2) during cooling. Because the gas is less dense it has a tendency to rise, resulting in higher concentrations of vesicles at the top of the flow. Analysis of primary and secondary porosity and permeability in hydrocarbon-bearing basalts from the Tertiary Liaohe basin, China, reveals that cavity volumes range from 10 to 50% and are widely interconnected (Chen *et al.* 1999). Clearly, the higher the initial gas content of the magma, the better the primary porosity is likely to be.

Permeability (k) porosity (ϕ) relationships have been investigated recently in vesicular basalts by Sarr & Manga (1999). Interestingly, measured values of these parameters did not follow according to empirical Kozney–Carman models of permeability in granular materials where $k \propto$ mean particle size (e.g. Dullien 1992). Instead, it was shown that the permeability of vesicular rocks is controlled not by particle (in this case bubble) size, but by the aperture spacing between the bubbles. Figure 2 shows a plot of porosity versus permeability for five samples analysed by Sarr & Manga (1999). With the exception of the high porosity (>40%) basaltic scoria (10^{-13} $m^2 < k < 10^{-11}$ m^2), the other lava types plot in distinct clusters, implying that the rock microstructure is governing both k and ϕ.

Flow porosity (Class F)

The most obvious discontinuity capable of exploitation by fluids in buried lavas and networks of sills is bedding planes. It is well known that much of the observed heterogeneity and anisotropy in the permeability of upper crustal rocks is due to contacts between planar layers (Manning & Ingebritsen 1999), which can be several orders of magnitude greater than the orthogonal permeability. Broadly planar contacts between individual lava flows can be laterally continuous on the scale of tens to hundreds of kilometres (e.g. Columbia River basalts, USA; Deccan Traps, India). The tops of lava flows are often regions of high vesicle concentration, thus providing a spatial link within the pile between Class F and Class D porosity features.

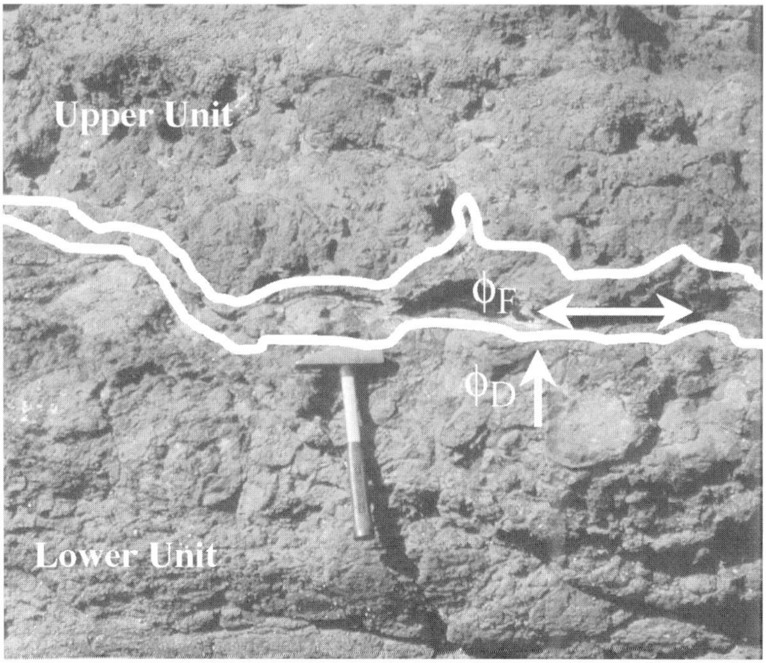

Fig. 3. Weathered contact (Red Bole) between two Tertiary basalt lava flows (Isle of Skye, Scotland). In the absence of any well-defined fracture or columnar jointing, one-dimensional flow porosity (ϕ_F) parallel to bedding and diffusion porosity (ϕ_D) domains in the vesicular top of the lower unit are indicated.

This relationship can be further exploited where secondary fracture porosity (tectonically induced) connects with gas cavities to improve reservoir quality (e.g. Chen *et al.* 1999). However, the surfaces of lava flows are especially prone to surface weathering (Fig. 3). These highly weathered zones are characterized by extensive alteration of basalt to a fine-grained mixture of clays that can extend tens of centimetres from individual bedding contacts, and could act to lower fluid transmissivity by reducing the flow porosity or flow channel. Lava tubes, a more exotic megapermeability feature in active lava fields such as Iceland and Hawaii, are unlikely to be preserved intact at depth.

Columnar joints provide another example of a primary igneous flow porosity related to initial cooling of magma (DeGraff & Aydin 1993). The joints develop layer by layer in response to thermal stresses building up behind the solidification front (Weaire & O'Carroll 1983), and occur in both basaltic and acidic rocks, often where magma interacts with ponded water (Lyle, 2000). Columnar joints can be discontinuous and locally developed within a flow, or pervasive and penetrative throughout. Clearly, a multi-sided pervasive joint network exposes a much greater surface area within a rock mass to a percolating fluid than a simple line discontinuity, resulting in a highly anisotropic porosity in the vertical direction (e.g. Fig. 4). Unfortunately, this geometry may lack the degree of connectivity required for high levels of permeability in tectonically fractured rocks (Shimo & Long 1987). In passing, it should be noted that the interaction of volcanic material with surface waters or hydrothermal fluids (in epithermal environments) can lead to localized hydrofracturing and brecciation, often accompanied by mineralization (Phillips 1972). The altered rock mass represents a zone of weakness that may preferentially accommodate strain during any later episode of tectonic deformation.

Volcanic rocks: granular

Granular volcanic rocks are emplaced at the Earth's surface by either violent gas-charged eruption or gravitational failure of a volcanic edifice. In many ways they are a more exotic kind of clastic sediment and amenable to all the tools of the trade employed by sedimentologists. There is a large database of technical literature on the various types and modes of emplacement

Fig. 4. Top surface of the *c.* 2–3 m thick Tajao phonolite ignimbrite, Tenerife, showing metre-sized, subpolygonal cooling cracks. These structures are good examples of Class F porosity in predominantly granular volcanic rocks.

of volcaniclastic and pyroclastic deposits and the interested reader is referred to relevant texts (e.g. Cas & Wright 1987).

Diffusive porosity (Class D)

A number of detailed studies have been made on the petrophysical properties (including porosity and permeability) of granular volcanic rocks. As with other granular materials (but *unlike* crystalline volcanic rocks), diffusive porosity in volcaniclastic materials is governed by the familiar Kozney–Carman relationship where the porosity is proportional to the particle size. Initial values of porosity in granular volcanic materials can exceed 50% (Williams & McBirney 1979 and Table 2).

Flow porosity (Class F)

As with clastic sediments, volcanic deposits can be bedded and internally layered on a range of scales. Graded bedding and cross-stratification within individual units are common, and may account for much of the 10^4 variation in permeability reported by Winograd (1971) from a single ash flow tuff. The most obvious difference between volcanic sediments and their clastic counterparts relates to the temperature of emplacement. For example, many pyroclastic flows and some air fall deposits are emplaced at the surface at temperatures in excess of several hundred degrees Celsius. These may undergo a series of syn- to post-depositional processes such as welding and compaction, that while analogous to diagenesis in sediments will reduce significantly any initial magmatic (Class D) porosity on timescales of hours to days (Cas & Wright 1987). In particular, zones of dense welding and compaction in some high-temperature ignimbrites can reduce the initial magmatic porosity by a factor of two (Fig. 5). However, welding can also result in the development of columnar jointing (Class F), implying that the primary petrophysical properties of some pyroclastic flows may evolve in scale rapidly with time. These rapid (days to years) temporal and scale variations in the primary petrophysical properties of igneous rocks are a common theme that sets them apart from sediments, where in general deposition and diagenetic changes take place on geological timescales.

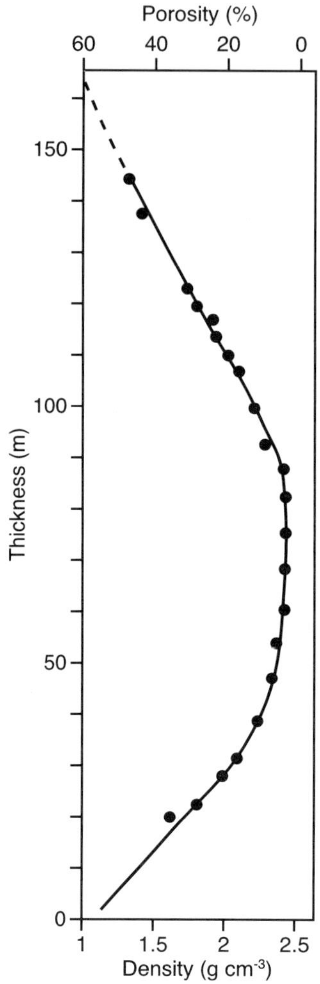

Fig. 5. Porosity–density versus depth variations in the Bishop Tuff welded ignimbrite, USA. Bulk density increases and primary porosity is reduced to a minimum in the central part of the deposit, corresponding to the zone of most intense apparent welding and compaction (after Ragan & Sheridan 1972).

Plutonic rocks

Plutonic rocks, cooling slowly inside the crust, develop patterns of jointing that differ from those formed in lavas and higher-level intrusions emplaced at or near the Earth's surface. The major porosity features in plutonic materials are fractures formed either during cooling or by later tectonic processes (e.g. Gerla 1988; Pollard & Aydin 1988). A strong influence on the style and extent of deformation is the temperature contrast between the intruding magma and surrounding country rock. For a given depth of emplacement, the temperature difference will be greatest where the invading magma is mafic in composition. The initial composition of the magma also plays an important role in secondary porosity development, with mafic plutonic rocks including gabbro and peridotite containing olivine and calcium-rich plagioclase feldspar being relatively more susceptible to the effects of weathering than most felsic plutonic rocks.

Diffusive porosity (Class D)

Primary grain-size porosity features in coarse-grained plutonic rocks are not as rare as one might think, given the crystalline nature of the material and mode of formation. For example, Norton & Knapp (1977) investigated the continuity, relative size distribution and location of the pore space in a porphyritic and equigranular granite. Measurements from the porphyritic sample suggested that the pore size is bimodally distributed between individual grains, with *c.* 30% of the pore volume located around pheoncrysts (Fig. 6a). In contrast, the equigranular rock shows a markedly different pore to grain-size distribution, with a higher cumulative proportion of pores larger than the largest grain size (Fig. 6b).

Other primary features are miarolitic cavities, gas escape structures similar to vesicles in volcanic rocks. The development and extent of miarolitic cavities in plutonic rocks will depend on the initial volatile content of the magma and the confining pressure at time of emplacement. High initial magma gas contents and shallow levels of emplacement thus favour cavity formation. In a recent study, Ohtani *et al.* (2001) describe a novel method using X-ray computed tomography (CT) to image the three-dimensional shape and distribution of miarolitic cavities in the Kakkonda granite, NE Japan. Estimated porosities are *c.* 0.9%. Exploration drilling has shown that distribution of strained and elongated cavities near the roof of the pluton apparently coincides with a region of elevated hydrothermal fluid flow, emphasizing the potentially important contribution of miarolitic cavities to the flow porosity.

An important feature in many granitic plutons is a primary magmatic flow fabric, defined by the alignment of early-formed minerals in the liquid during emplacement. The timing and tectonic significance of these fabrics still attracts considerable attention (Hutton 1988; Vernon 2000). Following Cloos (1925), the key point is that because they form before the magma has completely

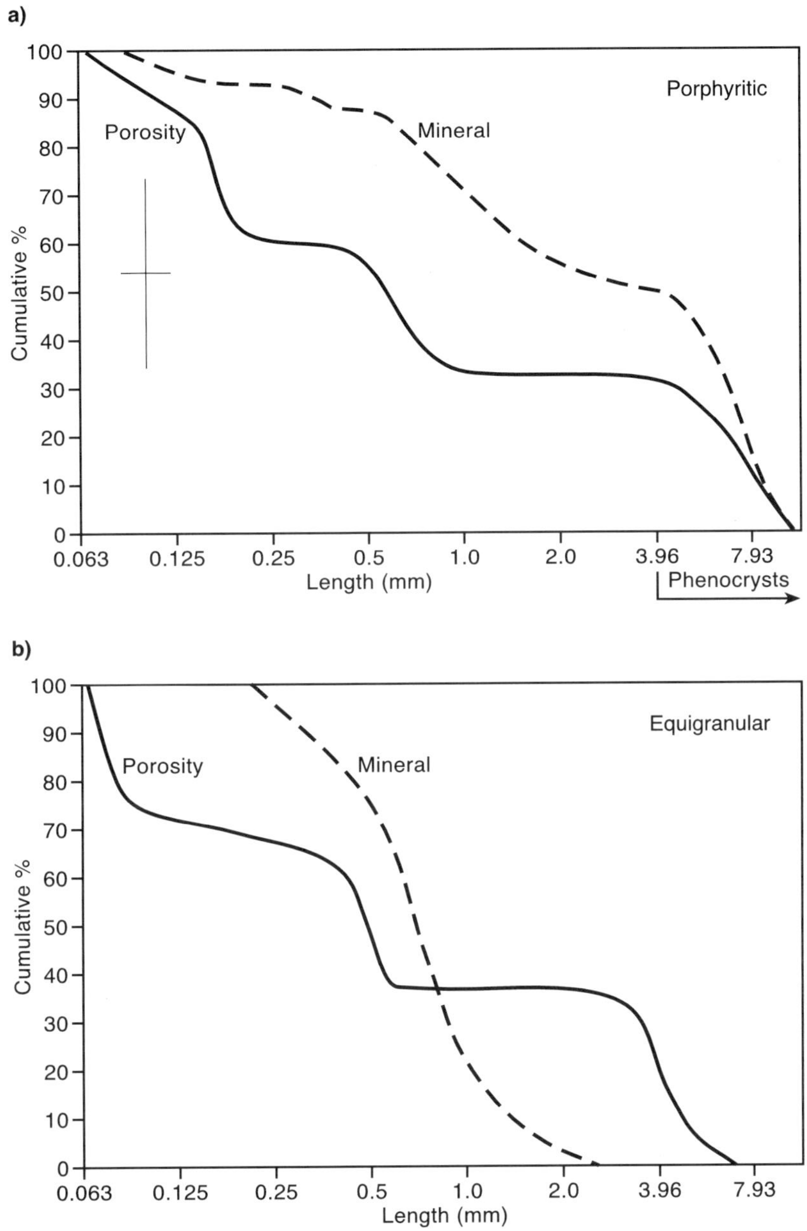

Fig. 6. Cumulative percentage porosity as a function of grain size for two granite samples (porphyritic and equigranular) from the Schultze pluton, Arizona: (**a**) Porphyritic granite, showing the convergence of both curves (mineral size and porosity) at the largest and smallest size intervals, suggesting that the pore volume is located bimodally around the phenocrysts (>4 mm), and the smallest grains (<0.1 mm). (**b**) Equigranular granite, showing only one domain of porosity associated closely with mineral size (after Norton & Knapp 1977).

solidified, they introduce an anisotropic 'grain' into the rock that can be exploited by later, more brittle processes associated with porosity development. This is shown in Figures 7 and 8, where a very strong, closely spaced joint pattern (Class F) in a granite pluton from the Andes of Peru has formed in the plane of the initial magmatic foliation (Class D). The rock has a flaggy

Fig. 7. Closely spaced sheet joints (generally <10 cm), in the Cordillera Blanca batholith, Peru, formed in the plane of an earlier synmagmatic foliation and 'embracing the flow fabric' as defined by Cloos (1925).

appearance (Fig. 7), with later joint development guided by the presence of planes of weakness defined by mineral alignments (principally biotite) in the rock mass (Fig. 8). Slickensides along the fracture openings show that some component of oblique (Mode II or III) failure was involved in joint formation (Petford & Atherton 1992). As with bedding planes, such discontinuities impart to the rock potential pathways for fluid and mass transfer.

Finally, initial magma composition plays a role in subsequent secondary porosity development in plutons via feldspar dissolution. In particular, the more An-rich cores of plagioclase crystals, along with some *k* feldspar, are prone to alteration by deuteric processes resulting in a secondary porosity commonly associated with kaolin mineralization (Bristow 1993; Psyrillos 2003). In some granites, dissolution porosity may exceed 20% (R. Maddox, pers. comm.). Mineral replacement processes that generate interconnected porosity can lead to reaction-enhanced permeability that may provide important local pathways for migrating fluids in metamorphic rocks (see Putnis 2002).

Flow porosity (Class F)

One of the earliest studies of fractures in plutonic rocks is the seminal work by Cloos (1925), followed soon after by the equally masterful *Structural Behaviour of Igneous Rocks* (Balk 1937). Cloos classified naturally occurring joint systems in granitic plutons into three major categories related to primary flow foliations common in most granitic plutons: (1) vertical cross-joints, at right angles to the magmatic flow direction, (2) longitudinal joints, parallel with the strike of the magmatic foliation, and (3) primary flat (sheet) joints that 'embrace' the flow fabrics (Cloos 1925 and Figs 7 and 8; see also Price & Cosgrove 1990, chap 3). Evidence that the cross-joints formed while the magma was still consolidating is seen in the form of late-stage aplite and pegmatite veins that commonly infill these joint sets (Fig. 9).

The veracity of cooling joints that form in the magmatic state is dependent strongly on the temperature contrast (ΔT) between the intruding magma and surrounding country rocks (Knapp & Norton 1981; Gerla 1988). The strain (e) due to contraction (shrinkage) in the cooling pluton is tensile, and can be estimated from $e = \alpha(\Delta T)$, where α is the thermal expansivity (see Koenders & Petford 2003 for a more through analysis). The steeper the thermal gradient, the greater the thermal stresses and more intensive the joint formation. It thus follows that cooling joints will be best developed in hot plutons emplaced at high crustal levels. This in turn implies a compositional control, as mafic magmas are hotter by up to several hundred degrees centigrade than their felsic counterparts. In fact, composite batholiths, emplaced at the same depth, show a range in cooling fractures that clearly relate to rock composition (Pitcher 1993), and possibly grain size as well (Sanders *et al.* 2003). Clearly, the maintenance of high permeability requires that the primary fractures remain open to fluid transport. Against this is field and experimental evidence showing that circulating fluids can reduce the permeability in fractured granites by sealing up the fracture network to values close to that of the intact rock (Moore *et al.* 1994; Morrow *et al.* 2001). However, primary fractures will impart zones of weakness in the rock mass that can be reworked by later tectonic deformation, and breaking fracture seals (Olsen *et al.* 1998).

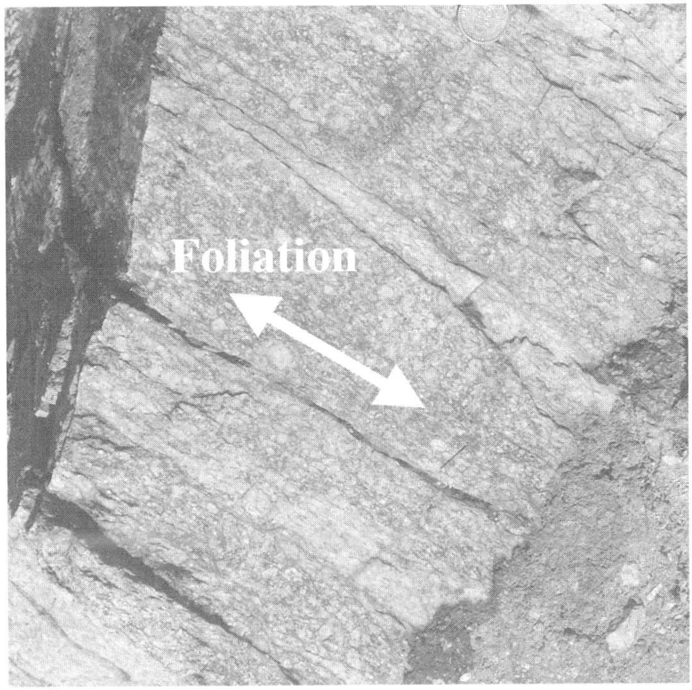

Fig. 8. Close-up section showing the initial magmatic foliation plane in the granite (Class D), overprinted by later sheet jointing (cf. Fig. 7).

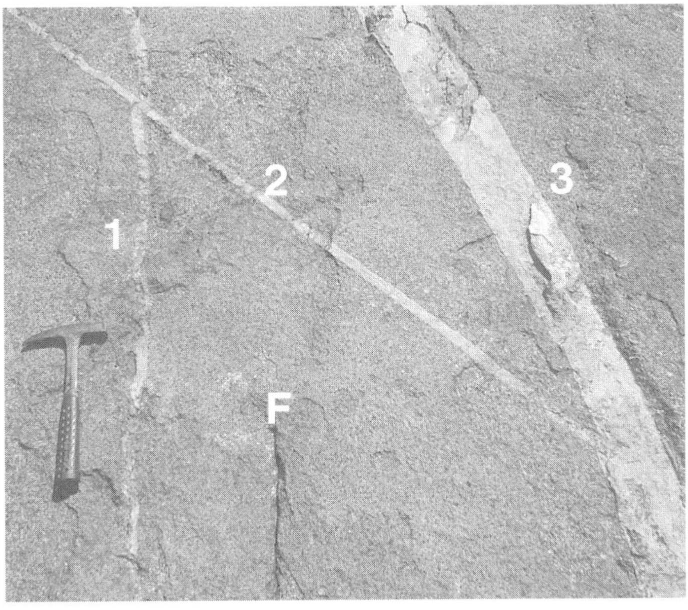

Fig. 9. Subvertical, late-stage (magmatic) cross-cutting aplite veins in the margin of the Cordillera Blanca batholith, Peru. Note that dilational (Mode I) failure is indicated at the intersection of the two narrower veins (numbered 1–3 in order of intrusion), and the unfilled fracture (F) parallel to 1, terminating in the lower half of the image.

Effects of weathering

Many of the primary igneous porosity and permeability features described above will be modified by weathering at the surface prior to reburial. Weathering will either enhance or reduce the ability of the rock to transmit fluids. Examples mentioned already include alteration of the upper surfaces of lava flows (Fig. 3) and feldspar dissolution porosity. An important point is that the weathering process is often acting upon an original primary igneous porosity feature relating directly to magma composition. Initial mineralogy is a prime example of this (cf. peridotite and granite weathering), but other examples include the etching out of pumice fragments in ignimbrites and other volcanic rocks, and the opening up of fracture seals infilled with secondary hydrothermal minerals prior to exposure. Indeed, intense (sub-tropical) weathering of primary magmatic fractures in granitic rocks now lying offshore in parts of SE Asia appears to have been a major contributing factor in the ability of these materials to act as hydrocarbon reservoirs after subsequent reburial (Dmitriyevskiy *et al.* 1993).

Discussion

Relative importance of flow, diffusion and residual porosity

Norton & Knapp (1977) give estimates of the diffusion and total porosities for a number of crystalline volcanic and plutonic rocks that allow an examination of the relative contribution made by each to the overall transport properties of the material. Some of their data relevant to this review are reproduced in Figure 10, which shows the laboratory measured diffusive porosity (ϕ_D) and the accompanying total porosity (ϕ_T) for rhyolites, dacites and three plutonic rocks (diorite, quartz monzonite and granite). While these values are averages of a relatively small number of analyses ($n < 10$), they nevertheless provide some useful insight into porosity development in these materials. For example, they show that while the diffusion porosity (Class D) is a small fraction of the total porosity (of the order 10^{-3} to 10^{-4}), it is largest in the two volcanic rocks (Fig. 10a), which also have largest total porosities. Norton & Knapp (1977) also give information on hydrothermally altered

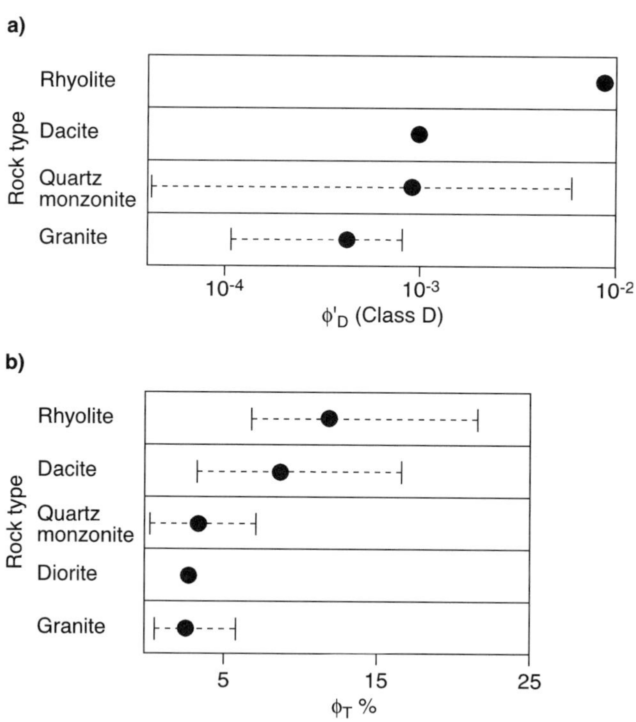

Fig. 10. Plots showing the average (filled circles) and range in values of diffusive (**a**) and total (**b**) porosity for selected volcanic (rhyolite and dacite) and plutonic (quartz monzonite, diorite and granite) rock types. Note the rhyolites and dacites have the highest average diffusive and total porosities (after Norton & Knapp 1977).

samples, showing that the altered rocks both have higher ϕ_D and ϕ_{total} compared to their unaltered equivalents. Using equation (4) and knowledge of the total rock porosity, it is possible to estimate the relative amounts of Class D to Class F porosity for the plutonic and volcanic rocks shown in Figure 10. For example, taking the total porosity of the quartz monzonite sample of *c.* 5% (Fig. 10b) and a diffusion porosity of *c.* 10^{-3}, then the flow porosity can be estimated from $\phi_F = \phi_{total} - \phi_D$ as ~0.049, easily the largest contributing porosity. Similar calculations can be made for the two volcanic rocks, which again show the dominance of ϕ_F on ϕ_{total}. However, this presents a misleading picture. Measured values of ϕ_F on a range of natural granites suggest a maximum range in flow porosity of 10^{-2}–10^{-8}. Taking an average of *c.* 10^{-5} for the flow porosity gives a new total porosity of just 0.1%. The remainder (from equation (4)), must be the residual porosity, ϕ_R, those pores not connected to ϕ_F or ϕ_D. Thus, following the earlier conclusion by Norton & Knapp (1977), provided the low values of flow porosity are typical ones, then residual porosity must comprise greater than 90% of the total porosity. One exception may be where plutonic rocks contain strong primary magmatic foliations (Figs 7 and 8). Information on metamorphic rocks (Manning & Ingebritsen 1999), has shown that diffusive porosities are increased parallel with bedding and foliation planes by between two and seven times, thus reducing the contribution to the total porosity made by unconnected pore space. This observation lends support to the assertion made earlier that synmagmatic fabrics in plutonic rocks may provide templates for regions of enhanced porosity in the solidified rock.

Permeability

Some estimates of primary permeability in igneous rocks are given in Table 2, based on a compilation of sources including core measurements, heat flow modelling and *in-situ* (reservoir scale) heat flow measurements. Secondary values (due to weathering) are shown in italics. The limited data suggest that Class F permeability may be higher in crystalline than in granular volcanic rocks, and that the effects of weathering can decrease the range in plutonic rocks. As noted earlier, it seems likely that domains of flow porosity govern the permeability of igneous rocks. However, it is unclear from the available data whether such domains (Table 2) occur on a scale large enough to affect the permeability of an entire reservoir. It should also be noted that laboratory measurements on core samples from the brittle upper crust (e.g. Huenges *et al.* 1997) may underestimate the true permeability by several orders of magnitude (Townend & Zobak 2000).

Preservation potential

To what extent is the primary porosity and permeability of igneous rocks preserved intact during burial? Most granular volcanic rocks are likely to suffer a fate similar to sediments, with a rapid drop-off in porosity and permeability at relatively shallow levels. However, the greater strength of igneous rocks is likely to make them more resistant to crustal loading and compaction, allowing the preservation of primary porosity and permeability to greater depths than their surrounding sediments (Schutter 2003a,b). The wide variation in permeability close to the surface in both continental and oceanic crust, and in general above the brittle–plastic transition (Fig. 1) requires that a large degree of heterogeneity (and associated anisotropy) is maintained in crystalline basement, a significant proportion of which may reflect inherent, primary igneous porosity and permeability.

Summary

Igneous rocks can develop a wide range of primary porosity and permeability features formed during cooling and emplacement that may contribute significantly to the bulk transport and storage properties of crystalline basement. Two broad classes of porosity, diffusive (Class D) and flow (Class F), define first-order petrophysical features of igneous materials, irrespective of composition or mode of emplacement. As identified in previous studies, megascopic fractures formed during cooling of magma either at the Earth's surface or at depth are probably the most important type of porosity feature. Both classes can be modified by weathering at the surface and tectonic deformation to improve their transport and storage properties through increased connectivity. Primary porosity and permeability in igneous rocks may vary strongly in time and scale due to thermal effects that induce changes analogous to diagenesis in sedimentary rocks on timescales as short as hours to weeks. As the absolute values and preservation potential of both classes are not yet well known, it may be unwise to pursue hydrocarbon exploration in an igneous environment by treating it as generic material with average properties.

I would like to thank JVPC (Japan Vietnam Oil Company) for providing the opportunity to work on problems associated with the extraction of oil from igneous rocks. S. Pugliese, K. McCaffrey, T. Murray (Midland Valley) and R. Maddox (Baker Atlas) have all contributed ideas that helped form the basis of this review (although any misconceptions are entirely my own). J. Imber is thanked for perceptive comments that helped improve the clarity of presentation.

References

BALK, R. 1937. *Structural Behaviour of Igneous Rocks*. Geological Society of America Memoirs, **5**.

BERGBAUER, S. & MARTEL, S. J. 1999. Formation of joints in cooling plutons. *Journal of Structural Geology*, **21**, 821–835.

BERGBAUER, S., MARTEL, S. J. & HIERONYMUS, C. F. 1998. Thermal stress evolution in cooling plutonic environments of different geometries. *Geophysical Research Letters*, **25**, 707–710.

BICKLE, M. J. & MCKENZIE, D. 1987. The transport of heat and matter by fluids during metamorphism. *Contributions to Mineralogy and Petrology*, **95**, 384–392.

BRACE, W. F. 1984. Permeability of crystalline rocks: new in situ measurements. *Journal of Geophysical Research*, **89**, 4327–4330.

BRISTOW, C. M. 1993. The genesis of china clays of Southwest England—a multistage history. *In*: MURRAY, H., BLUNDY, W. & HARVEY, C. (eds) *Kaolin Genesis and Utilisation*. Clay Minerals Society of America, Special Publications, **1**, 171–203.

BURGDORFF, K. & GOLDBERG, D. 2001. Petrophysical characterisation and natural fracturing in an olivine-dolerite aquifer. *Electronic Geosciences*, **6**, 3.

BURNHAM, C. W. 1979. The importance of volatile constituents. In: YODER, H. S. (ed.) *The Evolution of the Igneous Rocks*. Princeton University Press, New Jersey, 439–482.

CAS, R. A. F. & WRIGHT, J. V. 1987. *Volcanic Successions: Modern and Ancient*. Chapman & Hall, London.

CHEN, Z., YAN, H., LI, J., ZHANG, G., ZHANG, Z. & LIU, B. 1999. Relationship between Tertiary volcanic rocks and hydrocarbons in the Liaohe basin, People's Republic of China. *American Association of Petroleum Geologists, Bulletin*, **88**, 1004–1014.

CLAUSER, C. 1992. Permeability of crystalline rocks. *EOS*, **73**, 233, 237.

CLOOS, H. 1925. *Einfuhrung in die tektonische Behandlung magmatischer Erscheinungen*, 1. *Das Riesengebrige in Schlesien*. Gebr. Borntraeger, Berlin.

DAVID, C., PETFORD, N. & RISNES, R. 2001 (eds). Compaction, subsidence and mechanics of granular materials. *Physics and Chemistry of the Earth*, **26**, Nos. 1–2.

DEGRAFF, J. M. & AYDIN, A. 1993. Effect of thermal regime on growth increment and spacing of contraction joints in cooling basaltic lava. *Journal of Geophysical Research*, **98**, 6411–6430.

DMITRIYEVSKIY, A. N., KIREYEV, F. A., BOCHKO, R. A. & FEDOROVA, T. A. 1993. Hydrothermal origin of oil and gas reservoirs in basement rock of the South Vietnam continental shelf. *International Geology Review*, **35**, 621–630.

DULLIEN, F. L. A. 1992. *Porous Media—Fluid Transport and Pore Structure*, 2nd edn. Academic Press, San Francisco.

EVANS, D. D. & NICHOLSON, T. J. 1987. (eds) *Flow and Transport Through Unsaturated Rock*. Geophysical Monograph **42**, American Geophysical Union, Washington DC.

FISHER, A. T. 1998. Permeability within basaltic ocean crust. *Reviews of Geophysics*, **36**, 143–182.

FREEZE, R. A. & CHERRY, J. A. 1979. *Groundwater*. Prentice-Hall, Englewood Cliffs, NJ.

GERLA, J. P. 1988. Stress and fracture evolution in a cooling pluton: an example from Diamond Joe stock, western Arizona, USA. *Journal of Volcanology and Geothermal Research*, **34**, 267–282.

HUENGES, E., ERZINGER, J., KUCK, J., ENGESSER, B. & KESSELS, W. 1997. The permeable crust: geohydraulic properties down to 9101 m depth. *Journal of Geophysical Research*, **102**, 18,255–18,265.

HUTTON, D. H. W. 1988. Granite emplacement mechanisms and tectonic controls: inferences from deformation studies. *Transactions of the Royal Society of Edinburgh: Earth Sciences*, **79**, 245–255.

INGEBRITSEN, S. E. & MANNING, C. E. 1999. Geological implications of a permeability-depth curve for the continental crust. *Geology*, **27**, 1107–1110.

INGEBRITSEN, S. E. & SCHOLL, M. A. 1993. The hydrology of Kilauea volcano. *Geothermics*, **22**, 255–270.

KNAPP, R. B. & NORTON, D. 1981. Preliminary numerical analysis of processes related to magma crystallisation and stress evolution in cooling plutonic environments. *American Journal of Science*, **281**, 35–68.

KOENDERS, M. A. & PETFORD, N. 2002. Initial investigation into the effects of cooling on fracture development in tabular granitic plutons. *In*: PETFORD, N. & MCCAFFREY, K. (eds) *Hydrocarbon in Crystalline Rocks*. Geological Society, London, Special Publications, **214**, 143–150.

LANDES, K. K., AMORUSO, J. J., CHARLESWORTH, L. J., HEANY, F & LESPERANCE, P. J. 1960. Petroleum resources in basement rocks. *American Association of Petroleum Geologists Bulletin*, **44**, 1682–1691.

LYLE, P. 2000. The eruption environment of multi-tiered columnar basalt lava flows. *Journal Gelogical Society London*, **157**, 715–722.

MANGA, M. 1997. A model for discharge in spring-dominated streams and implications for the transmissivity and recharge of quaternary volcanics in the Oregon Cascades. *Water Resources Research*, **33**, 1813–1822.

MANNING, C. E. & INGEBRITSEN, S. E. 1999. Permeability of the continental crust: implications of geothermal data and metamorphic systems. *Reviews of Geophysics*, **37**, 127–150.

MCCAFFREY, K. J. W., LONERGAN, L & WILKINSON, J. J. (eds). 1999. *Fractures, fluid flow and mineralisation*. Geological Society, London, Special Publications, **155**.

MOORE, D. E., LOCKNER, D. A. & BYERLEE, J. D. 1994. Reduction in permeability of granite at elevated temperatures. *Science*, **265**, 1558–1561.

MORROW, C. A., MOORE, D. E. & LOCKNER, D. A. 2001. Permeability reduction in granite under hydrothermal conditions. *Journal of Geophysical Research*, **106**, 30,551–30,560.

NORTON, D. & KNAPP, R. 1977. Transport phenomena in hydrothermal systems: the nature of porosity. *American Journal of Science*, **277**, 913–936.

NORTON, D. & KNIGHT, J. 1977. Transport phenomena in hydrothermal systems: cooling plutons. *American Journal of Science*, **277**, 937–981.

ODLING, N. E. 1997. Fluid flow in fractured rocks at shallow levels in the Earth's crust: an overview. *In*: HOLNESS, M. B. (ed.) *Deformation-Enhanced Fluid Transport in the Earth's Crust and Mantle*. Chapman Hall, London, 289–322.

OHTANI, T., NAKANO, T., NAKASHIMA, Y. & MURAOKA, H. 2001. Three-dimensional shape analysis of miarolitic cavities and enclaves in the Kakkonda granite by X-ray computed tomography. *Journal of Structural Geology*, **23**, 1741–1751.

OLSON, M. P., SCHOLZ, C. H. & LEGER, A. 1998. Healing and sealing of a fault gauge under hydrothermal conditions: implications for fault healing. *Journal of Geophysical Research*, **103**, 7421–7430.

PETFORD, N. & ATHERTON, M. P. 1992. Granitoid emplacement and deformation along a major crustal lineament; the Cordillera Blanca, Peru. *Tectonophysics*, **205**, 171–185.

PHILLIPS, W. J. 1972. Hydraulic fracturing and mineralisation. *Journal of the Geological Society, London*, **128**, 337–359.

PITCHER, W. S. 1993. *The Nature and Origin of Granite*. Blackie Academic and Professional, London.

POLLARD, D. D. & AYDIN, A. 1988. Progress in understanding jointing over the last century. *Geological Society of America Bulletin*, **100**, 1181–1204.

PRICE, N. J. & COSGROVE, J. W. 1990. *Analysis of Geological Structures*. Cambridge University Press, Cambridge.

PSYRILLOS, A., MANNING, D. A. C., BURLEY, S. D. & FALLICK, A. E. 2003. Coupled evolution of a basin and high: kaolinization in the SW England granites is related to development of the Plymouth basin. *In*: PETFORD, N. & MCCAFFREY, K. (eds) *Hydrocarbon in Crystalline Rocks*. Geological Society, London, Special Publications, **214**, 175–196.

PUTNIS, A. 2002. Mineral replacement reactions: from macroscopic observations to microscopic mechanisms. *Mineralogical Magazine*, **66**, 689–708.

RAGAN, D. H. & SHERIDAN, M. F. 1972. Compaction of the Bishop Tuff, California. *Geological Society of America, Bulletin*, **83**, 95–106.

ROBERTS, S., SANDERSON, D. J. & GUMIEL, P. 1998. Fractal analysis of Sn-W mineralisation from central Iberia: insights into the role of fracture connectivity in the formation of an ore deposit. *Economic Geology*, **93**, 360–365.

SANDERS, C. A. E., FULLARTON, L. & CLAVET, S. 2003. Modelling fracture systems in extensional crystalline basement. *In*: PETFORD, N. & MCCAFFREY, K. (eds) *Hydrocarbon in Crystalline Rocks*. Geological Society, London, Special Publications, **214**, 219–234.

SANFORD, W. E. 1997. Correcting for diffusion in carbon-14 dating of groundwater. *Groundwater*, **35**, 357–361.

SARR, M. O. & MANGA, M. 1999. Permeability-porosity relationship in vesicular basalt. *Geophysical Research Letters*, **26**, 111–114.

SCHUTTER, S. R. 2002a. Occurrences of hydrocarbons in and around igneous rocks. *In*: PETFORD, N. & MCCAFFREY, K. (eds) *Hydrocarbon in Crystalline Rocks*. Geological Society, London, Special Publications, **214**, 35–68.

SCHUTTER, S. R. 2002b. Hydrocarbon occurrence and exploration in and around igneous rocks. *In*: PETFORD, N. & MCCAFFREY, K. (eds) *Hydrocarbon in Crystalline Rocks*. Geological Society, London, Special Publications, **214**, 7–33.

SEGALL, P. & POLLARD, D. D. 1983. Joint formation in granitic rock of the Sierra Nevada. *Geological Society of America Bulletin*, **94**, 563–575.

SHIMO, M. & LONG, J. 1987. A numerical study of transport parameters in fracture networks. *In*: EVANS, D. D. & NICHOLSON, T. J. (eds) *Flow and Transport Through Unsaturated Rock*. Geophysical Monograph 42, American Geophysical Union, Washington DC, 12–132.

TOWNEND, J. & ZOBAK, M. D. 2000. How faulting keeps the crust strong. *Geology*, **28**, 399–402.

VERNON, R. H. 2000. Review of microstructural evidence of magmatic and solid state flow. *Electronic Geosciences*, **5**, 2.

WEAIRE, D. & O'CARROLL, C. 1983. A new model for the Giant's Causeway. *Nature*, **302**, 240–241.

WILLIAMS, H. & MCBIRNEY, A. R. 1979. *Volcanology*. Freeman Cooper, San Francisco.

WINOGRAD, I. J. 1971. Hydrogeology of ash-flow tuff: a preliminary statement. *Water Resource Research*, **7**, 994–1006.

ZHANG, X. & SANDERSON, D. J. 1996. Effects of stress on the two-dimensional permeability tensor of natural fracture networks. *Geophysics Journal International*, **125**, 912–924.

ZHU, W. & WONG, T-F. 1999. Network modeling of the evolution of permeability and dilatancy in compact rock. *Journal of Geophysical Research*, **104**, 2963–2971.

Fracture formation and evolution in crystalline rocks: Insights from attribute analysis

K. J. W. McCAFFREY,[1] J. M. SLEIGHT,[1,3] S. PUGLIESE[2] & R. E. HOLDSWORTH[1]

[1] *Reactivation Research Group, Department of Geological Sciences, University of Durham, Durham DH1 3LE, UK (e-mail: k.j.w.mccaffrey@durham.ac.uk)*

[2] *CEESR, School of Earth Sciences and Geography, Kingston University, Kingston upon Thames, Surrey KT1 2EE, UK*

[3] *Present address: Shell UK Exploration and Production, 1 Altens Farm Road, Nigg, Aberdeen AB9 2H7, UK*

Abstract: Fractures are ubiquitous in crystalline rocks and control the strength and geophysical and fluid transport characteristics of the Earth's upper crust. A quantitative description of fracture attributes may constrain models of fracture formation and evolution. In this study, fracture attributes collected from one-dimensional samples across exposures of typical crystalline rocks show comparable variability in fracture size and spacing to sedimentary rocks. Vein thickness and fracture aperture data show predominately power-law distributions. Vein and fracture spacing data are best described by exponential distributions with negative slopes and appear to vary with composition in intrusive rocks. The fracture systems exhibit a range of anti-clustered to clustered patterns, and densities are an order of magnitude higher for joints compared to veins. Fracture clustering data can be used in conjunction with the spatial distributions to provide information on the controlling processes of fracture spacing. We suggest that exponential spacing distribution is produced as a sampling effect for both periodic-spaced and clustered fracture sets. In the examples given here, thermal stress-related joint patterns are distinguishable from tectonic-related fractures in plutonic rocks and fracture density and clustering is increased towards a major reactivated basement fault.

Much of the continental and oceanic crust is composed of crystalline rocks that may play host to significant resources, e.g. minerals, water and hydrocarbons. The nature and evolution of fracture systems influences directly the fluid flow and mechanical properties of crystalline rocks and hence controls the formation and exploitation of these resources (McCaffrey *et al.* 1999). To fully understand the nature and evolution of fracture systems, it is first necessary to describe their three-dimensional (3-D) geometry, spatial organization, kinematics and nature of infills from crustal to microscopic scales. Fracture attribute analysis provides a methodology for collecting and analysing quantitative data taken from sub-samples through fracture systems. The method combines discontinuity spacing analysis (Priest & Hudson 1981) with geometrical, displacement and textural information on fracture systems (Price 1966). Fracture attribute analysis provides a quantitative assessment of: (1) the scaling properties up- or down-scale from that of the sample; (2) the 3-D fracture system geometry; (3) the fracture sealing history; (4) controls on the evolution of the fracture system. Fracture attribute data are now used routinely in tectonic models of sedimentary reservoirs (e.g. Gillespie *et al.* 2001); however, there are few published equivalent studies for crystalline rocks.

In this study, we report three case studies where fracture attribute analysis methods have been applied to crystalline rocks. The data sets were collected from veins, faults and joints formed in a range of common igneous and high-grade metamorphic rocks. All of the data sets presented here have been collected from outcrop traverses carried out over comparable scales. Taken together the data are typical of a range of lithologies found in many basement terrains. We describe the data sets that have been collected and systematic variations in fracture aperture/thickness and spatial attributes. We discuss models of fracture formation for the individual examples and demonstrate how the results from fracture attribute analysis provide insights into the processes of fracture formation, fluid transport in crystalline rocks and reactivation of pre-existing fracture systems.

From: Petford, N. & McCaffrey, K. J. W. (eds) 2003. *Hydrocarbons in Crystalline Rocks*. Geological Society, London, Special Publications, **214**, 109–124. 0305-8719/03/$15

Table 1. *A summary of sample lines and fracture attribute data used in this study*

Locality	Data set identifier	Host lithology	Fracture type	Trend	N	Line length (m)	Th or Ap	Exponent	Strain %	% alteration	Sp	Exponent	C_v	FD
							Size attributes				*Spatial attributes*			
Cornwall, UK														
Goonbarrow	GT1	granite	greisen	134-h	87	16.95	PL	0.55	—	0.23	—	—	—	—
Goonbarrow	GT1	granite	greisen	134-h	87	16.95	—	—	—	—	Exp	0.0081	1.77	5.13
Goonbarrow	GT2	granite	Qtz-tm vs	122-h	78	40.79	PL	0.91	0.01	—	Exp	0.0014	1.27	1.91
Goonbarrow	GT2g	granite	greisen	122-h	69	40.79	PL	0.93	—	0.06	Exp	0.0012	1.24	1.69
Cligga	CT1	granite	Qtz-tm vs	143-h	47	17.78	PL?	1.05	0.01	—	Exp	0.0032	0.74	2.64
Cligga	CT1g	granite	greisen	143-h	51	17.78	Exp	0.44	—	0.23	Exp	0.0039	0.66	2.87
Park	PT1	Bodmin granite	Qtz-tm vs	140-h	63	25.71	PL	0.69	0.03	—	Exp	0.0031	1.79	2.45
Vung Tau, Vietnam														
Big Mountain Q	BMQ1a	andesite	Joint	350-h	108	15.42	—	—	—	—	Exp	0.007	0.92	7.07
Big Mountain Q	BMQ1c	andesite	Cal vs	350-h	47	15.42	PL	1.64	<0.01	<0.01	Exp	0.027	0.96	3.04
Big Mountain Q	BMQ2m	microgranite	Joint	080-h	102	8.57	—	—	<0.01	0.01	Exp	0.012	0.90	11.90
Big Mountain Q	BMQ3a	andesite	Joint	230-h	122 (51)	2.88	PL?	1.60	0.02	0.07	Exp	0.068	0.71	45.83
Intersection point	IP1m	microgranite	Joint	004-h	89	5.69	—	—	—	—	Exp	0.013	1.00	15.64
Intersection point	IP2m	microgranite	Joint	Vert.	34	0.95	—	—	—	—	Exp	0.042	1.19	36.84
Intersection point	IP1d	diorite	Joint	352-h	76	14.12	—	—	—	—	Exp	0.005	1.01	5.38
Long Hai	LH1m	monzogranite	Joint	080-h	131 (52)	8.05	PL	0.82	0.02	<0.01	Exp	0.014	1.06	16.27
Long Hai	LH2m	monzogranite	Joint	174-h	112	8.11	—	—	<0.01	0.13	Exp	0.007	1.34	18.81
Jesus Mountain	JM1a	andesitic tuff	Joint	132-h	116 (36)	1.65	PL	1.67	0.02	<0.01	Exp	0.073	0.91	70.30
Møre Trøndelag Fault Complex, Norway														
Verran Fault (30)	47	gneiss	fracture	145-h	94	1.570	—	—	—	—	Exp	0.072	1.30	59.87
Verran Fault (28)	137	gneiss	fracture	135-h	80	1.317	—	—	—	—	Exp	0.061	1.10	60.74
Verran Fault (475)	28a	gneiss	fracture	315-h	35	1.418	—	—	—	—	Exp	0.025	1.26	24.68
Verran Fault (500)	28b	gneiss	fracture	280-h	52	1.910	—	—	—	—	Exp	0.035	0.73	27.23
Verran Fault (600)	28c	gneiss	fracture	330-h	38	2.743	—	—	—	—	Exp	0.013	1.09	13.85
Verran Fault (1926)	132a	gneiss	fracture	305-h	34	4.541	—	—	—	—	Exp	0.007	1.11	7.49
Verran Fault (1300)	133	gneiss	fracture	145-h	52	6.924	—	—	—	—	Exp	0.007	1.24	7.51
Verran Fault (500)	157	gneiss	fracture	315-h	117	5.154	—	—	—	—	Exp	0.024	1.00	22.70
Verran Fault (475)	164	gneiss	fracture	325-h	125	7.130	—	—	—	—	Exp	0.013	0.81	17.53
Verran Fault (-3)	130	gneiss	fracture	313-h	67	1.241	—	—	—	—	Exp	0.059	1.21	53.99
Verran Fault (20)	139	gneiss	fracture	136-h	24	0.592	—	—	—	—	Exp	0.041	1.50	40.54
Verran Fault (-3)	161	gneiss	fracture	140-h	89	2.070	—	—	—	—	Exp	0.053	1.60	43.00
Verran Fault (-800)	140	gneiss	fracture	310-h	12	4.127	—	—	—	—	Exp	0.002	1.11	2.91
Verran Fault (140)	48g	amphibolite	fracture	160-h	37	1.442	—	—	—	—	Exp	0.029	1.32	25.66
Verran Fault (28)	137	cataclasite	fracture	128-h	22	0.553	—	—	—	—	Exp	0.043	1.32	39.78

Abbreviations: N—number of fractures in sample; Th—thickness distribution; Ap—aperture distribution; Sp—spacing distribution; C_v—coefficient of variation; FD—fracture density.

Fracture attribute analysis

We use the term 'fracture' to describe any, generally planar, filled or unfilled discontinuity found in a rock volume. Commonly these are joints, faults and veins. A 'fracture system' is a coherent set of fractures that formed under the same stress regime. A 'fracture network' is a system in which the fractures show linkage at the scale of observation. A fracture 'attribute' is any property or parameter that may be quantified, e.g. dip, strike, spacing, fault throw and infill. Parameters may be given for a single fracture or considered as a bulk attribute computed for an entire sample, e.g. fracture density or strain. After collection, attribute data are analysed using fracture population methods based on the common statistical distributions (see below). The methods used to collect fracture attribute data are conveniently divided into three categories that relate to how a rock volume is sampled.

One-dimensional sampling methods

Any fracture data collected on a linear traverse (also known as a line section or a scan line), across outcrops, represents a one-dimensional (1-D) sample through a rock volume. Such data are often routinely collected during resource exploration in the form of boreholes or drill core logs. The position along the sample line and the attributes for individual fractures are recorded successively (see Table 1 for common attributes). Attribute collection from 1-D samples is simple but can be time consuming and the method produces results that are easily comparable between different data sources. Where a fracture system is truly 3-D in character, 1-D sampling may give a poor approximation of the spatial properties (Gillespie *et al.* 1993). Beacom *et al.* (2001) show that fracture system characteristics may be obtained from the spatial attributes in 1-D samples if the fractures have relatively simple geometries.

Two-dimensional sampling methods

Field outcrops or geological maps provide the other readily available source for fracture and fault data, and are two-dimensional (2-D) surfaces through a rock volume. Fracture data determined from 2-D sampling of rock volumes may be depicted on a fracture map (horizontal planes) or cross-section (vertical or inclined planes). Fracture maps permit an analysis of connectivity, orientation and density; spatial properties that are important controls on fluid flow (Odling 1997; Ryan *et al.* 2000).

Three-dimensional sampling methods

Full three-dimensional (3-D) characterization of fracture attributes is achieved primarily by producing maps from closely spaced 2-D samples or slices, followed by interpolation between the samples to reconstruct the 3-D fracture geometry. Examples include use of 3-D seismic data, serial section reconstruction of hand specimens (e.g. Pugliese & Petford 2001) or X-ray computed tomographic (CT) scanning if the fractures show compositional changes that may be imaged (Carlson *et al.* 1999). Despite the high costs and scale limitations of some of these methods, the results are potentially powerful because the full 3-D fracture geometry may be extracted. To extrapolate these 3-D methods from small scale (hand specimen or 3-D seismic data set) to a basin or regional scale requires a detailed knowledge of the scaling of fracture attributes. This can only be achieved using 1-D and 2-D sampling at the larger scale in conjunction with the 3-D models.

Fracture population analysis

Most natural populations of fractures yield skewed distributions of their commonly measured attributes (e.g. spacing, length, displacement). The common statistical distributions that have been used to describe these populations range from the normal distribution (unskewed) to the increasingly skewed lognormal, exponential and power-law distributions (Gillespie *et al.* 1993, 1999). The simplest qualitative test to decide which of these distributions is appropriate for a given set of fracture attributes is to plot the data on a cumulative frequency (number) versus attribute size graph (herein called a 'population plot') (Villemin & Sunwoo 1987; Childs *et al.* 1990) (Fig. 1). Population plots are generated by ranking the attribute data in descending order and then plotting attribute against the cumulative frequency. Depending on which combination of logarithmic and linear axes is used, the most common distributions may be distinguished (Fig. 1). Distributions may be quantitatively compared by using distribution curves fitted using least squared regression analysis. Data should be converted to the same units to ensure comparability between the best-fit distributions.

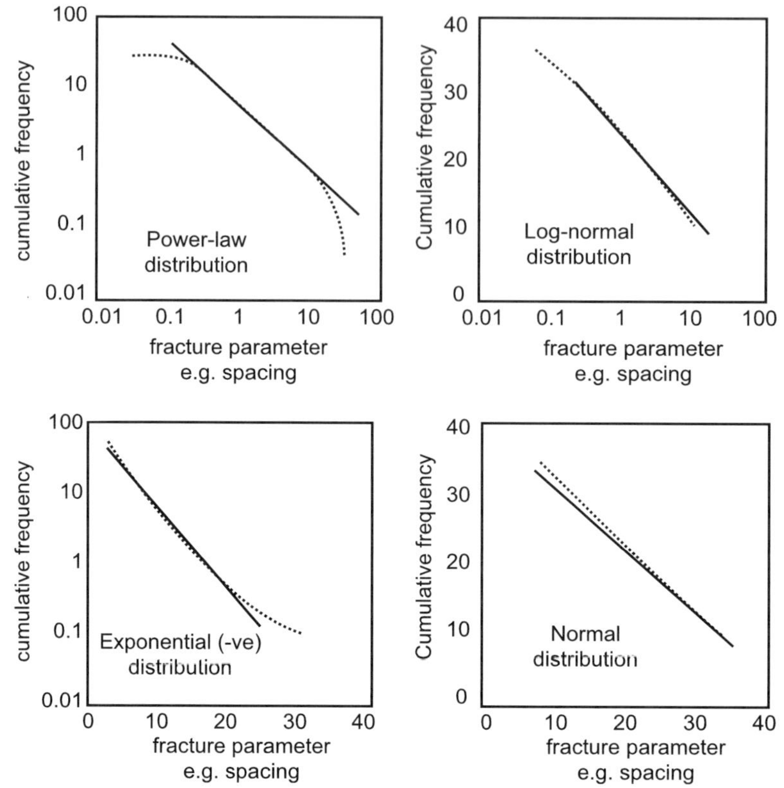

Fig. 1. Typical population plots for fracture attribute distributions. Solid line ideal distribution, dotted line typical of a 'real' data set.

Bulk attributes

Where thickness or aperture data are available for tensile fractures, 1-D strain can simply be calculated as the percentage sum of the thickness or aperture divided by the sample line length.

In this study, such data were not always available or the strain was of negligible amount (less than 0.01%). Where the width of planar alteration zones adjacent to fractures has been measured, the total has been worked out as percentage of the sample line length in a similar way to strain (see Table 1).

In order to provide a more complete spatial analysis, two further attributes were calculated for each data set. Fracture density (FD) (the number of fractures per unit distance) allows simple comparisons to be made between data sets. Relative fracture density between samples may also be assessed visually from the population plots, because each sample line is normalized with regard to the sample line length. The coefficient of variation (C_V), defined as the ratio of the standard deviation to the mean, is another useful parameter that measures the degree of spatial clustering or anti-clustering in each data set (Gillespie *et al.* 1999, 2001). If $C_V = 1$ then the distribution is identical to a Poisson (random) distribution (Cox & Lewis 1966), whereas if $C_V < 1$ the fractures are anti-clustered (more regular) and if $C_V > 1$ the fractures are clustered (less regular).

Fractal analysis

Certain fracture attributes such as fault throw, fault length, vein thickness and some fracture spacing data have been shown to be best described by a power-law distribution (Childs *et al.* 1990; Scholz & Cowie 1990; Velde *et al.* 1990; Marrett & Allmendinger 1991; McCaffrey *et al.* 1993; Gillespie *et al.* 1993). An important property of a power-law distribution is that it is defined without reference to a characteristic scale, whereas other common distributions have inherent scale dependence (e.g. a mean at a certain size value). The term 'fractal' has been applied to any fracture attribute that follows a power-law distribution (Turcotte 1992); however, as pointed out by Bonnet *et al.* (2001), 'fractals' should be restricted to a description of the spatial organization of fractures because it is a geometric concept. Fracture systems, in general, have been shown to display a wide variation in their size and spatial properties, due not only to the range of settings and host rocks in which they form, but also to the methodologies employed during sampling and analysis (Gillespie *et al.* 1993, 1999; Bonnet *et al.* 2001).

Data set locations

Mineralized fractures in the St Austell, Bodmin and Cligga Head granites, Cornwall

The Cornubian batholith was intruded during the latest Variscan orogenic events in SW England (Willis-Richard & Jackson 1989). The batholith formed the heat source that drove the Sn–W and base metal mineralization system (Dines 1956; Moore & Jackson 1977; Jackson *et al.* 1989; Floyd *et al.* 1993). Individual plutons have undergone locally intensive, late-stage alteration in the form of tourmalinization, greisenization (quartz–sericite alteration) and kaolinization. In order to investigate the spatial distributions of the mineral veins, the fracture attributes (vein thickness and spacing) were sampled at three localities: (1) Goonbarrow china clay pit, St Austell granite; (2) Park china clay pit, Bodmin granite, and (3) the Cligga Head granite (Fig. 2a). In all localities, 1-D traverses were taken at high angles to the ENE–WSW striking and moderately NNW- or steeply SSE-dipping main structures that trend parallel to the main regional Sn–W mineralization lode trend (Dines 1956). A genetic relationship between zones of kaolinization and pre-existing hydrothermal veins was proposed (Jackson *et al.* 1989); however, more recent studies now support a model that suggests that meteoric fluids were channelled along the earlier veins causing clay formation (Psyrillos *et al.* 1998).

Igneous lithologies from the Vung Tau peninsula, Vietnam

In order to investigate fracture system attributes in a variety of igneous rocks, Jurassic- to Cretaceous-aged components of the Kontum basement block of central and SE Vietnam (Rangin *et al.* 1995) were studied (Fig. 2b). This was instigated in the Vung Tau peninsula because hydrocarbons have been discovered in similar Cretaceous intrusive rocks offshore. The intrusive suite at Vung Tau comprises basalt, diorite, granodiorite, andesite, monzogranite and rhyolite. Volcanic rocks are generally either andesitic or rhyolitic tuffs containing varying amounts of ellipsoidal lapilli, plagioclase megacrysts and small subangular xenoliths. The majority of joints, faults, dykes and intrusive contacts in the Vung Tau peninsula are steeply dipping, and strike NE–SW, north–south and east–west. These fracture orientations are consistent with regional north–south and NE–SW fracture trends that have been identified in the Kontum block (Rangin *et al.* 1995). Most of the joints are barren and have small apertures (1–3 mm) but north–south and east–west striking joints are filled with carbonate or silica minerals. North–south, NE–SW and east–west joint sets are characterized by alteration halos in their wall-rocks, suggesting fluids have been transported along these structures. The thickness and spacing attributes of these structures in different lithologies were measured along 1-D samples taken across coastal and quarry exposures (Fig. 2b). Only the joint and vein data are sufficiently numerous for population studies; however, not all joints yielded measurable apertures and wall-rock alteration, and therefore data sets are only presented where numbers permit.

Granitic gneiss lithologies in the Møre Trøndelag, Norway fault system

The ENE–WSW trending Møre Trøndelag Fault Complex (MTFC) in central Norway (Fig. 2c) is a 10–20 km wide, steeply dipping zone of fault-related deformation that extends offshore to coincide with the southern margin of the Møre basin and the northern margin of the Viking graben in the North Sea (Doré *et al.* 1997). The study investigated the variation in fracture attributes in the vicinity of a major fault zone in an area chosen because of the uniform Palaeoproterozoic granitic gneiss host rocks. These have been highly reworked during Scandian (Siluro-Devonian) deformation and metamorphism. The MTFC comprises a series of Early Devonian ductile shear zones that are overprinted by multiple brittle reactivations in the Permo-Carboniferous, Mesozoic and Cenozoic (Doré *et al.* 1997; Watts, unpublished PhD thesis) with offsets varying from strike–slip through oblique–slip to dip–slip at different structural levels. An extensive data set of 1-D and 2-D fracture characteristics has been collected and analysed from the MTFC (Sleight, unpublished

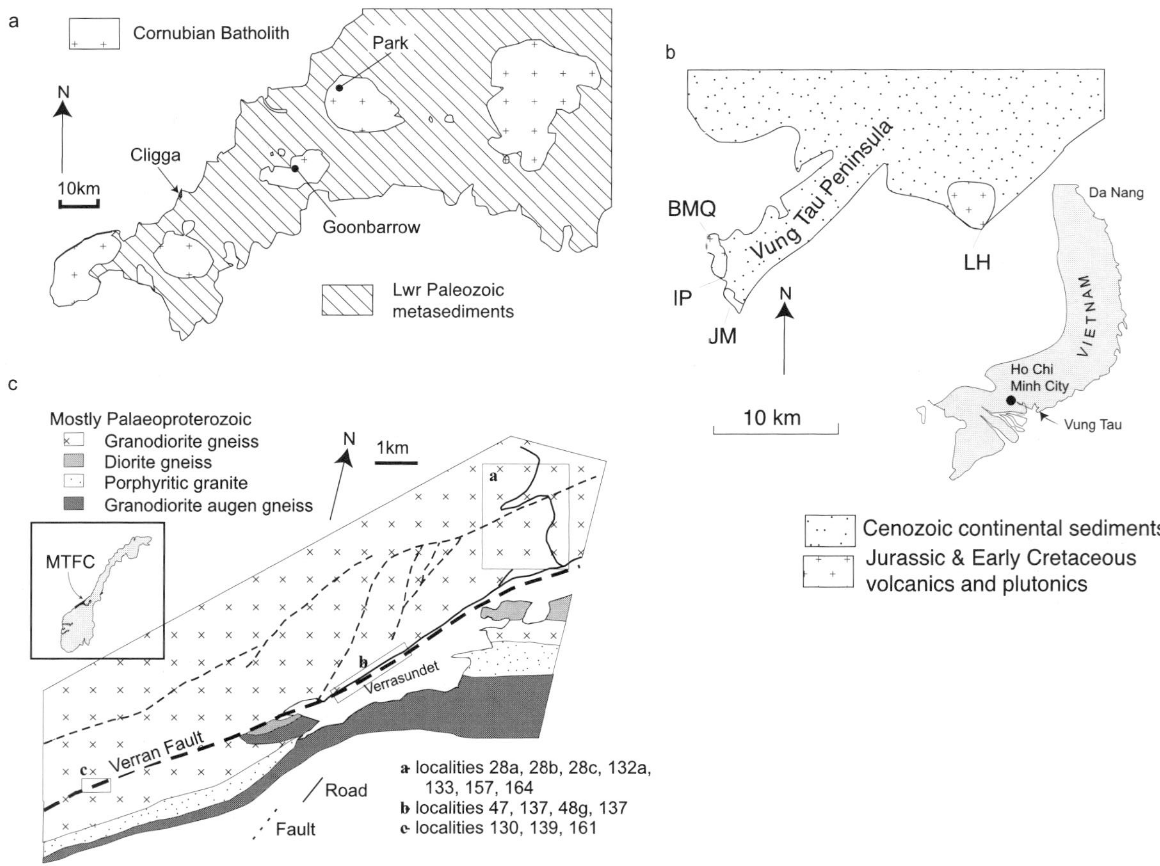

Fig. 2. Location maps for the data sets presented in this study: (**a**) Goonbarrow and Park china clay pits and Cligga Head granite, SW England. (**b**) Vung Tau peninsula, Vietnam: BMQ, Big Mountain quarry; IP, Intersection point; JM, Jesus Mountain; LH, Long Hai. (**c**) Sample localities relative to the Møre Trøndelag Fault Complex (MTFC), Norway.

PhD thesis). Here, we report the results of 1-D outcrop transects measured at distances up to 2 km from the Verran Fault, which is the southern strand of the MTFC (Fig. 2c). Up to 500 m from the Verran fault plane (VFP), fracture orientations show four groups: ENE–WSW—parallel to the fault plane, east–west, NNW–SSE and north–south. Data sets presented here have been collected along 1-D transects orientated perpendicular to the MTFC (and VF) trend, and therefore predominantly sample the ENE–WSW trending fractures which are parallel to the main fault trend. Many of the fractures show infills of epidote-rich cataclasite, calcite and zeolite mineralization, and minor pseudotachylite, chlorite and clay-rich gouges indicating that fracture formation was related to brittle displacement on the MTFC.

Fracture data sets

Details of sample lines and a summary of the fracture attributes for each sample line are given in Table 1. Representative population plots from each data set are given in Figures 3–7. In each plot, cumulative frequency values have been divided by sample line length to ensure comparability between the different data sets.

Cornwall—vein thickness data

Thickness attribute data are shown plotted on population plots with calculated best-fit curves (Fig. 3). Most thickness data are best described by a power-law relationship. Greisen thicknesses

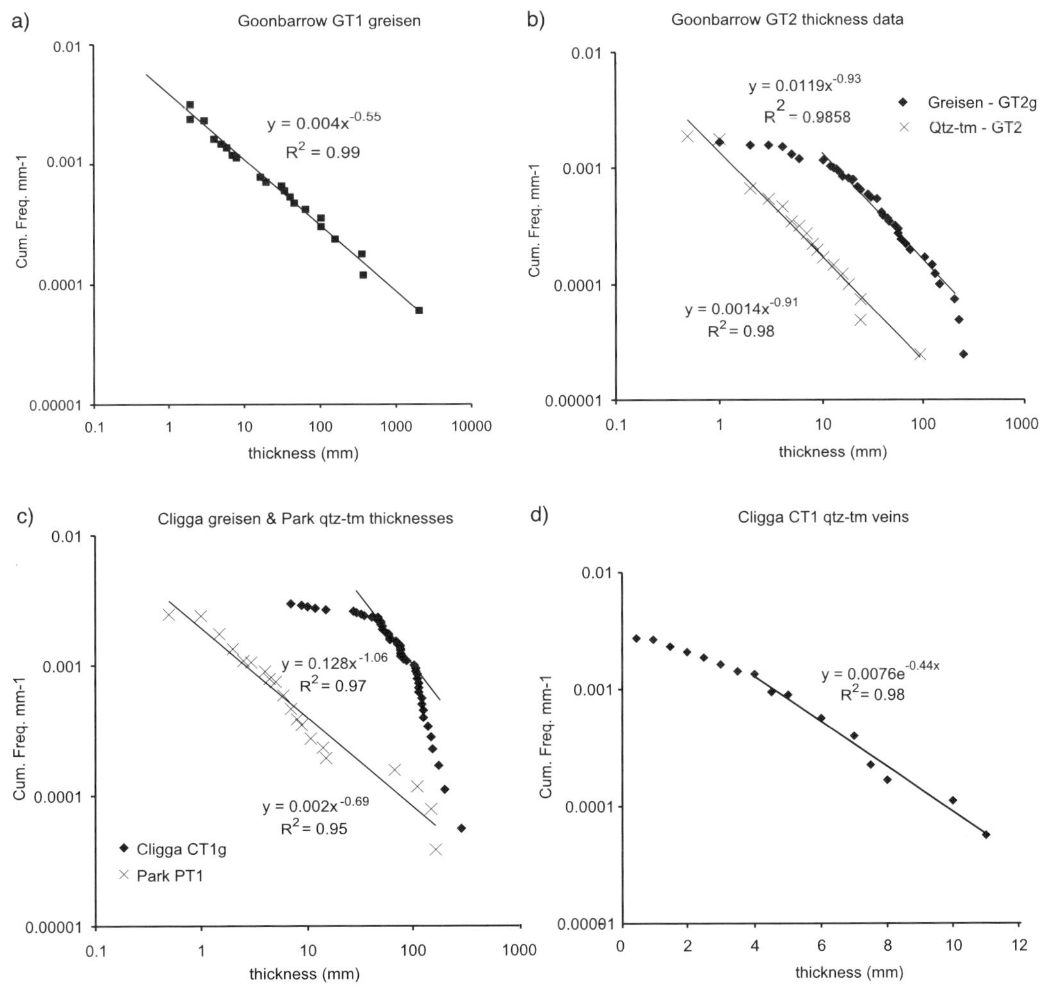

Fig. 3. Population plots for vein and greisen thickness data from SW England.

(Fig. 3a–c) show good (curve spanning three orders of magnitude) to poor (half an order of magnitude span) accordance with a power law distribution. The left-hand portion of the curves for GT2g and CT1g is flatter than the calculated best-fit curve (Fig 3b, c) for the main part of the data. This is often attributed to an under-sampling or truncation effect (Pickering *et al.* 1995); however, this is thought unlikely to apply to these data because the change of slope occurs in the range 10–15 mm thickness that is well above any outcrop-scale resolution threshold. A steeper right-hand end of the curve also shown by GT2g and CT1g (Fig. 3c, d) has been attributed to censoring, a 'cut effect' due to the 1-D sample lines not intercepting the thickest part of the veins (Pickering *et al.* 1995). Goonbarrow and Park quartz–tourmaline vein thicknesses (Fig. 3b, c) show typical power-law distributions, whereas Cligga Head quartz–tourmaline vein thicknesses are better described by an exponential (log/linear axes) distribution (Fig. 3d). Calculated strain values are low, ranging from 0.01% to 0.03%, and greisen alteration (equivalent in this case to total greisen thickness divided by sample line length) varies from 0.06% to 0.23% (Table 1).

Cornwall—vein spacing data

The distribution of spaces between adjacent greisen and quartz–tourmaline veins for both Cligga and Goonbarrow (GT2) show good agreement with exponential distributions (Fig. 4a, b). Where veins and greisen were measured in the same transect (Fig. 4a, d), the obvious similarity in distribution confirms the close spatial relationship between the two structures observed in the field. The spacing distributions of quartz–tourmaline veins from Park and Goonbarrow T1 are shown on both exponential (Fig. 4c) and power-law plots (Fig. 4d) because the sampled distribution lies somewhere between these two models. In both cases, the concave upwards nature of the right-hand end of the curve (Fig. 4c) implies that there is an excess of larger spaces in the sample than would be expected

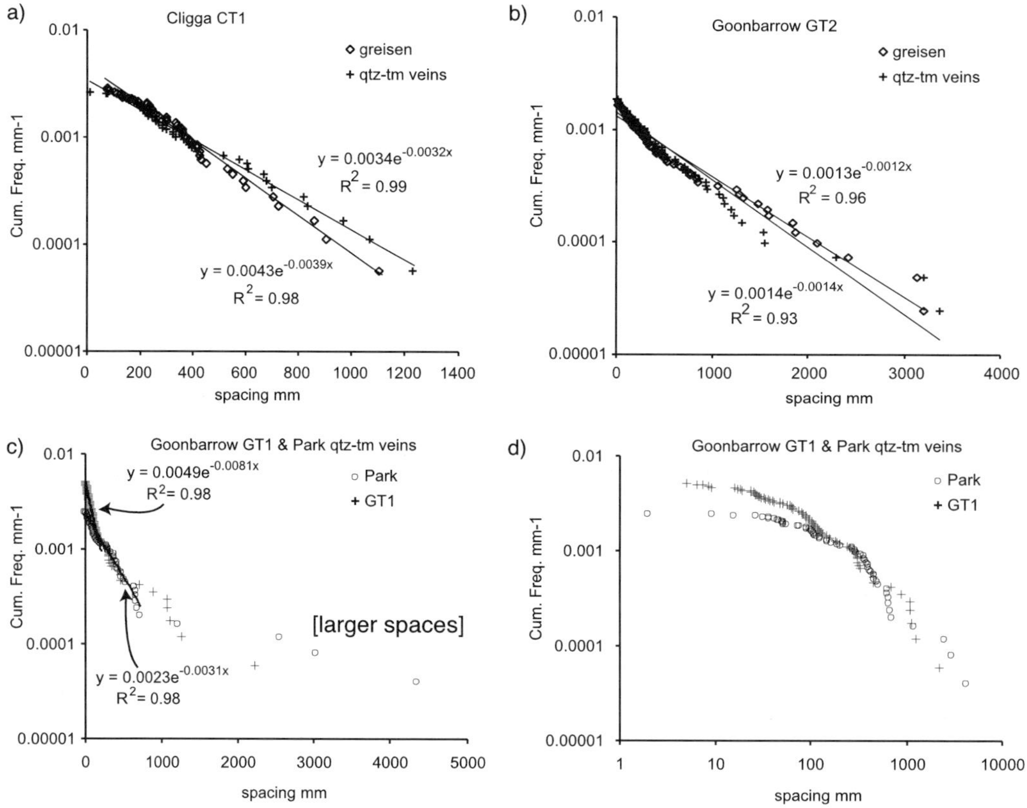

Fig. 4. Vein and greisen spacing distributions from SW England.

relative to a fitted exponential distribution. These data are also shown on a power-law plot which suggests that the spacing distributions, whilst being more clustered than an exponential distribution, are not clustered enough to be regarded as a power-law (Fig. 4d).

The C_v for each data set distinguished different clustering properties (Table 1). Goonbarrow (GT1) and Park (PT1) yielded C_v that are 1.77–1.79, i.e. highly clustered, whereas Goonbarrow (GT2) is closer to a random distribution (1.24–1.27), and Cligga (0.66–0.74) anti-clustered. Fracture density varies from 1.69 m^{-1} to 5.13 m^{-1}.

Vung Tau, Vietnam—fracture populations

Calcite vein thickness and joint aperture data follow power-law distributions with exponents in the range 0.8–1.64 (Table 1, Fig. 5). The limited scale ranges for most data sets reflect the low overall strain (less than 0.1%); however, joint apertures for Long Hai (LH1m, Fig. 5b) do span two orders of magnitude. More complete availability of joint spacing data means that the fracture spacing distributions are more robust. They are all best described by exponential distributions with negative slopes (Fig, 6, Table 1) with exponents for the joint data sets ranging from 0.005 to 0.073. The calcite vein distribution has an exponent of 0.027. Most C_v values are close to 1.00 with the Long Hai monzogranite (LH2m) yielding the most clustered value of 1.34. Fracture densities varied from a 3 m^{-1} to a maximum of over 70.3 m^{-1} for the andesitic tuff.

Møre-Trøndelag, Norway—fracture spacing

Fracture spatial attributes collected from the MTFC adjacent to the VFP are listed in Table 1 and representative exponential population plots are shown in Figure 7. Most data sets show good correspondence to the exponential distribution, with exponents that range from 0.006 to 0.720. Sample line lengths are shorter on average than for the Cornwall or Vietnam data sets; however, the fracture density is generally higher (7–61 m^{-1}). The C_v values for the MTFC fracture spacing distributions in the vicinity of the Verran fault have values that range between 0.73 and 1.6 with an average C_v of 1.18. Fracture data sets collected from amphibolite and cataclasite show similar distribution type to those collected from granitic gneiss (Fig. 7).

Comparative analysis between fracture data sets

The fracture attribute data collected from the three studies, although collected for different purposes, were all collected by the 1-D line sample method and thus the distribution types are comparable. All data sets reported here show consistent size (vein thickness or joint aperture) distributions that are best described by power-law distributions. The average power-law exponent for the vein thickness distributions from the Cornish samples was 0.83; however, the other vein data set (calcite veins in andesite from Vung Tau) yielded a higher exponent value (1.64). Joint apertures in the Vietnamese igneous rocks yielded poorly constrained power-law

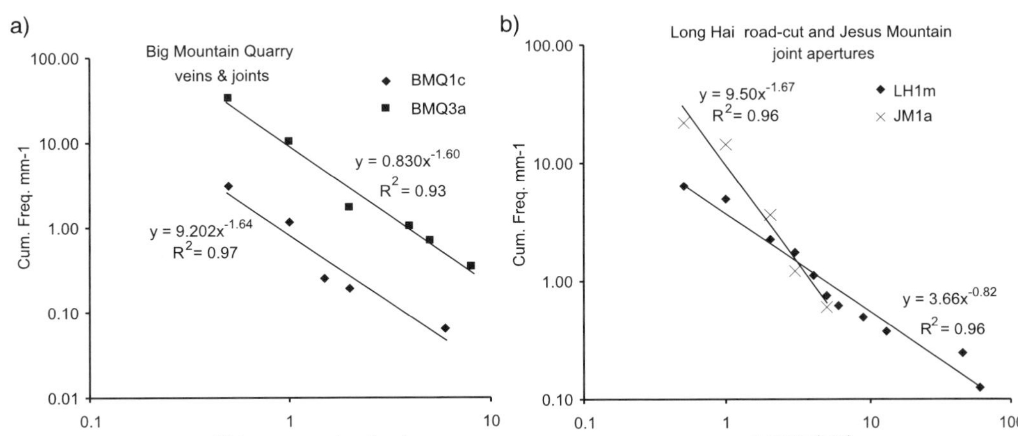

Fig. 5. Vein thickness and joint aperture distributions in igneous rocks from the Vung Tau region, Vietnam.

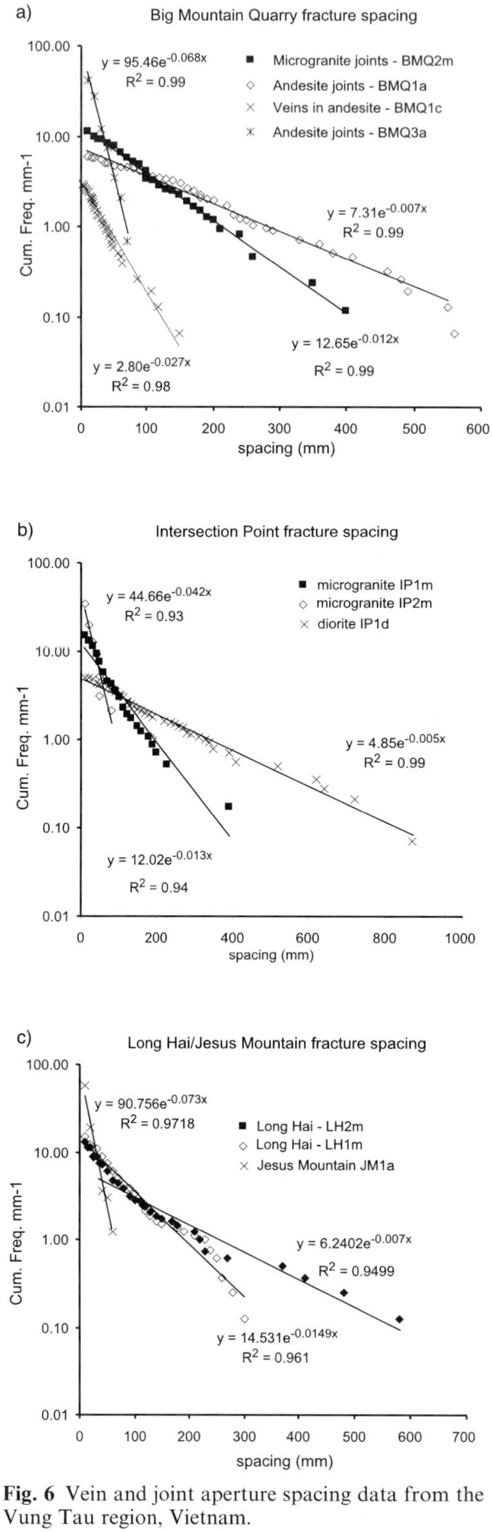

Fig. 6 Vein and joint aperture spacing data from the Vung Tau region, Vietnam.

distributions with average slopes of *c.* 1.36. Overall, strain values were low for the vein systems (~0.1%) and negligible for the joints.

The spacing distributions for all data sets are best described by an exponential distribution with a negative slope. The exponent for the veins was between 0.0031 and 0.014, whereas for joints the exponent ranged from 0.005 to 0.068. The fracture density values were generally lower for veins ($<3\,m^{-1}$) and higher for joints ($>5\,m^{-1}$). The C_v values varied from 0.66 to 1.79, with an average of 1.13. There was no relationship between the number of fractures and the exponent either for the power-law or exponential distributions.

Discussion

Fracture size data

The vein thickness data from Cornwall and Vung Tau are consistent with those reported for vein systems in other igneous lithologies (McCaffrey *et al.* 1993; Braithwaite *et al.* 2001; Gudmundsson *et al.* 2001). A power-law exponent of 0.8 was given as the mode for 1-D samples through veins developed in a range of metamorphic and sedimentary lithologies (Gillespie *et al.* 1999). Fault displacement populations measured commonly follow a power-law distribution with an exponent in the range 0.4–1.0 (Childs *et al.* 1990; Walsh *et al.* 1994; Nicol *et al.* 1996). Joint aperture distributions generally yield higher power-law exponents of *c.* 1.4 (Barton & Zoback 1992). The results from this study show that vein thickness and joint aperture distributions in crystalline rocks yield similar power-law distributions to those in other lithologies with joints displaying higher exponents.

Fracture spacing data

Compared to fracture size distributions, spacing data generally display more variability in the types of distribution and the range of exponent values (Gillespie *et al.* 1993). In igneous rocks, vein spacing distributions reported by Braithwaite *et al.* (2001) show non-power-law spacing distributions. For sedimentary and metasedimentary rocks, the observed spacing distribution appears to vary as a function of the type of fracturing, the host-rock lithology and the conditions of deformation (Gillespie *et al.* 1999, 2001). Power-law fracture spacing distributions, representing the most highly clustered of the range of possible distributions,

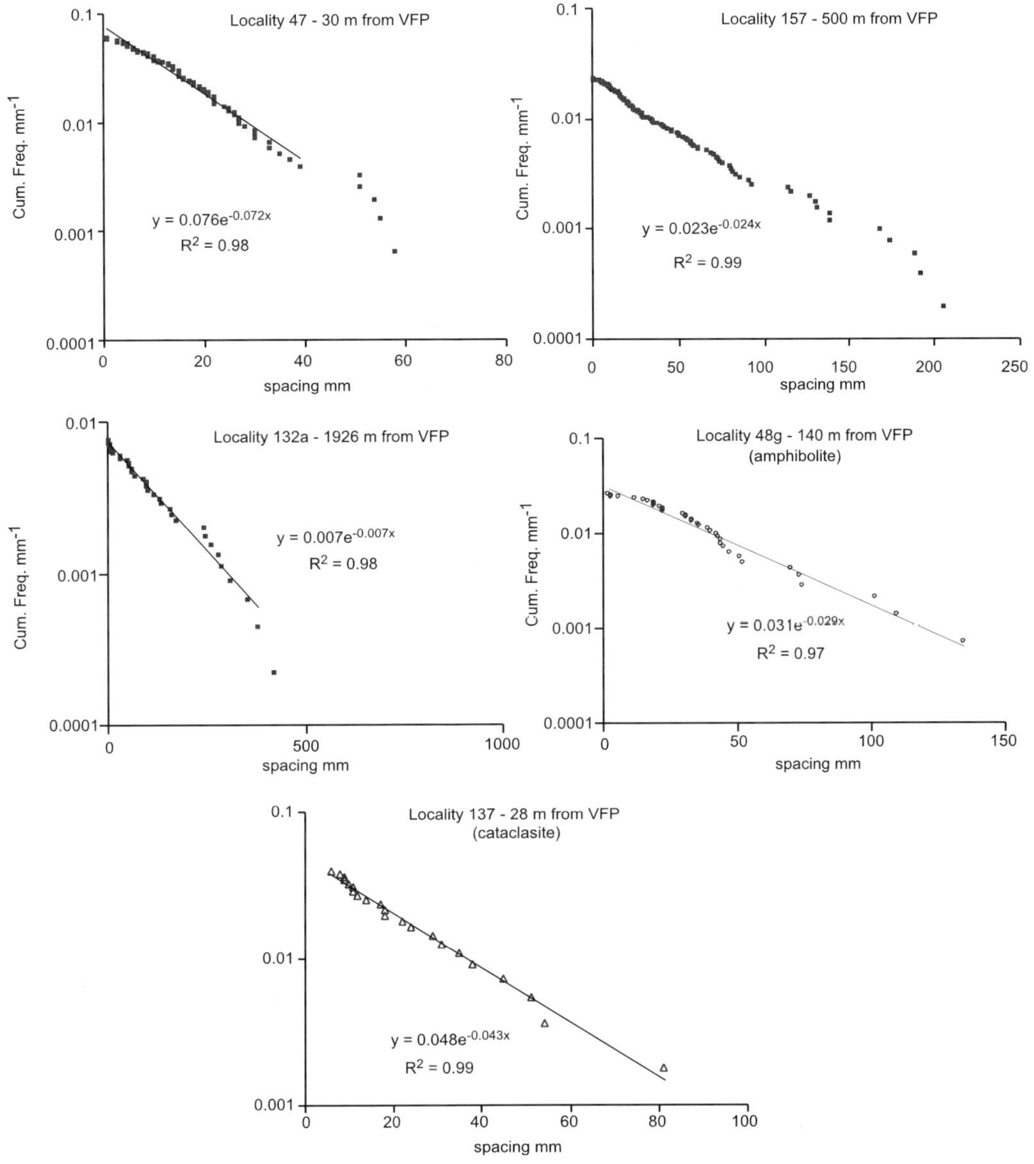

Fig. 7. Representative fracture spacing data from the Møre Trøndelag Fault complex, Norway.

have been obtained for tectonic faults (Gillespie *et al.* 1993), some non-strata bound hydrothermal vein systems (Gillespie *et al.* 1999) and metamorphic veins (Manning 1994). Less clustered distributions or more random distributions are characteristic of joint systems giving exponential, gamma or lognormal distributions (Huang & Angelier 1989; Rives *et al.* 1992; Pascal *et al.* 1997). Regular spacing (anti-clustered) distributions are shown by fractures developed in well-bedded sedimentary rocks with the average spacing being related to bed thickness (Ladeira & Price 1981; Narr & Suppe 1991; Bai & Pollard 2000).

Mechanisms envisaged for generating fractures generally produce *either* scale-dependent, anti-clustered (normal or log-normal) distributions where internal effects (thermal, chemical or mechanical) predominate, *or* scale-independent, clustered (power-law) distributions where rupture processes due to external tectonic forces predominate (see Bonnet *et al.* 2001 for review).

Neither of these mechanisms should produce the exponential distributions found in this study that are more typical of a random (Poisson) point process (Cox & Lewis 1966; Gillespie *et al.* 1993). Here, we suggest that since a mechanism or process that generates purely random fractures has not yet been proposed, then the exponential distributions are likely to be artefacts or pseudo-random in nature. These pseudo-random distributions could arise in one of two ways: (1) Gillespie *et al.* (1993) showed that regularly spaced joint sets with different orientations and densities when sub-sampled in 1-D may produce an exponential distribution; (2) it is also known that sampling and finite range effects may result in departures from an ideal power-law that could resemble a lognormal or exponential distribution (Pickering *et al.* 1995; Main *et al.* 1999). A full analysis of this problem requires a modelling solution and is not developed further here. Nevertheless, in the three examples given below, we suggest that the exponential spacing distributions, when supplemented with other fracture attributes or an analysis of variations between samples, can point to underlying controls in the fracturing process.

Vein formation in Cornwall

Formation of the greisen and hydrothermal vein systems in the Cornubian granites has been attributed to a variety of processes (Halls 1987; Jackson *et al.* 1989). These are: (1) hydrofracturing due to increased fluid pressures; (2) hydraulic shock due to decompression; (3) thermal stresses forming contraction joints in a cooling pluton; (4) active faulting. Mechanisms (1) to (3) are due to the formation and dissipation of internal stresses and might be expected to give rise to more anti-clustered distributions, whereas mechanism (4) should give rise to a highly clustered distribution. Modelling of the thermal stresses formed in cooling plutons by Jackson *et al.* (1989) suggests that mineralized contraction joints would only form in small granite bodies. The fracture spatial data reported here for the larger plutons (St Austell, Bodmin) all show evidence for clustering (distributions that are more clustered than exponential with C_v values 1.24–1.79), whereas the smaller pluton (Cligga Head granite) yielded exponential distributions with C_v values (0.66–0.74) indicating a tendency towards anti-clustering. Hence, the fracture spatial data, although of a preliminary nature, support the Jackson *et al.* (1989) interpretation that thermal contraction joints controlled mineralization processes in the smaller pluton. In

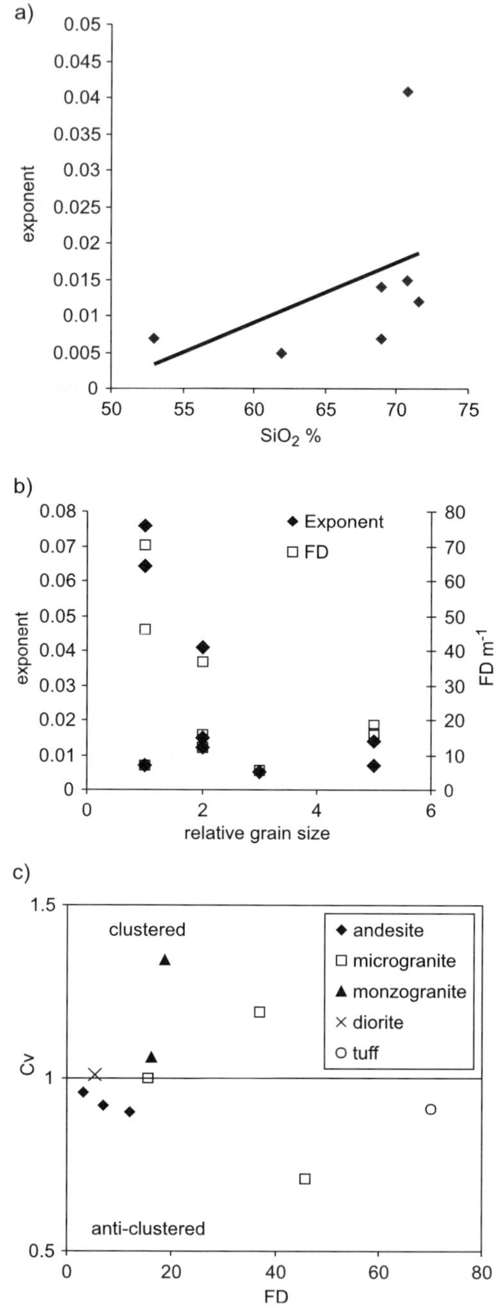

Fig. 8. Variation in fracture spatial properties in igneous lithologies from the Vung Tau region, Vietnam: (**a**) Exponent from spacing distribution versus SiO_2 content for intrusive lithologies. (**b**) Exponent and fracture density versus relative grain size: 1, andesite; 2, microgranite; 3, diorite; 5, monzogranite. (**c**) Coefficient of variation (C_v) versus fracture density (FD) for different lithologies (see text for discussion).

the larger plutons, the thermal-induced stresses produced fractures that were presumably additionally influenced by tectonic stresses.

Lithological variations in Vietnam

Fracture population data have been presented from a variety of igneous lithologies at Vung Tau, Vietnam. Little comparative data exist on the fracture attributes of different igneous lithologies, despite their importance as crystalline aquifers and hydrocarbon reservoirs (Stearns & Friedman 1972; Parnell 1988). Our data all yielded exponential joint spacing distributions and, where constrained, power-law joint aperture distributions. There is a weak relationship between exponent and the silica content of the intrusive rocks (Fig. 8a). This would suggest that the proportion of small to large spaces is higher for high silica rocks compared to more basic rocks. There are also weak relationships between grain size and the distribution exponent or fracture density (Fig. 8b). These relationships are not sufficiently constrained to provide any empirical statements about the relationships between composition and fracturing, but are interesting enough to warrant further investigation.

The relative spatial characteristics of the different lithologies may be summarized and compared on a plot of C_v versus fracture density (FD) (cf. Beacom *et al.* 2001). This plot is of use when fracture density variations associated with major fault structures can be eliminated. Different lithologies tend to plot in distinct areas on this plot (Fig. 8c), e.g. andesites plot in the low-density, slightly anti-clustered region, whereas the microgranite samples show medium fracture densities and the tuff the highest density. These exponential distributions are characterized by C_v values close to 1 in most cases, and probably arise from interference of several joint sets with periodic distributions but different orientations and densities, as in the Gillespie *et al.* (1993) model discussed earlier. It is notable that the one data set (LH2m) that displayed clustering ($C_v = 1.34$) was taken across a NE–SW orientated sinistral fault exposed on the beach at Long Hai. This suggests that clustering of fractures associated with tectonic processes is discernible in the fracture attribute data.

Systematic variations in fracture population adjacent to the VFP, MTFC

The data presented here permit investigation of the fracture attributes of a major basement fault structure. Beacom *et al.* (2001) have shown that both compositional variation and the host rock anisotropy variation can influence the spatial and orientation characteristics of subsequent fault reactivation. The data sets presented here have been collected from homogeneous granitic gneiss, with the exception of localities 48g and 137 that were collected within amphibolite and cataclasite, respectively. The relatively constant host rock composition and fabric means that any systematic variation in fracture properties is likely to have a tectonic origin. All data sets are best described by an exponential spacing distribution (Table 1). The position of each sample line relative to the Verran fault plane is shown in Figure 9 plotted against the exponents from the fracture spacing distributions and fracture density values. An expected increase in fracture density towards the fault plane is shown, but also noticeable is a clear increase in the exponent value towards the centre of the fault. The data permit a definition of an inner fault zone (<100 m from the fault plane with high exponent values >0.04), an outer damage zone (up to 750 m from fault plane, with exponents 0.01–0.04) and the background (>750 m from fault plane, with exponents <0.01). Both the cataclasite and the amphibolite samples have slightly lower values than would be expected for their distance from the fault plane, illustrating that lithological variation, where present, is significant. The data demonstrate clearly that fracture characteristics vary systematically with distance to the major structure.

The exponential spacing distributions that have been obtained from data sets collected within the MTFC appear to be at odds with fracture mechanisms, and suggest that fault-related fractures in homogeneous rocks should be clustered rather than random (see above discussion). We can address this by examining the relative variations in fracture spacing attributes. Higher exponent values at the centre of the VF infers that the data set is more dominated by small spaces relative to large spaces compared to background levels outside the fault damage zones. This suggests that the fractures are more clustered within the fault than outside it. This is supported by higher C_v values within the inner fault damage zones (average 1.33) compared to outside (average 1.07), and also qualitative observations and strain profiles that suggest more clustering closer to the fault planes (Sleight, unpublished PhD thesis). We therefore suggest that 1-D samples through the granitic gneisses yield exponential distributions of fractures in background areas away from the fault because

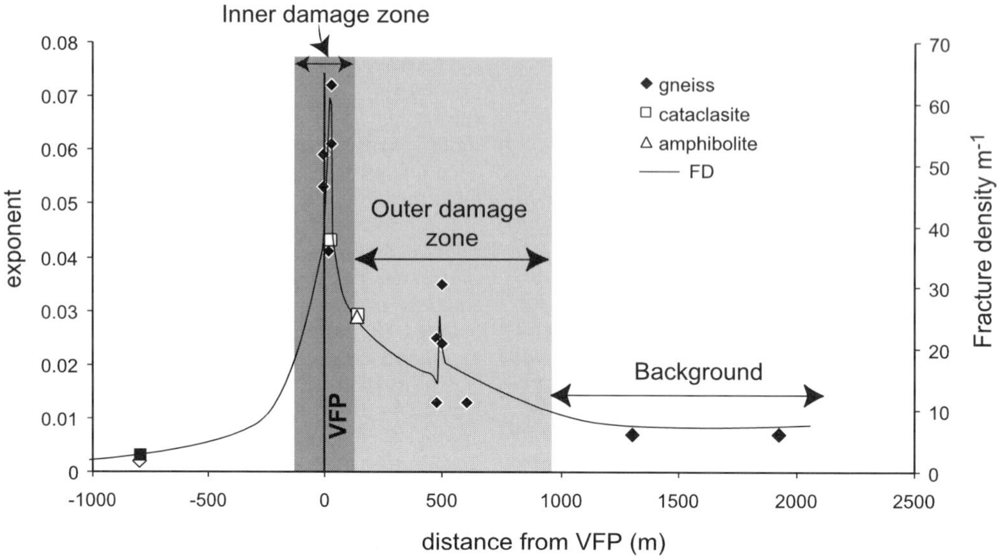

Fig. 9. Variation of the exponent from the fracture spacing distribution and fracture density with distance from the Verran Fault plane (VFP).

of the effect of multiple orientations of fracture. Within the fault damage zone (<600 m from VFP), fracture clustering caused by faulting is present and can be detected in the fracture attributes, but the clustering is never strong enough to yield a power-law spacing distribution. This may be due to the controlling influence of pre-existing fracture patterns that existed prior to the initiation of the MTFC. Basement rocks ubiquitously contain old fracture systems with characteristic spatial patterns, and any new fault system superimposed is likely to inherit some of the older fracture attributes. This has implications for the remote detection of fault systems from fracture patterns. The attributes should show different spatial characteristics when compared with similar faults developed in cover sequences without pre-existing structures.

Conclusions

(1) Outcrop traverses or 1-D methods of collecting fracture attribute data from sedimentary rocks have been applied successfully to fracture systems formed in crystalline rocks.

(2) Fracture attribute data from typical crystalline rocks from three locations studied show comparable variability in their fracture size and spacing distributions to sedimentary rocks.

(3) Vein thickness and fracture aperture data show predominately power-law distributions with exponents of 0.8 and 1.2, respectively. Vein spacing data are best described by exponential distributions with negative slopes. C_v values indicate a range of anti-clustered to clustered patterns, and fracture densities are an order of magnitude higher for joints compared to veins.

(4) Exponential spacing distributions may result from 1-D sampling of both multiply orientated fracture sets with periodic distributions, and clustered distributions showing finite range effects. Used in conjunction with measures of fracture clustering (C_v), the spacing data can be indicative of the mode of origin.

(5) Fracture spatial attributes in igneous rocks may vary with composition; they can potentially distinguish thermal stress related joints from tectonic related fractures in igneous rocks. Fractures increase in density and clustering towards a major reactivated basement fault structure; however, the spatial attributes of crystalline basement fractures appear to be inherited by younger systems.

The authors wish to acknowledge L. Watts for discussion of MTFC, W. Cox for assistance with data collection and Kingston University for travel assistance. T. Aoyama of JVPC is thanked for supporting the study in Vung Tau, and J. Howe at ECC is thanked for access to Goonbarrow and Park china clay pits. JMS wishes to thank Statoil UK for support.

References

BAI, T. & POLLARD, D. D. 2000. Closely spaced fractures in layered rocks: initiation mechanism and propagation kinematics. *Journal of Structural Geology*, **22**, 1409–1425.

BARTON, C. A. & ZOBACK, M. D. 1992. Self-similar properties of macroscopic fractures at depth in crystalline rocks in the Cajon Pass Scientific drillhole. *Journal of Geophysical Research—Solid Earth*, **97**, 5181–5200.

BEACOM, L. E., HOLDSWORTH, R. E., MCCAFFREY, K. J. W. & ANDERSON, T. B. 2001. A quantitative study of the influence of pre-existing compositional and fabric heterogeneities upon fracture-zone development during basement reactivation. *In*: HOLDSWORTH, R. E., STRACHAN, R. A., MAGLOUGHIN, J. F. & KNIPE, R. J. 2001. *The Nature and Tectonic Significance of Fault Zone Weakening*. Geological Society, London, Special Publications, **186**, 195–211.

BONNET, E., BOUR, O., ODLING, N. E., DAVY, P., MAIN, I. G., COWIE, P. & BERKOWITZ, B. 2001. Scaling of fracture systems in Geological Media. *Reviews of Geophysics*, **39**, 347–383.

BRAITHWAITE, R. L., CARGILL, H. J., CHRISTIE, A. B. & SWAIN, A. 2001. Lithological and spatial controls on the distribution of quartz veins in andesite- and rhyolite-hosted epithermal Au–Ag deposits of the Hauraki Goldfield, New Zealand. *Mineralium Deposita*, **36**, 1–12.

CARLSON, W. D., DENISON, C. & KETCHAM, R. A. 1999. High-resolution X-ray computed tomography as a tool for visualization and quantitative analysis of igneous textures in three dimensions. *Electronic Geosciences*, **4**, 3.

CHILDS, C., WALSH, J. J. & WATTERSON, J. 1990. A method for estimating the density of fault displacements below the limits of seismic resolution in reservoir formations. *In*: BULLER, A. T., BERG, E., HJELMELAND, O., KLEPPE, J., TORSÆTER, O. & AASEN, J. O. (eds) *North Sea Oil and Gas Reservoirs—II*. Graham and Trotman, London, 309–318.

COX, D. R. & LEWIS, P. A. 1966. *The Statistical Analysis of Series of Events*. Methuen, London.

DINES, H. G. 1956. *The metalliferous mining region of south-west England*. Great Britain Geological Survey Memoir.

DORÉ, A. G., LUNDIN, E. R., FICHLER, C. & OLESEN, Ø. 1997. Patterns of basement structure and reactivation along the NE Atlantic margin. *Journal of the Geological Society, London*, **154**, 85–92.

FLOYD, P. A., EXLEY, C. S. & STYLES, M. T. 1993. *Igneous rocks of south-west England*. The Geological Conservation Review Series, **5**. Chapman & Hall, London & New York.

GILLESPIE, P. A., HOWARD, C., WALSH, J. J. & WATTERSON, J. 1993. Measurement and characterization of spatial distributions of fractures. *Tectonophysics*, **226**, 113–141.

GILLESPIE, P. A., JOHNSTON, J. D., LORIGA, M. A., MCCAFFREY, K. J. W., WALSH, J. J. & WATTERSON, J. 1999. Influence of layering on vein systematics in line samples, *In:* MCCAFFREY, K. J. W., LONERGAN, L. & WILKINSON, J. J. (EDS) *Fractures, fluid flow and mineralization*. Geological Society, London, Special Publications, **155**, 35–56.

GILLESPIE, P. A., WALSH, J. J., WATTERSON, J., BONSON, C. G. & MANZOCCHI, T. 2001. Scaling relationships of joint and vein arrays from The Burren, Co. Clare, Ireland. *Journal of Structural Geology*, **23**, 183–201.

GUDMUNDSSON, A., BERG, S. S., LYSLO, K. B. & SKURTVEIT, E. 2001. Fracture networks and fluid transport in active fault zones. *Journal of Structural Geology*, **23**, 343–353.

HALLS, C. 1987. A mechanistic approach to the paragenetic interpretation of mineral lodes in Cornwall. *Proceedings of the Ussher Society*, **6**, 548–554.

HUANG, Q. & ANGELIER, J. 1989. Fracture spacing and its relation to bed thickness. *Geological Magazine*, **126**, 355–362.

JACKSON, N. J., WILLIS-RICHARDS, J., MANNING, D. A. C. & SAMS, M. S. 1989. Evolution of the Cornubian Ore Field, Southwest England: Part II. Mineral deposits and Ore forming processes. *Economic Geology*, **84**, 1101–1133.

LADEIRA, F. L. & PRICE, N. J. 1981. Relationship between fracture spacing and bed thickness. *Journal of Structural Geology*, **13**, 179–183.

MAIN, I. G., LEONARD, T., PAPASOULIOTOS, O., HATTON, C. G. & MEREDITH, P. G. 1999. One slope or two? Detecting statistically significant breaks of slope in geophysical data, with application to fracture scaling relationships. *Geophysical Research Letters*, **26**, 2801–2804.

MANNING, C. E. 1994. Fractal clustering of metamorphic veins. *Geology*, **22**, 335–338.

MARRETT, R. & ALLMENDINGER, R. W. 1991. Estimates of strain due to brittle faulting: sampling of fault populations. *Journal of Structural Geology*, **13**, 735–738.

MCCAFFREY, K. J. W., JOHNSTON, J. D. & FEELY, M. 1993. Use of fractal statistics in the analysis of Mo–Cu mineralization at Mace Head, County Galway. *Irish Journal of Earth Sciences*, **12**, 139–148.

MCCAFFREY, K. J. W., LONERGAN, L. & WILKINSON, J. J. 1999. *Fractures, fluid flow and mineralization*. Geological Society, London, Special Publications, **155**.

MOORE, J. MCM. & JACKSON, N. 1977. Structure and mineralization in the Cligga granite stock, Cornwall. *Journal of the Geological Society, London*, **133**, 467–480.

Narr, W. & Suppe, J. 1991. Joint spacing in sedimentary rocks. *Journal of Structural Geology*, **13**, 1037–1048.

NICOL, A., WALSH, J. J., WATTERSON, J. & GILLESPIE, P. A. 1996. Fault size distributions—are they really power-law? *Journal of Structural Geology*, **18**, 191–197.

ODLING, N. E. 1997. Scaling and connectivity of joint systems in sandstones from western Norway. *Journal of Structural Geology*, **19**, 1257–1271.

PARNELL, J. 1988. Migration of biogenic hydrocarbons into granites: a review of hydrocarbons in British plutons. *Marine and Petroleum Geology*, **5**, 385–396.

PASCAL, C., ANGELIER, J., CACAS, M.-C. & HANCOCK, P. L. 1997. Distributions of joints: probabilistic modelling and case study near Cardiff (Wales, U.K.). *Journal of Structural Geology*. 19, 1273–1284.

PICKERING, G., BULL, J. M. & SANDERSON, D. J. 1995. Sampling power-law distributions. *Tectonophysics*, **284**, 1–20.

PRICE, N. J. 1966. *Fault and joint development in brittle and semi-brittle rocks*. Pergamon Press, New York.

PRIEST, S. D. & HUDSON, J. A. 1981. Estimation of discontinuity spacing and trace length using scanline surveys. *International Journal of Mechanics, Mining Science and Geomechanics Abstracts*, **18**, 183–197.

PSYRILLOS, A., MANNING, D. A. C. & BURLEY, S. D. 1998. Geochemical constraints on kaolinization in the St Austell Granite, Cornwall, England. *Journal of the Geological Society, London*, **155**, 829–840.

PUGLIESE, S. & PETFORD, N. 2001. Reconstruction and visualisation of melt topology in veined microdioritic enclaves. *Electronic Geosciences*, **6**, 2

RANGIN, C., HUTCHON, P., LEPICHON, X., BELLON, H., LEPVRIER, C., ROQUES, D., HOE, N. D. & QUYNH, P. V. 1995. Cenozoic deformation of central and southern Vietnam. *Tectonophysics*, **251**, 179–196.

RIVES, T., RAZACK, M., PETIT, J.-P. & RAWNSLEY, K. D. 1992. Joint spacing: analogue and numerical simulations. *Journal of Structural Geology*, **14**, 925–937.

RYAN, J. L., LONERGAN, L. & JOLLY, R. J. 2000. Fracture spacing and orientation distributions for two-dimensional data sets. *Journal of Geophysical Research—Solid Earth*, **105,** 19305–19320.

SCHOLZ, C. H. & COWIE, P. 1990. Determination of total strain from faulting using slip measurements. *Nature*, **346**, 837–839.

SLEIGHT, J. M. 2001. *Fracture characteristics from two reactivated basement fault zones: Examples from Norway and Scotland.* Unpublished PhD thesis, University of Durham, UK.

STEARNS, D. W. & FRIEDMAN, M. 1972. Reservoirs in fractured rock. *In*: *Fracture Controlled Production.* American Association of Petroleum Geologists, Reprint Series, **21**, 174–198.

TURCOTTE, D. L. 1992. *Fractals and chaos in geology and geophysics.* Cambridge University Press, Cambridge.

VELDE, B., DUBOIS, J., TOUCHARD, G. & BADRI, A. 1990. Fractal analysis of fractures in rocks: the Cantor's Dust method. Tectonophysics, **179**, 345–252.

VILLEMIN, T. & SUNWOO, C. 1987. Distribution logarithmique self-similaire des rejets et longeurs de failles: exemples de Bassin Houllier Lorrain. *Comptes rendus de l'Académie des sciences, Série IIa*, **305**, 1309–1312.

WALSH, J. J., WATTERSON, J. & YIELDING, G. 1994. Determination and interpretation of fault size populations: procedures and problems. *In*: AASEN, J. O., BERG, E., BULLER, A. T., HJELMELAND, O., HOLT, R. M., KLEPPE, J. & TORSÆTER, O. (eds) *North Sea Oil and Gas Reservoirs—III.* Graham and Trotman, London, 141–155.

WATTS, L. M. 2001. *The Walls Boundary fault zone and the Møre Trøndelag fault complex: a case study of two reactivated fault zones.* Unpublished PhD thesis, University of Durham, UK.

WILLIS-RICHARD, J. & JACKSON, N. J. 1989. Evolution of the Cornubian Ore Field, Southwest England: Part I. Batholith modelling and Ore deposition. *Economic Geology*, **84**, 1078–1100.

Characterization of rough-walled fractures in crystalline rocks

S. R. OGILVIE*, E. ISAKOV, C. W. TAYLOR & P. W. J. GLOVER

Department of Geology and Petroleum Geology, University of Aberdeen, Aberdeen AB24 3UE, UK

**Present address: Statoil Research Centre, Rotvoll, Trondheim, Norway*

Abstract: The full characterization of the rough surfaces of fractures and their resulting apertures is an important step in the drive toward an improved understanding of the factors which control fluid flow through rocks. This is crucial in igneous and metamorphic rocks, since fractures in these rocks may form the only significant pathways for fluid migration. Here we describe a three-pronged approach for the full characterization of rough fracture surfaces in a selection of crystalline rocks using a suite of software developed in house. Firstly, profiling is carried out using an optical method, which converts images of epoxy fracture surfaces covered with dyed water into topographies using the Lambert–Beer Law. Many hardware and software (OptiProf™) developments give this method the upper hand over previous attempts at spectrophotometric analysis. It is not possible to profile every fracture surface, therefore numerical modelling of fluid flow is carried out using synthetic fractures with rough fracture surfaces that are representative of the natural rock fractures. ParaFrac™ allows the analysis and parametrization of fracture surfaces and apertures. SynFrac™ enables the numerical synthesis of fracture surfaces and apertures with prescribed basic parameters. Both procedures take full account of the complex matching properties of the fracture surfaces as a function of wavelength, as well as anisotropy within the properties defining the fracture surfaces and their resulting aperture. They have been rigorously tested on a large suite of synthetic fractures as well as real rock fractures. These tests have allowed relationships between the standard deviation of surface asperity heights, the fractal dimension and the matching parameters to be related to the resulting aperture of the fractures.

Fractures are ubiquitous discontinuities in the Earth's crust and understanding their influence upon fluid flows has much practical benefit in the petroleum (Jones *et al.* 1998), water (Gudmundson 2000), geothermal (Barton 1998) and nuclear industries (Moreno *et al.* 1985). This is of particular importance in tight and crystalline reservoirs, as flow occurs mainly in fractures (Sahimi 1993; Abelin *et al.* 1994; Hakami & Larsson 1996).

Recently, the influence of fracture surface roughness upon fluid flow has received great attention, as this feature introduces much deviation from the parallel-plate model (Local Cubic-Law) for predicting fluid flow through rock fractures (e.g. Iwano & Einstein 1995; Glover *et al.* 1998a, b; Meheust & Schmittbuhl 2000; Renshaw *et al.* 2000). The scale invariance of such surfaces, important in modelling their effect upon fluid flow in crystalline hydrocarbon reservoirs, has been described in a variety of crystalline rocks including marble (Poon *et al.* 1992), granite (Brown 1988), gabbro (Durham & Bonner 1995) and basalt (Plouraboué *et al.* 2000).

Multi-fracture reservoir models oversimplify fracture parameters and would clearly benefit from the incorporation of realistic fracture roughnesses. To do this, we must be able to (1) measure the roughness of a range of fracture surfaces in nature, (2) analyse the characteristic features of the measured surfaces, and (3) create synthetic fractures numerically that share these characteristic features. In order to improve our estimates of the influences of rough fractures on mechanical and transport properties, we employ this three-stage approach aided by a suite of in house software (Fig. 1). Firstly, we describe an improved optical method for the profiling of rough fracture surfaces in a variety of crystalline rocks aided by OptiProf™ software (Isakov *et al.* 2001a; Ogilvie *et al.* 2002a, b). This forms the basis for the parametrization of fracture surfaces using ParaFrac™ software to provide the basic statistical parameters for the creation of synthetic models (Isakov *et al.* 2001b). These suites of numerical fractures are generated by a new, powerful and flexible method implemented in SynFrac™ software. The synthetic fractures can then be used for modelling fluid flow using the Local Cubic Law, solution of the Reynolds equation, or solution of the Navier–Stokes equation (e.g. Brown *et al.* 1995; Oron & Berkowitz 1998) (Fig. 1). The synthetic numerical fractures are also used for investigations of the statistical properties of rock fractures. In particular, the mean fracture aperture dependence on the properties of bounding surfaces is obtained.

From: PETFORD, N. & MCCAFFREY, K. J. W. (eds) 2003. *Hydrocarbons in Crystalline Rocks*. Geological Society, London, Special Publications, **214**, 125–141. 0305-8719/03/$15

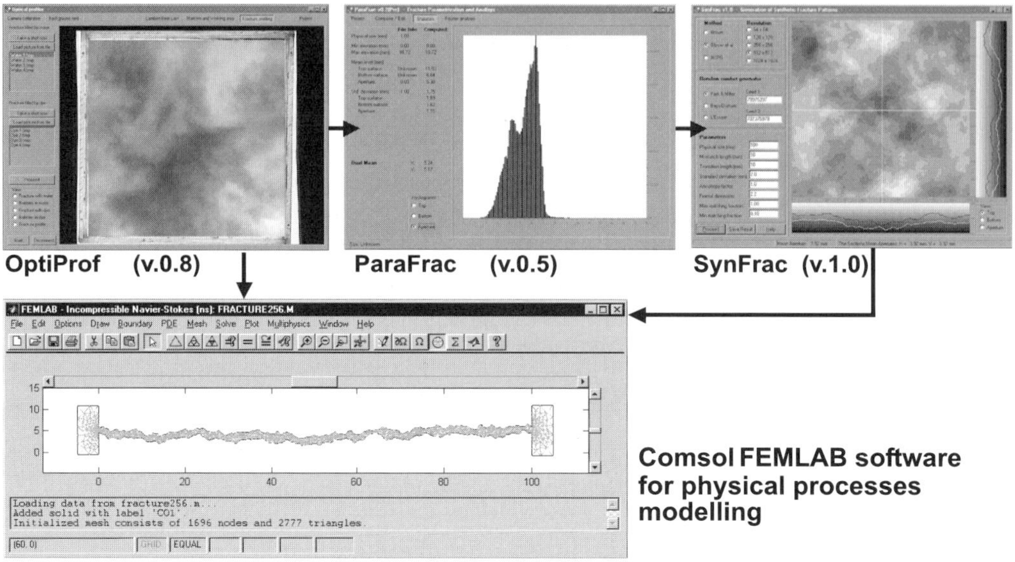

Fig. 1. The software framework for characterization of rough fractures in crystalline rocks. OptiProf^{TM} provides considerable control over the imaging process and calculates the final surface topographies. This data can be input into Femlab^{TM} physical process modelling software, however, this is more likely to be obtained from numerical models created using ParaFrac^{TM} and SynFrac^{TM} software and tuned to the profiling data.

Fracture surface profiling

The majority of fracture surface profiling is carried out by mechanical or optical techniques (see Adler & Thovert 1999; Devili *et al.* 2001). Mechanical stylus profilometers, popularly used for surface characterization of real rock fractures in the laboratory (e.g. Poon *et al.* 1992; Glover *et al.* 1998b), or in the field (e.g. Power & Tullis 1992; Schmittbuhl *et al.* 1993) work by tracking a stylus across the rough surface with surface elevations measured on a grid at various spacings. The vertical resolution of this type of profilometer is ~10 μm (Glover *et al.* 1998b). However, because mechanical profiling measures single profiles, one at a time, the horizontal resolution in one direction (*c.* 0.02 μm) is usually considerably better than that in the other direction (greater than 1000 μm). Furthermore, complete coverage of the surface is not possible at high resolution because (1) it takes too long, and (2) there are problems aligning the multiple profiles accurately.

Optical methods generally involve casting of real rock fractures and the point-wise measurement of fracture topographies using the Lambert–Beer Law (Persoff & Pruess 1995; Hakami & Larsson 1996; Brown *et al.* 1998; Yeo *et al.* 1998; Amundsen *et al.* 1999; Renshaw *et al.* 2000). This contribution describes the development of an improved optical method, which provides a fuller and more formalized framework for measuring the surfaces and apertures of fractures than previous attempts (Isakov *et al.* 2001a; Ogilvie *et al.* 2002a).

This involves (1) advances in high fidelity polymer model (HFPM) preparation, (2) significant improvements in hardware image capture, lighting and firmware image capture (the capturing of images from the camera through to the final stored images), (3) robust methodologies for reliably calibrating the fluids used in the imaging process, and (4) improved control over the imaging process by using OptiProf^{TM} software (Fig. 1) to correct for technical difficulties and calculate the final measured topography of the surface by calibrating the image produced to dye thickness. This is a fast and high-resolution process, allowing high resolutions (15 μm for 10 × 10 cm area) to be attained in the mean plane of the fracture surface using a high-resolution digital camera.

Methods and measurements

Sample descriptions

A suite of three crystalline rocks (granite, syenite and gabbro) was trimmed to form blocks with 120 × 120 × 100 mm nominal dimensions. Two parallel grooves were created on the opposite

Table 1. *Petrophysical data for intact samples*

	Rock type	Sample location	Grain size	$Hg\phi$	S_{WI} (%)	$He\phi$ (%)	K_L (m^2)
A	granite	Norway	coarse	0.5	99.2	4.45	3.94×10^{-6}
B	syenite	Sweden	coarse	1.2	94.5	0.2	3.94×10^{-7}
C	gabbro	South Africa	fine	0.31	98.8	0.47	3.94×10^{-6}

Notes: $Hg\phi$ = mercury injection porosity, S_{WI}, irreducible water saturation derived from mercury porosimetry; $He\phi$, helium porosity; K_L, equivalent liquid (Hassler-sleeve) gas permeability.

faces of each block and mode I fractures were carefully propagated using a bolster chisel from one groove to the other.

These samples were chosen due to their very low matrix permeabilities (Table 1). They also have very low (helium and mercury technique) porosities, from 0.2% to 4.5% (Table 1). Sample B (syenite) consists of coarse, labradorite laths, which define a parallel, anisotropic fabric. It splits more easily than the others, producing a smooth surface with lower fractal dimension (Table 2).

High fidelity polymer model (HFPM) preparation

Each fracture half was moulded with Dow Corning Silastic® E RTV rubber, which is a flexible,

Table 2. *Rock fractures tested with the new method*

Parameter	A	B	C
Surface parameters			
Standard deviation (upper), $_U\sigma_s$ (mm)	1.97	1.89	1.82
Standard deviation (lower), $_L\sigma_s$ (mm)	2.07	2.03	2.07
Variance (upper), $_U\sigma_s^2$ (mm^2)	3.88	3.57	3.31
Variance (lower), $_L\sigma_s^2$ (mm^2)	4.28	4.12	4.28
Fractal dimension (upper), $_UD_f$ (-)	2.25	2.17	2.25
Fractal dimension (lower), $_LD_f$ (-)	2.16	2.18	2.22
Anisotropy in fractal dimension (upper), $_UA_{sD}$ (-)	0.88	1.04	1.42
Anisotropy in fractal dimension (lower), $_LA_{sD}$ (-)	0.86	0.97	1.41
Physical size, L (mm)	95.9	96.8	100
Measurement points per fracture size (-)	512	480	500
Resolution (μm)	190	200	200
Fracture parameters			
Mis-match wavelength (ML)	4.5	2.5	2.3
Transition length (TL)	40	90	72
Maximum matching fraction (MaxMF)	0.98	0.99	0.99
Minimum matching fraction (MinMF)*	−0.02	−0.06	0
Standard deviation, σ_a	0.65	0.42	0.64
Fractal dimension, D_f	2.64	2.69	2.78
Anisotropy in fractal dimension of the aperture, $A_a\sigma$	1.02	1.07	0.76
Arbitrary parameters			
Arithmetic mean aperture $\langle z_a \rangle_a$ (mm)	1.71	0.82	1.40
Harmonic mean aperture $\langle z_a \rangle_h$ (mm)	0	0	0
Geometric mean aperture $\langle z_a \rangle_g$ (mm)	0	0	0
Dual mean of fracture aperture in x-direction (mm)	1.33	0.66	1.11
Dual mean of fracture aperture in y-direction (mm)	1.34	0.70	1.22

Notes: Harmonic and geometric means of the fracture are 0 because at least one touching point exists, where the aperture is 0. A, granite; B, syenite; C, gabbro.

* The theory presumes the minimum value for the Minimum Matching Fraction (MinMF) to be zero. Negative values were found in the process of measurements and characterization may mean the theory does not reflect fracture features perfectly. Authors have no reasonable explanations to these negative values.

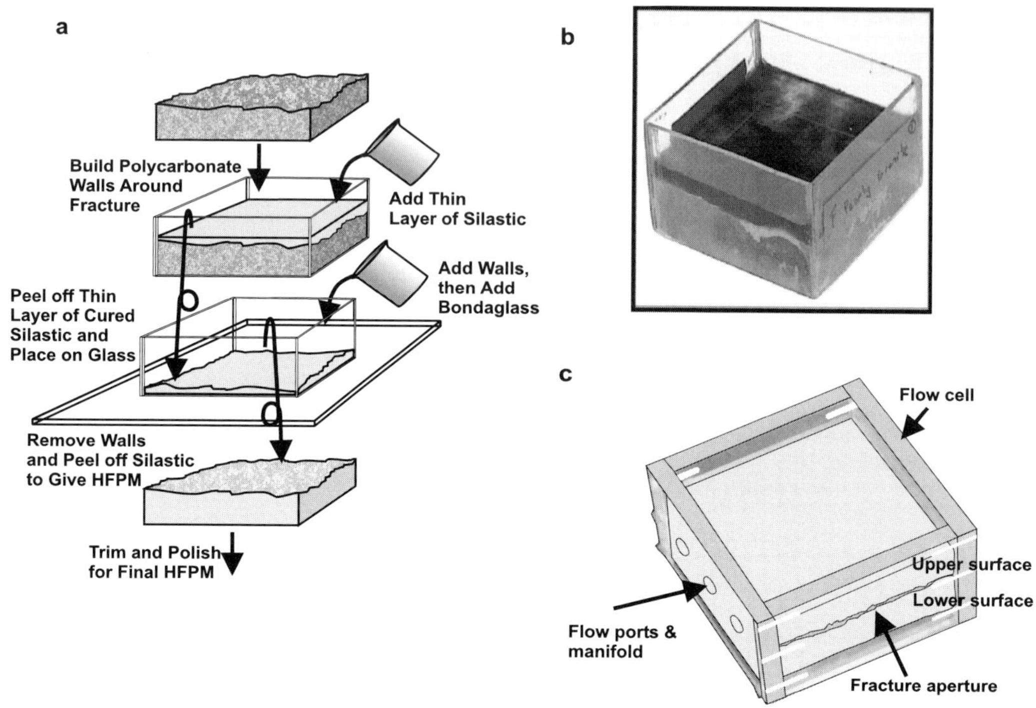

Fig. 2. (**a**) The procedure for creating high fidelity polymer models (HFPMs). (**b**) HFPM with built-up sides for point-wise determination of fracture surface topography. (**c**) Two surfaces married and housed in a cell for fluid flow experiments.

mould-making rubber for intricate detail moulds. The rubbery nature of the final mould allows the fracture surface to be removed from the mould without damage to either surface. Any plucking of mineral grains from the fracture surface (e.g. Yeo *et al.* 1998) was minimized by spraying it with aerosol glue (SprayMount®), a few minutes before moulding. Once set, the mould was filled with *c.* 400 cm^3 of clear casting resin (Fig. 2). When fully set, the casting resin was removed from the mould, trimmed to $100 \times 100 \times 30$ mm nominal dimensions, and polished on all surfaces except that of the rough surface. The original rock surfaces are reproduced to within 1 μm as illustrated by comparison of SEM images of a real rock surface of sample A (granite) and its cast (Fig. 3). These models are therefore referred to as high fidelity polymer models (HFPMs). However, reproduction in certain areas (i.e. centre top in Fig. 3b) was affected by (1) dust and other particles, (2) bubbles and (3) minerals dislodged during the casting process (Yeo *et al.* 1998). Polycarbonate walls were attached to the sides of each HFPM in preparation for digital optical imaging (Fig. 2b). The fracture surfaces were also mated and inserted into a cell for fluid flow experiments (Fig. 2c) (Ogilvie *et al.* 2001). These results will be reported in a later contribution.

Digital optical imaging

The HFPM was placed on a light box under a digital colour camera (640×480 pixels, 8-bit grey-scale depth) and imaged, first containing distilled water, and then containing dyed water (Fig. 4). When filled with water, the resin surfaces were lightly scrubbed using a small brush to remove water bubbles. This method has a lateral resolution of 15 μm, which can be improved if higher resolution cameras are used. It has a height resolution of *c.* 30 μm but could be smaller if 16-bit or 24-bit imaging hardware was used. The ratio of the intensity of light for a given pixel at a given location on the fracture between the images containing dye and those containing water is related to the thickness of fluid covering the rough surface, i.e. the Lambert–Beer Law, $I_x = I_0 \, e^{-KT}$, where I_x is the intensity of the transmitted light, I_0 is the intensity of the incident light, K is a material dependent property describing the efficiency with which dyed water adsorbs light, and T is the thickness

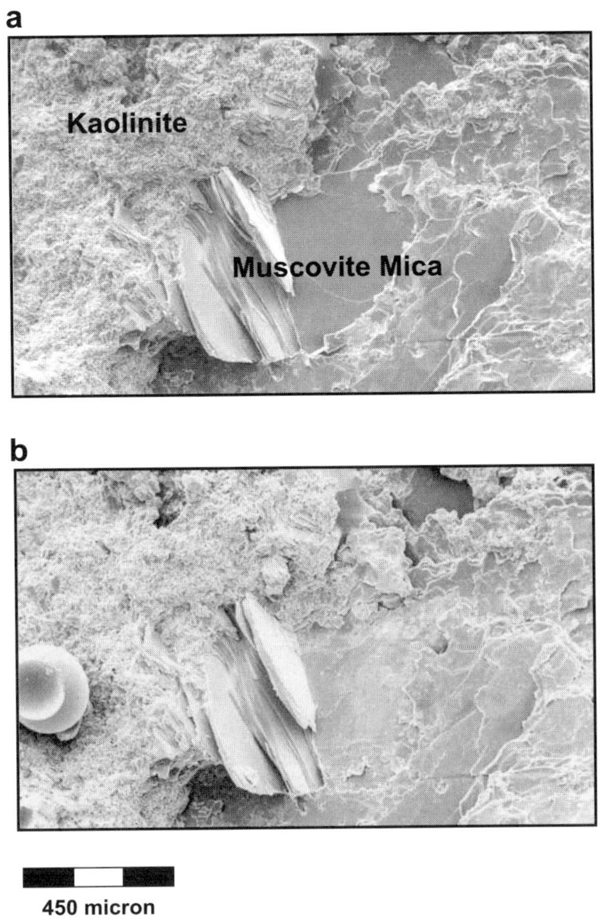

Fig. 3. The quality of reproduction of the fracture surfaces of sample A (granite) by HFPMs. (**a**) SEM backscattered image of the surface of the original rock showing muscovite mica growth in two directions to the right of the field of view and kaolinite to the left. (**b**) The exactly corresponding area of the resulting HFPM. A gas bubble in (**b**) usefully distinguishes between the two images.

of the material through which the light has passed.

Fluid calibration

Calibration was carried out for distilled water and dyed water (0.5 g/l Dylon® black) using a wedge vial device (aperture varies linearly from 0.00 mm to ~17.5 mm), providing a measurement of the light extinction properties of the dye and to calculate the fluid thickness. This allowed the Lambert–Beer law to be applied to the measured data to convert intensities into accurate surface heights instead of arbitrarily scaling the relative surface height measurements.

The wedge was filled with each of the fluids, and imaged multiple times, producing 8-bit greyscale images, intensities varying from 0 to 255. The mean (stacked) image was calculated to remove possible dynamic fluctuations in the incident light intensity or in the sampled video stream. A Clearfield equalization was then performed, i.e. subtraction of background image (without subject) from that of the subject to remove spatial variations in incident light intensity. Optical similarity between the HFPM material and this device is not required. This is because the ratio of intensity of light passing through the subject filled with water to the intensity of light passing the subject filled with dyed water depends upon depth and light-absorbing properties of the dye only. The resulting image represents the ratio of the intensities recorded when the wedge is full of dyed water and undyed water, and is dependent upon the fluid

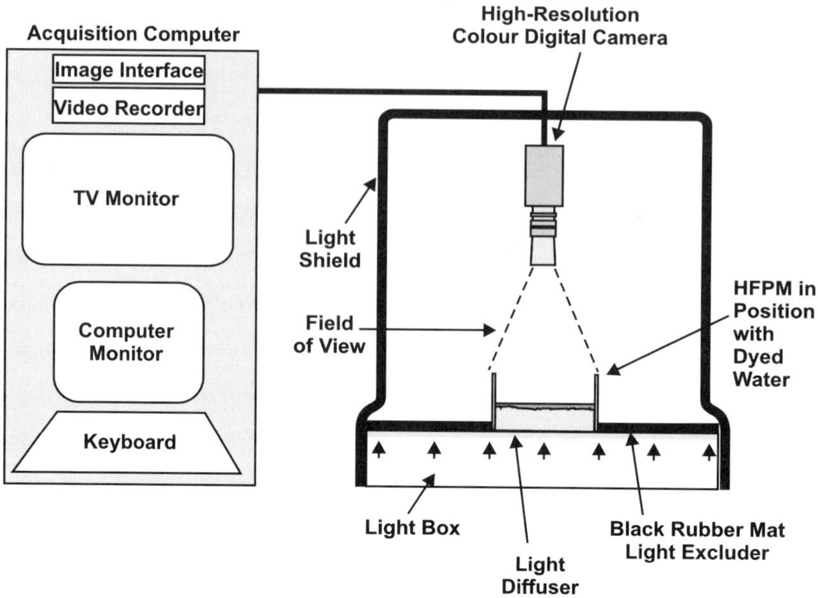

Fig. 4. Digital optical imaging set-up. This set-up is arranged so that the intensity of light captured by the camera can be accurately measured. Hence, the camera was set to manual exposure control, manual light colour balance, and a high quality manual aperture and focus zoom lens was used. The whole arrangement was shielded from ambient and reflected light, using black-out curtains and a thick rubber mat that covered all parts of the light box except that directly under the subject.

thickness if the type and concentration of the dye remains constant. This image was analysed in SigmaScan Pro 5™ to obtain the intensity (0 to 255) distributions for the wedge. The calibration curve of intensity as a function of fluid thickness is shown in Figure 5. This curve is linear on the semi-log scales used in this diagram, thereby conforming to the Lambert–Beer Law as expected. From the gradient of the calibration plot, we can derive a value for K of the dye solution, $K = 0.136\,\mathrm{mm}^{-1}$. This is input into the respective GUI in OptiProf™ software.

Conversion to fracture surface topographies

A function fitted to the wedge-derived calibration data provided us with a conversion from

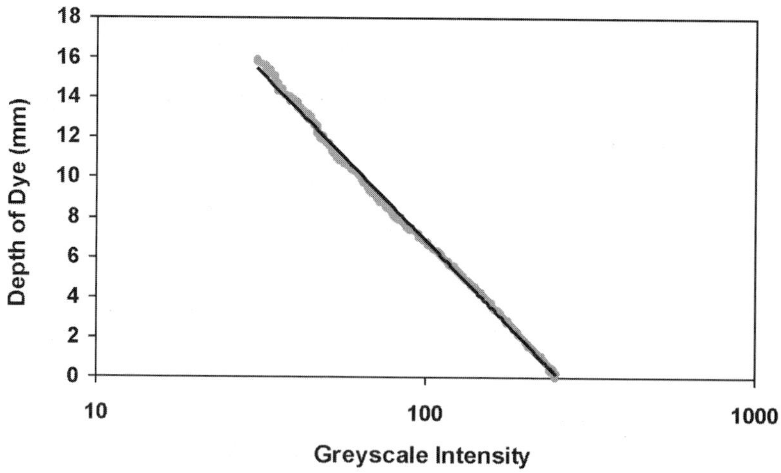

Fig. 5. Fluid calibration. The calibration curve, which is linear thereby obeying the Lambert–Beer Law.

intensity ratio to dye thickness that allowed each pixel of the fracture intensity ratio map to be converted to a thickness of dye below the fluid surface. The resulting data were transformed to provide a fully determined topography for each surface using OptiProf^{TM}. Isakov *et al.* (2001a) combined the data from each of two surfaces to provide an aperture map of the fracture. In each case, the arithmetic mean aperture was greater than expected as a result of (1) the difficulty in positioning and marrying the surfaces, and (2) differences in the slope of both surfaces with respect to fluid level. Since then, OptiProf^{TM} software has been further developed to position and marry the fracture surfaces more accurately for the creation of the aperture maps and for subsequent statistical analyses by ParaFrac^{TM}. The upper fracture surface of each sample (1) was firstly filled with water and profiled using the procedure described in Isakov *et al.* (2001a), the HFPM fixed in position using markers before being filled with dyed water for profiling. The fracture profile is calculated by OptiProf^{TM}. The lower surface (2) was then firstly filled with dyed water (to enable the software to compute the elevation map for positioning purposes) and placed under the camera in the field of view. Surface 1 was loaded into the 'positioning' section but flipped vertical and reversed so that it could be married to surface 2.

To obtain accurate matching, the modified software (from that used in Isakov *et al.* 2001a and in Ogilvie *et al.* 2002a) has image contrast and step options. The software computes the map elevations for the two surfaces simultaneously and displays the topographic maps of both surfaces in one field of view in two colours. Changing the values in the 'step' tab enables us to contour the dyed water intensities into appropriate contours for matching. There is also an 'adjust slope' tab, which enables more accurate matching of slightly sloping surfaces. Once matching is achieved, fine positioning is undertaken by selecting more detailed areas of the fracture surface to ensure accurate matching. Surface 2 is now profiled in the same way as surface 1 and OptiProf^{TM} calculates the surface topography and aperture maps using the approach already described.

Technical difficulties

There are several technical difficulties, which must be overcome in the imaging process if the fracture surface roughnesses are to be profiled accurately and the variable apertures calculated correctly (Detwiler *et al.* 1999; Isakov *et al.* 2001a). These, together with their solutions provided by software (OptiProf^{TM}) and hardware

Table 3. *Technical difficulties encountered during the imaging process and their solutions, which are incorporated into OptiProf^{TM} software*

Technical difficulty	Solutions
1. Fluid level control: each of the fractures must be filled with dyed and undyed water up to exactly the same arbitrary level.	A horizontal datum line is marked onto opposite walls surrounding the fracture. Lining these up and filling to this level removes the parallax errors.
2. Lateral alignment: the imaged surface must be in exactly the same *xy* position for imaging with dyed water and undyed water even though it must be moved for replacing the fluids.	Four fixed reference points within the software are set over point marks that are etched into the top of the walls surrounding the fracture These are used to realign the HFPM when removed and refilled.
3. Dynamic noise in the imaged light intensity: from the light source and video stream. Leads to variations in brightness of the imaged intensities.	Taking multiple images of the HFPM with each fluid in place, and averaging the result pixel by pixel.
4. Static noise in video signal: a stripe effect on an image of a uniform field, which varies with camera lens aperture.	The variation in the sensitivity of the CCD between each pixel on its surface was removed by calibrating each pixel of the CCD individually, for every aperture.
5. Non-uniformity of the light source.	Each of the averaged intensity images from the measurements on the undyed and dyed water were subjected to a clearfield equalization to remove variations due to changes in the incident light source.
6. Bubbles and dust in the fluids: particles and bubbles in the fluids are mobile if the fluid is perturbed.	The software compares multiple images with bubbles and dust in different locations and recognizes characteristics which move. These are removed from the relevant images prior to averaging.
7. Opaque particles in the HFPM: small, uncommon and obvious in the final image as thin low intensity spikes.	Recognized and removed with the affected pixel being reduced to the weighted mean of the surrounding eight pixels.

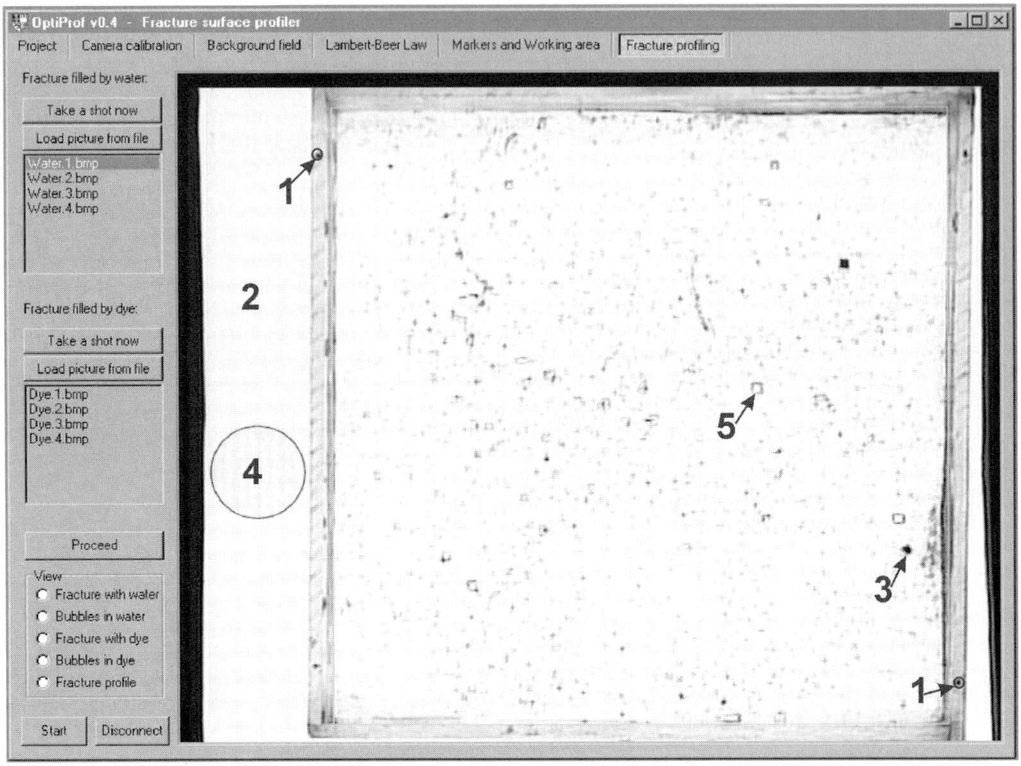

Fig. 6. An image of the HFPM during measurement by fracture surface profiler software (OptiProf^{TM}) showing the different technical difficulties encountered: 1, usage of location marks on walls to lock HFPM in position during imaging; 2, dynamic noise across whole image; 3, particles in the HFPMs; 4, static distortions giving regular structure in video signal; 5, bubbles and dust in fluids.

developments are listed in Table 3 and some are illustrated in Figure 6.

Rough fracture parametrization

This section describes the procedure for creating synthetic fractures using ParaFrac^{TM} and SynFrac^{TM} software developed in house. Flow modelling on a suite of such fractures allows the mean flow behaviour to be judged, which is representative of that type of fracture. This enables us to recognize and account for particular topographic geometries, which may be unrepresentative of the suite in general. Furthermore, the scatter in the flow modelling results represents the range of expected values for all fractures with the given geometrical parameters. The fracture parameters chosen to represent the complexity of fracture surfaces are listed below—some of which are defined for the first time. These are determined from profiling data and used in the creation of the synthetic fractures for comparison with real rock fracture data.

Fracture parameters

We classify the fracture parameters into those parameters associated with individual surfaces, parameters that are only defined when using two surfaces to make a fracture, and arbitrary parameters.

Surface parameters. (1) Standard deviation σ_s (or variance σ_s^2) of asperity heights on each fracture surface. This is a measure of the roughness of the surface asperities (i.e. the difference between the peaks and troughs).

(2) The fractal dimension D_f of each fracture surface. This is a measure of the scaling behaviour of the surface, and contains information regarding the relative positions of asperities of different sizes on the surface. This parameter is calculated from the log–log slope of the power density spectrum of the surface as a function of wavelength (e.g. Power & Tullis 1992). For natural fracture surfaces, it falls in the range of

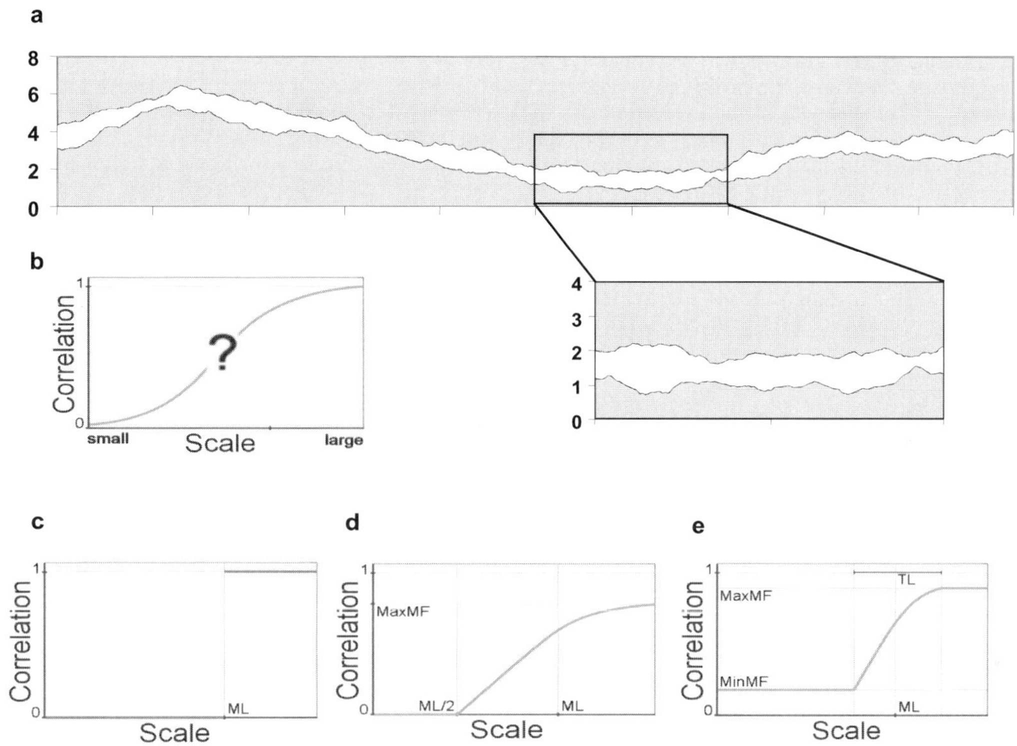

Fig. 7. Approaches in the matching of fracture surfaces. (**a**) Upper: well matched at large scale; lower: independent behaviour at small scale; (**b**) variation in matching behaviour between scales; (**c**) the classic approach used by Brown (1995); (**d**) the approach used by Glover *et al.* (1998a, b); (**e**) the approach developed in this work to more accurately reflect matching behaviour in natural fractures.

2 to 3 with smaller values representing smoother surfaces. In fact, Durham & Bonner (1995) found this to be a useful parameter to distinguish fracture surface roughnesses in different rock types: gabbro, $D_f = 2.2$, granite, $D_f = 2.4$.

(3) The anisotropy of fractal dimension of the surface A_{sD}, which allows the surface to have different fractal dimensions in different directions across the surface.

(4) The resolution of the measurement in the fracture plane. This is expressed in measurement points per fracture size, and may be any value for measured fractures depending upon the measurement technique, but is a binary multiple for synthetic fractures (ranging from 256 $\times$ 256 pixels to 1024 $\times$ 1024 pixels in this paper).

Fracture parameters. (5) The matching characteristics. Once the fracture surfaces have been separated for profiling, it is important to fit them back together again for numerical modelling using the correct 'matching' approaches. Rough fractures are matched to some degree at long wavelength and relatively unmatched at short wavelength (Power & Tullis 1995) (Fig. 7a). In fact, Brown & Scholz (1986) and Power & Tullis (1992) found that surfaces are well matched above the scale of a few millimetres. In between, the degree of matching varies, but currently it is unknown how it behaves for individual rock types (Fig. 7b). The Brown (1995) approach uses the mis-match wavelength (ML), and it represents the wavelength lower than which there is no correlation between two fracture surfaces, and higher than which there is complete matching (Fig. 7c). Glover *et al.* (1998a, b) realized that there is a smooth transition between independent and dependent behaviour and also introduced the maximum matching fraction (MaxMF) (Fig. 7d) to define the degree of matching reached using the fracture model used. In order to take more account of the variation in real rock fractures we define a (1) mis-match wavelength (ML), (2) the size of a transition zone around the mis-match wavelength: the transition length (TL), (3) a minimum matching fraction (MinMF), and (4) a maximum matching fraction (MaxMF) (Fig. 7e).

(6) Standard deviation σ_a (or variance σ_a^2) of the aperture defined by the two fracture surfaces. This parameter is a measure of the complexity of the aperture (i.e. the difference between the constrictions and wide portions of the aperture) irrespective of the position of the asperities on the surface. It is commonly non-Gaussian for rock fractures apertures (e.g. Gentier 1986; Gale 1987; Power & Tullis 1992).

(7) The fractal dimension of the aperture D_f (e.g. Cox & Wang 1993). This parameter is a measure of the scaling behaviour of the aperture, and contains information regarding the relative positions of asperities of different sizes on the surface. This parameter can be calculated from the log–log slope of the power density spectrum of the aperture as a function of wavelength. However, the aperture distribution character is not exactly fractal, especially at the low spatial frequencies.

(8) The anisotropy of fractal dimension of the aperture $A_{a\sigma}$. This parameter results from any anisotropy in the standard deviations of the fracture surfaces.

Arbitrary parameters. (9) The arithmetic mean height of each surface $\langle z_a \rangle_a$. This occurs at the peak of the probability distribution of surface heights. This is an arbitrary parameter if a fracture surface is used singly, and only important if two surfaces are used together, as the relative arithmetic mean height of each surface will then control the resulting aperture, with the minimum difference between the arithmetic mean height of each surface being non-zero and controlled by the scenario where the surfaces just touch.

(10) The mean aperture. This depends upon the mean surface heights of the two surfaces used to define the fracture aperture. The drawback with popularly used geometric mean apertures $\langle z_a \rangle_g$ (e.g. Brown 1987), and harmonic mean apertures $\langle z_a \rangle_h$ (e.g. Oron & Berkowitz 1998) as measures of hydraulic aperture of a fracture is that they collapse to zero if any touching point exists. We therefore introduce a Dual Mean approach (Isakov *et al.* 2001a); the arithmetic mean of the geometric mean apertures along all profiles in the direction of presumed fluid flow through the fracture. It has a physical basis, and is sensitive to anisotropy in the plane of the fracture; i.e. it has different values in different directions through the fracture. We use the dual mean in the two cartesian directions in the plane of the fracture x and y, and give them the symbols $\langle z_a \rangle_{dx}$ and $\langle z_a \rangle_{dy}$, respectively.

Fracture parametrization

The two profiled surfaces comprising the fracture are loaded into ParaFrac™ software (Fig. 1). These surfaces have been carefully positioned and married using OptiProf™ software as described, so the top surface can be cropped in ParaFrac™ and the same cropping values can be applied to the bottom surface. Now, the fracture surfaces and the aperture can be parametrized. This enables (1) the calculation of basic statistical parameters for each surface and the resulting aperture, (2) the display and Gaussian/non-Gaussian fitting of probability distributions for surface heights and apertures, (3) the calculation and display of power spectral density plots of the surfaces and apertures together with the calculation of their respective fractal dimensions, and (4) the calculation and display of power spectral density ratio plots of the fracture for the derivation of matching parameters.

Basic statistics. This section takes the individual surfaces and the resulting aperture data, and for each calculates and displays the probability density plot of surface heights and the fracture aperture, respectively. These plots can be used to judge the normality of the distribution. The statistical data produced for real rock fractures are summarized in the first section of Table 2.

Fourier analysis. This section takes the individual surfaces and the resulting aperture data, and for each uses Fast Fourier Transforms (FFTs) to calculate and display the power spectral density (PSD) of the surfaces and their resulting aperture as a function of wavenumber on log–log scales (where the wavenumber k, the wavelength λ and the frequency ν of the Fourier components are related by $k = 1/\lambda$ and $\nu = 2\pi/\lambda$). Linear regression to the full PSD for the surfaces, and to the linear high wavenumber portion of the aperture allows the fractal dimensions of the surfaces and the aperture to be calculated. Anisotropy in the fractal dimensions of surfaces and the resulting apertures are obtained by unwrapping the surface (or aperture) in the appropriate direction prior to application of the FFT. The PSDs and their fitted regressions are displayed.

Matching parameters. This section calculates the ratio of the PSDs from the aperture with the sum of the PSDs of the two surfaces composing the fracture and plots it as a function of wavenumber on a log–log scale. This parameter is the

PSD Ratio (PSDR), where,

$$\mathrm{PSDR} = \frac{\mathrm{PSD\,(aperture)}}{\mathrm{PSD\,(upper\,surface)} + \mathrm{PSD\,(lower\,surface)}}$$

At large wavenumbers, k (i.e. high frequencies and small wavelengths), the PSDR tends to unity if the surfaces are completely independent (Glover *et al.* 1998b). If the PSDR is less than unity, it shows that there is some matching occurring at the highest wavenumbers available in the dataset. It can be seen, therefore, that the PSDR at the highest wavenumber of the dataset $(\mathrm{PSDR})_{k\,\max} = (1 - \mathrm{MinMF})$, and hence MinMF can be obtained. As matching of the two surfaces occurs, the PSDR drops to values below unity, but never below zero. This is because there is increasing correlation between the two fracture surfaces that results in loss of power of the Fourier components of the aperture. Consequently, the PSDR at the lowest wavenumber of the dataset $(\mathrm{PSDR})_{k\,\min} = (1 - \mathrm{MaxMF})$, and hence MaxMF can be obtained. We define the mis-matching wavelength (ML) for the system as the wavelength λ_{ML} (represented by the wavenumber k_{ML}, where $\lambda_{\mathrm{ML}} = 1/k_{\mathrm{ML}}$), which lies equidistant between the wavelength at which minimum matching occurs and that at which maximum matching occurs in the dataset. The transition length (TL) is defined as the difference in wavelength between that at which maximum matching occurs and that at which minimum matching occurs, and corresponds to the width (expressed in wavelength) of the transition zone.

Creation of synthetic fractures

SynFrac™ software, allows the numerical synthesis of fracture surfaces and apertures with eight prescribed basic parameters to be carried out (Fig. 1). It can provide fractures calculated using the Brown (1995) method, Glover *et al.* method (1998a, b), or our improved method of controlling the degree of fracture surface correlation with wavelength (Isakov *et al.* 2001b). Fracture surface generation is carried out using spectral synthesis (Saupe 1988) on a grid up to 1024 × 1024 pixels and at any physical scale. The program implements three different types of high-quality random number generation methods, allowing suites of physically distinct fractures to be created which all share the same basic parameters, allowing them to be used in statistically rigorous modelling studies.

Two random number seeds control the actual topographies of the two fracture surfaces, and therefore control the resulting aperture. Hence, we can create a suite of synthetic fractures, say 20, using 20 sets of two random numbers and one set of geometrical parameters derived from a natural fracture. The resulting 20 synthetic fractures will each share the same geometrical parameters as the original natural fracture, but they will be different physically.

Fracture synthesis procedure

The spectral synthesis method involves defining a symmetric matrix containing Fourier components. These Fourier components are calculated to obey the various parameters for the fracture. Each component has two parts, the amplitude and the phase. The amplitude scales with a power law that contains the fractal dimension information, and any information about the relative anisotropy of surface heights. The phase part is controlled by random numbers, which depend in their turn upon the two original random number seeds and the matching parameters.

The first step is to generate two matrices where each point in each matrix corresponds to that in the final matrix of Fourier components. These two matrices contain random numbers that are partially correlated to some degree. The degree of partial correlation depends upon the matching parameters. This step is not necessary for the Brown (1995) approach, as one surface can be generated with one set of random numbers. He then generated a second fracture surface using (1) a different set of random numbers for the Fourier components corresponding to wavelengths that were less than the mis-match wavelength (i.e. where the surfaces were independent), and (2) the same random numbers that were used to generate the first fracture for the Fourier components corresponding to wavelengths that were greater than the mis-match wavelength (i.e. where the surfaces were perfectly matched). The implementation of the Glover *et al.* (1998a, b) method required the use of two different sets of random numbers for the wavelengths that are less than half the mis-match wavelength, but the generation and use of partially correlated random numbers for wavelengths above this value. To do this, they linearly mixed the two random number sets using a linear weighting, which varied from zero at half of the mis-match wavelength to some fraction less than unity representing the maximum matching fraction at the largest wavelength contributing to the fracture. While this procedure does produce a partially correlated set of numbers, they cannot be considered to be truly random because they lose their uniform distribution over the interval zero to unity. Hence, there is a fundamental problem in mixing random numbers this way. We have overcome this problem by implementing a position-swapping algorithm that enables a given mixing of two uniformly distributed random number data sets to be attained while retaining a uniform distribution in the final mixed and partially correlated random number data set. This is an elegant solution, but one that requires significant CPU time. Consequently, we use the improved method for creating partially correlated random number data sets to both the new matching approach and our improved implementation of the Glover *et al.* (1998a, b) approach.

When all the Fourier components are known and arranged in a 2-D complex and symmetric matrix, they are submitted to a 2-D Fast Fourier Transform (FFT), the real part of which is the fracture surface with a mean value of zero. It only remains then to scale the surface to the required physical size, to scale the asperities to the size defined by the standard deviation of surface heights, and to shift the mean level of the fracture surface to whatever is required.

Results and discussion

Our optical method involves new developments that make existing concepts (i.e. the spectrophotometric analysis of epoxy casts) into a very robust method for determining rough fracture surfaces with respect to a fidelity of reproduction to within 1 μm, improvements in hardware and firmware image capture, and the usage of calibration devices enabling the Lambert–Beer law to be applied to the measured data to convert intensities into accurate surface heights. This method has a decent lateral resolution, which was 15 μm for our camera/imaging set-up (for an area of 10 × 10 cm) and *c.* 30 μm for a maximum topography of *c.* 17.5 mm. Advances in the multiple capture of images from the video stream have improved the quality of the initial images as well as making advanced image analysis options possible on the captured image data.

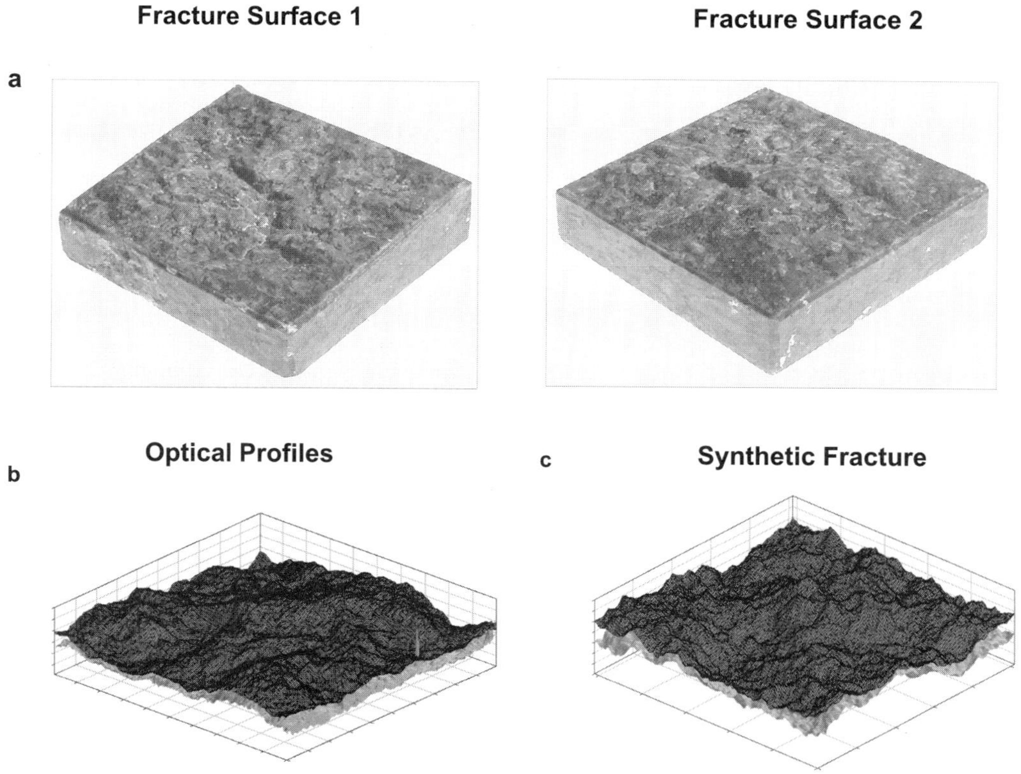

Fig. 8. Profiled and numerical fracture results for Sample B (syenite). (**a**), Rough fracture surface 1 & 2; (**b**), surface 1 & 2 combined to show the fracture aperture; (**c**) synthetic fracture.

OptiProf^{TM} software developed in house was used for the analysis of the image data to correct for and negate problems with the image capture process. A full comparative analysis of the sources of error in the measurements, together with the scope for their reduction, is presented such as that carried out by Detwiler *et al.* (1999). Techniques that are newly implemented are multiple imaging, clearfield equalization, stacking, software keying, bubble detection, static detection, individual pixel calibration, and precise filling.

The optical profiles for both surfaces of the three crystalline rock samples are accurate representations of the fracture surfaces for each sample (Fig. 8). There are, however, limitations in terms of the number of profiles which can be produced. Therefore, combinations of numerical fractures with the same basic geometry, but with different physical topographies, are generated using SynFrac^{TM} software. The fracture parameters analysed by ParaFrac^{TM} for this process from the respective real rock fracture surfaces are listed in Table 2.

About 619 synthetic rough fractures (100 × 100 mm) in rocks were created (Fig. 8) as a function of fractal dimension (from 2 to 2.4), standard deviation of bounding surfaces (from 0.01 mm to 5 mm), mis-match length (from 1 mm to 50 mm), and minimum and maximum matching fractions (from 0% to 20% and from 80% to 100%, respectively). For each set of parameters a suite of 10–30 fractures was created. Each fracture was analysed to ascertain whether the resulting synthetic fractures had parameters, which matched the synthesizing parameters. In this way we verified the synthesizing algorithms and their implementation in the software.

The real rock aperture surface height distributions have log-normal shapes (Fig. 9). This is a common situation for rough fracture apertures (e.g. Power & Tullis 1992). The mean arithmetic apertures of the resulting fractures were obtained

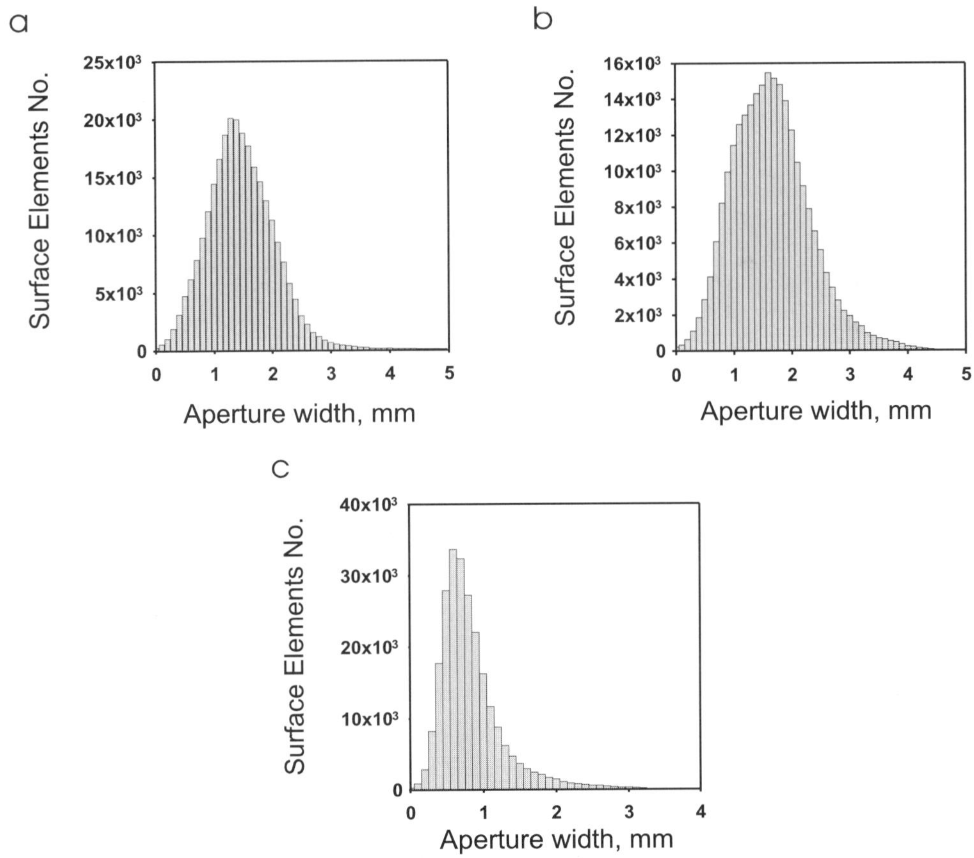

Fig. 9. Basic statistics (from ParaFrac^{TM}) for rough rock fracture apertures; all apertures approximate log-normal distributions. (**a**) Granite; (**b**) syenite; (**c**) gabbro.

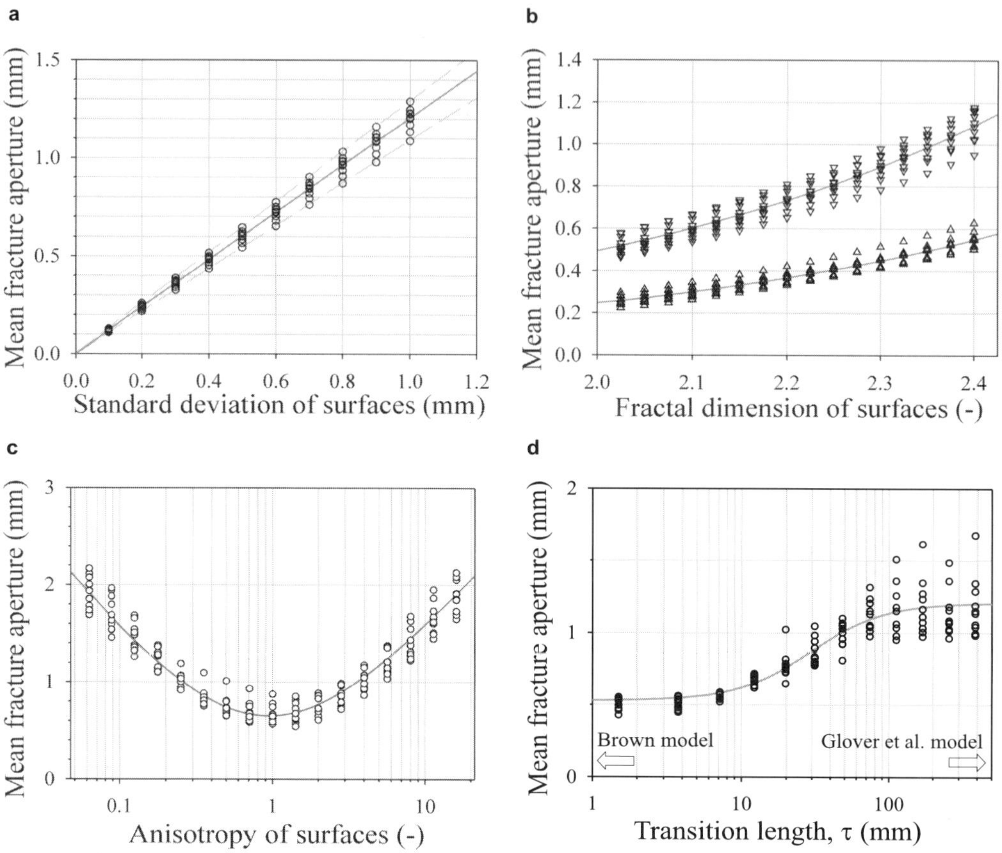

Fig. 10. (**a**) Mean synthetic fracture aperture as a function of the standard deviation of surfaces. Fractal dimension 2.2, mis-match length 10 mm, transition length 20 mm. (**b**) Mean synthetic fracture aperture as a function of the fractal dimension of surfaces. Mismatch length 10 mm, transition length 20 mm, standard deviation of surfaces 0.3 mm (△) and 0.6 mm (▽). (**c**) Mean synthetic fracture aperture as a function of the anisotropy of surfaces. Fractal dimension 2.1, mis-match length 10 mm, transition length 20 mm, standard deviation of surfaces 0.5 mm. (**d**) Mean synthetic fracture aperture as a function of the transition length. Fractal dimension 2.2, mis-match length 10 mm, transition length 20 mm, standard deviation of surfaces 0.5 mm.

for each suite of fractures, and have been examined as a function of surface asperity distribution (standard deviation), fractal dimension, anisotropy, and mis-match parameters (Fig. 10). Dependence of mean fracture aperture on the standard deviation of bounding surfaces is shown in Figure 10a (isotropic fractal dimension 2.2, mis-match length $\lambda_c = 10$ mm, transition length $\tau = 20$ mm). The scattering in the mean fracture aperture increases as the mean fracture aperture increases, so the relative scattering values are constant.

The fracture aperture depends non-linearly upon the fractal dimension (Fig. 10b). Two data sets are presented in this figure, symbols △ and ▽ correspond to surface standard deviations of 0.3 mm and 0.6 mm, respectively. The change of the standard deviation causes a proportional change of fracture aperture again. The fracture aperture increases with fractal dimension as the roughness of fracture surfaces increases. This is clear for the real rock fractures (Table 2) and for synthetic fractures. It was found that isotropic fractures have the least mean aperture, if all other parameters remain constant (Fig. 10c, fractal dimension 2.1, standard deviation 0.5 mm). The mean aperture increases as anisotropy appears.

The Brown (1995) model underestimates fracture aperture, and the Glover *et al.* (1998a, b) model tends to overestimate it as they use restricted parameters to control the fracture

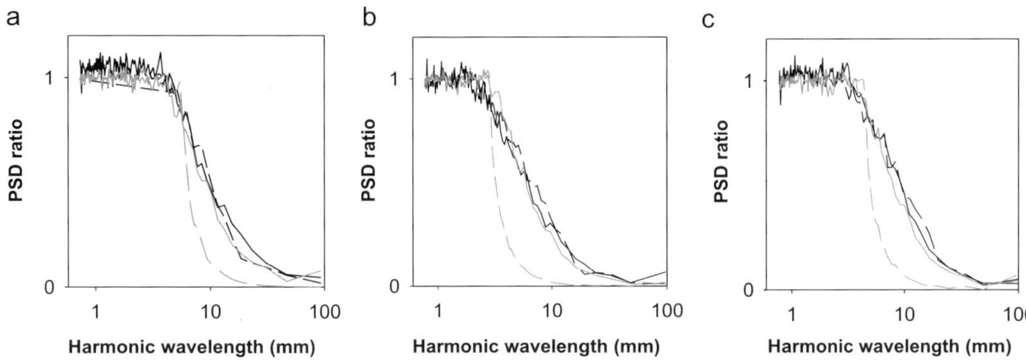

Fig. 11. Power spectral density (PSD) ratio plots for derivation of fracture parameters. (**a**) Granite; (**b**) syenite; (**c**) gabbro. Natural fracture (——), new model (- - -), Glover *et al.* (1998a, b) model (——), Brown (1995) model (- - -).

aperture. We therefore use four parameters to describe the matching characteristics of a fracture; the mis-match wavelength (ML), and maximum-matching fraction (MaxMF) of Glover *et al.* (1998a, b), and two additional parameters, a minimum matching fraction (MinMF), and a transition length (TL). The new definition of mis-match wavelength is a measure of the wavelength halfway between the largest wavelength at which minimum matching occurs and the smallest wavelength at which maximum matching occurs. The transition length (TL) is introduced, which describes how fast in wavelength space the matching develops. Consequently, the new definition of mis-match wavelength (ML) will always produce an ML greater than the Brown (1995) or Glover *et al.* (1998) values.

Variation of the transition length parameter yields smooth transition between Brown (1995) and Glover *et al.* (1998a, b) methods. The variation of the mean fracture aperture during this transition is shown in Figure 10d. The method of Brown (1995) gives the smallest values of fracture aperture and can underestimate it. The mean aperture of the synthetic fracture increases as the transition length increases up to 80 mm. As the transition length become comparable to the size of whole fracture (100 mm), further increasing of the transition length does not much affect the fracture properties. A large transitional length causes considerable scattering of mean aperture values, because the correlation between long-wave harmonics with highest amplitude becomes random. The fractured syenite has the greatest transition length.

In addition to determination of fractal dimension, we introduce the use of Fourier analyses as a means of analysing the matching behaviour of fracture surfaces. Power spectral density ratios (PSDRs) from the aperture and the sum of the PSDs of the two surfaces composing the fracture, plotted as a function of wavelength for the three real rock fracture apertures, are shown in Figure 11. It is clear from this analysis that the Glover *et al.* (1998a, b) model and our new model more closely reflect the complex matching properties of real rock fractures than the Brown (1995) approach and are likely to be more accurate predictors of fracture apertures in rough fractures.

Conclusions

Fracture surface roughness has a major control upon the flow of fluids through rock fractures in the lithosphere. However, the parallel-plate model assumed in large-scale multi-fracture models is insufficient to account for 'roughness' in crystalline rock reservoirs. This work describes new approaches in the characterization of 'roughness', as part of the drive to replace this model with one that fully accounts for rough surfaces at a variety of scales (e.g. Ge 1997).

This is a three-stage approach, aided by specially developed software. Improved optical profiling of fracture topographies from three different crystalline rocks has been carried out by OptiProf^{TM}, the images of which are imported into ParaFrac^{TM} and then statistically analysed to provide the input parameters for SynFrac^{TM} to create suites of synthetic fractures tuned to the real rock fractures. Outside the scope of this paper is the combination of these results with actual flow experiments and modelling (e.g. Nicholl *et al.* 1999; Ogilvie *et al.* 2001).

A new definition of mean aperture is used (the dual mean) which removes the difficulty of zero calculated aperture for rock apertures that touch at a single point, but is physically reasonable.

Relations between the fractal dimension, matching parameters and the resulting aperture were established. It was found that the resulting aperture increases linearly as the standard deviation of the fracture surfaces increases. The resulting aperture depends non-linearly on the fractal dimension and anisotropy of bounding surfaces. It was also found that the Brown (1995) model of rough fractures essentially underestimates the fracture aperture, while the Glover *et al.* (1998a, b) model slightly over predicts it. Our new model, with improved synthesis methods and two additional parameters (minimum matching fraction and transition length), is most flexible, allowing the range of fracture instances between the Brown (1995) and Glover *et al.* (1998a, b) models in order to reflect features of natural fractures in the best way.

This work was funded by the Natural Environmental Research Council of the UK, as part of the Micro-to-Macro Thematic Programme ongoing in the UK.

References

ABELIN, H., BIRGERSSON, L., WIDEN, H., AGREN, T., MORENO, L. & NERETNIEKS, I. 1994. Channelling experiments in crystalline fractured rocks. *Journal of Contaminant Hydrology*, **15**, 129–158.

ADLER, P. M. & THOVERT, J. F. 1999. *Theory and Applications of Transport in Porous Media: Fractures and Fracture Networks*. Kluwer Academic Publishers, Dordrecht.

AMUNDSEN, H., WAGNER, G., OXAAL, U., MEAKIN, P., FEDER, J. & JOSSANG, T. 1999. Slow-two-phase flow in artificial fractures: Experiments and simulations. *Water Resources Research*, **35**, 2619–2626.

BARTON, C. A. 1998. Reservoir scale fracture permeability in the Dixie Valley, Nevada, geothermal field. *SPE 4731* presented at *SPE/ISRM Eurorock '98* held in Trondheim, Norway, 8–10 July, 1998.

BROWN, S. R. 1987. Fluid flow through rock joints: The effect of surface roughness. *Journal of Geophysical Research*, **92**, 1337–1347.

BROWN, S. R. 1988. Correction to 'A note on the description of surface roughness using fractal dimension'. *Geophysical Research Letters*, **15**, 286.

BROWN, S. R. 1995. Simple mathematical model of a rough fracture. *Journal of Geophysical Research*, **91**, 5941–5952.

BROWN, S. R. & SCHOLZ, C. H. 1986. Closure of rock joints. *Journal of Geophysical Research*, **91**, 4939–4948.

BROWN, S. R., CAPRIHAN, A., & HARDY, R. 1998. Experimental observation of fluid flow channels in a single fracture. *Journal of Geophysical Research*, **103**, 5125–5132.

BROWN, S. R., STOCKMAN, H. W. & REEVES, S. J. 1995. Applicability of the Reynolds equation for modelling fluid flow between rough surfaces. *Geophysical Research Letters*, **22**, 2537–2540.

COX, B. L. & WANG, J. S. Y. 1993. Single fracture aperture patterns: Characterization by slit island analysis. *In*: *High Level Radioactive Waste Management: Proceedings of the Fourth International Conference, Las Vegas, NV*, 26–28 April, **2**, American Society of Civil Engineers, New York, 2053–2060.

DETWILER, R. L., RUSSEL, L., PRINGLE, S. E., SCOTT, E. & GLASS, R. J. 1999. Measurement of fracture aperture fields using transmitted light: An evaluation of measurement errors and their influence on simulations of flow and transport through a single fracture. *Water Resources Research*, **35**, 2605–2617.

DEVILI, K., BABADAGLI, T. & COMLEKCI, C. 2001. A new computer-controlled surface-scanning device for measurement of fracture surface roughness. *Computers and Geosciences*, **27**, 265–277.

DURHAM, W. B. & BONNER, B. P. 1995. Closure and fluid flow in discrete fractures. *In*: Myer, L., Cook, N., Goodman, R. & Tsang C. (eds) *Fractured and Jointed Rock Masses*, Balkema, Rotterdam, 441–447.

GALE, J. E. 1987. Comparison of coupled fracture deformation and fluid flow models with direct measurements of fracture pore structure and stress flow properties. *Procedures of the 28th U.S. Symposium on Rock Mechanics*, 1213–1222.

GE, S. 1997. A governing equation for fluid flow in rough fractures. *Water Resources Research*, **33**(1) 53–61.

GENTIER, S. 1986. *Morphologie et comportement hydromechanique d'une fracture naturelle dans un granite sous contrainte normale*. PhD Dissertation, Université d'Orléans, Orléans, France.

GLOVER, P. W. J., MATSUKI, K., HIKIMA, R. & HAYASHI, K. 1998a. Synthetic rough fractures in rocks. *Journal of Geophysical Research*, **103**, 9609–9620.

GLOVER, P. W. J., MATSUKI, K., HIKIMA, R. & HAYASHI, K. 1998b. Fluid flow in synthetic rough fractures and application to the Hachimantai geothermal HDR test site. *Journal of Geophysical Research*, **103**, 9621–9635.

GUDMUNDSON, A. 2000. Active fault zones and groundwater flow. *Geophysical Research Letters*, **27**(18), 2993–2996.

HAKAMI, E. & LARSSON, E. 1996. Aperture measurements and flow experiments on a single natural fracture. *International Journal of Mechanics, Mineral Sciences and Geomechanical Abstracts*, **33**, 395–404.

ISAKOV, E., OGILVIE, S. R., TAYLOR, C. W. & GLOVER, P. W. J. 2001a. Fluid Flow through rough fractures in rocks. I: High resolution aperture determinations. *Earth and Planetary Science Letters*, **191**, 267–282.

ISAKOV, E., GLOVER., P. W. J. & OGILVIE, S. R. 2001b. Use of synthetic fractures in the analysis of natural fracture apertures. *Proceedings of the 8th European Congress for Stereology and Image Analysis, Image Analysis and Stereology*, **20**(2) Suppl. 1, Sept., 366–371.

IWANO, M. & EINSTEIN, H. H. 1995. Laboratory experiments on geometric and hydromechanical characteristics of three different fractures in granodiorite. *In*: Fuji, T. (ed.) *International Congress on Rock Mechanics, Proceedings*, **2**, Toyko, Japan, 743–750.

JONES. G., FISHER, Q. J. & KNIPE, R. J. 1998. *Faulting, Fault Sealing and Fluid Flow in Hydrocarbon Reservoirs*. Geological Society, London Special Publications, **147**.

MEHEUST, Y. & SCHMITTBUHL, J. 2000. Flow enhancement of a rough fracture. *Geophysical Research Letters*, **27** (18), 2989–2992.

MEHEUST, Y. & SCHMITTBUHL, J. 2001. Geometrical heterogeneities and permeability anisotropy of rough fractures. *Journal of Geophysical Research*, **106**, 2089–2102.

MORENO, L., NERETNIEKS, I. & ERIKSEN, T. 1985. Analysis of some laboratory tracer runs in natural fissures. *Water Resources Research*, **21**, 951–958.

NICHOLL, M. J., RAJARAM, H., GLASS, R. J. & DETWILER, R. 1999. Saturated flow in measured aperture fields. *Water Resources Research*, **35**(11), 3361–3373.

OGILVIE, S. R., ISAKOV, E., GLOVER, P. W. J. & TAYLOR C. W. 2001. Use of image analysis and finite element analysis to characterise fluid flow in rough rock fractures and their synthetic analogues. *Proceedings of the 8th European Congress for Stereology and Image Analysis, Image Analysis and Stereology*, **20**(2) Suppl. 1, Sept., 504–509.

OGILVIE, S. R., ISAKOV, E., TAYLOR, C. W. & GLOVER, P. W. J. 2002a. A new high resolution optical method for obtaining the topography of fracture surfaces in rocks. *Image Analysis and Stereology*, **21**(1) 61–66.

OGILVIE, S. R., ISAKOV, E. & GLOVER, P. W. J. 2002b. Advances in the characterization of rough fractures in hydrocarbon reservoirs. *First Break*, **20**(4) April, 233–239.

ORON, A. P. & BERKOWITZ, B. 1998. Flow in rock fractures: The local cubic law assumption re-examined. *Water Resources Research*, **34**, 2811–2825.

PERSOFF, P. & PRUESS, K. 1995. Two-phase flow visualization and relative permeability measurement in natural rough-walled rock fractures. *Water Resources Research*, **35**, 1175–1186.

PLOURABOUÉ, F., KUROWSKI, P., BOFFA, J.-M., HULIN, J.-P. & ROUX, S. 2000. Experimental study of the transport properties of rough self-affine fractures. *Journal of Contaminant Hydrology*, **46**, 295–318.

POON, C. Y., SAYLES, R. S. & JONES, T. A. 1992. Surface measurement and fractal characterization of naturally fractured rocks. *Journal of Physics D: Applied Physics*, **25**, 1269–1275.

POWER, W. L. & TULLIS, T. E. 1992. The contact between opposing fault surfaces at Dixie Valley, Nevada, and implications for fault mechanics. *Journal of Geophysical Research*, **97**, 15425–15435.

POWER, W. L. & TULLIS, T. E. 1995. Review of the fractal character of natural fault surfaces with implications for friction and the evolution of fault zones. *In*: BARTON, C. C. & LA POINTE, P. R. (eds) *Fractals in the Earth Sciences*, Plenum Press, New York.

RENSHAW, C. E., DADAKIS, J.-S. & BROWN, S. R. 2000. Measuring fracture apertures: A comparison of methods. *Geophysical Research Letters*, **27**, 289–292.

SAHIMI, M. 1993. Flow phenomena in rocks: from continuum models to fractals, percolation cellular automata, and simulated annealing. *Reviews in Modern Physics*, **65**, 1393–1534.

SAUPE, D. 1988. Algorithms for random fractals. *In*: Peitgen H.-O. & Saupe, D. (eds) *The science of fractal images*. Springer-Verlag, New York, 71–136.

SCHMITTBUHL, J., GENTIER, S. & ROUX, S. 1993. Field measurements of the roughness of fault surfaces. *Geophysical Research Letters*, **20**(8), 639–641.

YEO, I. W., DE FREITAS, M. H. & ZIMMERMAN, R. W. 1998. Effect of shear displacement on the aperture and permeability of a rock fracture. *International Journal of Rock Mechanics and Mineral Science Abstracts*, **35**, 1051–1070.

Thermally induced primary fracture development in tabular granitic plutons: a preliminary analysis

M. A. KOENDERS & NICK PETFORD

Centre for Earth and Environmental Science Research, Kingston University, Kingston-upon-Thames, Surrey KT1 2EE, UK

Abstract: We present an analytical model that predicts some of the mechanical effects associated with the intrusion and subsequent cooling of a rectangular intrusion emplaced at a uniform temperature into elastic continental crust. Assuming an idealized geometry and initial conditions, we recover the temperature field and subsequent strain field as a function of both position and time. The strain field is particularly relevant as it provides information on the *primary* (cooling-related) fracture formation pattern and direction within and immediately surrounding the pluton. We find a large strain jump across the pluton–country rock contact, implying that fracture formation should be maximized at the edges and corners of the intrusion. The direction of the fractures is predominantly vertical within the pluton centre, but becomes progressively more inclined towards the pluton margin and into the adjacent country rock. Fracture orientation may depend critically on the geometry of the intrusion, in particular the ratio of the longest to shortest dimension L_1/L_2.

It has long been known that plutonic rocks, cooling slowly beneath the Earth's surface, develop distinctive patterns of jointing caused mostly by stresses related to heat loss accompanying the phase transition from liquid to solid (e.g. Balk 1937; Norton & Knight 1977; Knapp & Norton 1981; Segall & Pollard 1983; Gerla 1988; Velde *et al.* 1991; Bergbauer *et al.* 1998). Fractures considered to have formed while the pluton was solidifying were first referred to by Cloos (1925) as cross joints, since they formed perpendicular to magmatic flow fabrics or flow lines. These joints are predominantly vertical Mode 1 (tensional) joints (Price 1966; Pollard & Aydin 1988). In granitic plutons, cross joints are commonly infilled either with late-stage magmatic products such as aplites or pegmatites (Balk 1937), or hydrothermal minerals including quartz, epidote, chlorite and muscovite (Segall & Pollard 1983). These observations are important, as they confirm both the magmatic provenance of the jointing and its ability to provide pathways for magmatic or externally derived fluids to enter and circulate through the cooling rock. Where these fluids are hydrothermal in origin, the associated mineralization can result in economically important ore deposits (Lindgren 1907; Sams & Thomas-Betts 1988). A further motivation for studying the formation of primary joints in plutons is that fractured and hydrothermally altered crystalline basement composed of granitic material is a potentially important type of hydrocarbon reservoir (Dmitriyevskiy *et al.* 1993; Schutter 2003), and that primary fracture sets may contribute significantly to their porosity and connectivity (see Sanders *et al.* 2003).

In this short contribution, we present the initial results of an analytical investigation into the effects of intruding a hot pluton into cooler country rocks. Our objective was to predict the magnitude and orientation of thermally induced *primary* fractures as a guide to help constrain the identification of these structures in the field. The mechanical model we present is successful in that it predicts the local fracture orientation and stress field relevant to an ideal tabular pluton. The results form part of an ongoing study, and a more detailed analysis will be published elsewhere.

Mechanical (analytical) model for cooling fracture formation

As an example of primary fracture (hence permeability) development, we consider the specific case of a cooling tabular pluton emplaced instantaneously into continental crust. The mechanical effects around a cooling rectangular inclusion, including the temperature profile, the strain field and the likely fracture pattern, are calculated analytically. Earlier studies by Knapp & Norton (1981), Segall & Pollard (1983) and Bergbauer & Martel (1999) have all demonstrated convincingly that thermal stresses play an important role in forming joints in cooling plutons. An important finding (see also Gerla 1988) is that the initial geometry of the pluton is critical to the subsequent development of the stress field, hence joint pattern, that develops in the vicinity of the cooling pluton.

From: PETFORD, N. & MCCAFFREY, K. J. W. (eds) 2003. *Hydrocarbons in Crystalline Rocks.* Geological Society, London, Special Publications, **214**, 143–150. 0305-8719/03/$15

Analysis

Our thermal model begins with an isothermal, tabular-shaped intrusion (McCaffrey & Petford 1997; Cruden & McCaffrey 2001), emplaced instantaneously into country rock at a uniform dimensionless temperature at time $t = 0$. If we assume for simplicity that the mechanical and thermal coefficients and constants in the intrusion and country rock are everywhere the same (see Heuze 1983 for typical values), then the problem can be scaled according to geometrical quantities only, in effect the ratio of the longest and shortest dimensions L_1/L_1. This useful and important result is discussed in more detail in a later section.

To begin, it is necessary to solve the relevant energy balance relating to magma intrusion. This can be done easily by using the familiar heat conduction equation (Carslaw & Jaeger 1959), which in two dimensions reads:

$$\frac{\partial^2 T}{\partial x^2} + \frac{\partial^2 T}{\partial y^2} = \frac{1}{\kappa}\frac{\partial T}{\partial t}, \quad (1)$$

where κ is the thermal diffusivity. For the temperature field to be solved as a function of time (t), the initial temperature distribution and the boundary conditions are required. While acknowledging that advection can play an important role in heat loss from intrusive bodies, for this investigation the key point is that the temperature field gives rise to a strain field that can result in primary fracturing as cooling takes place. Following Landau & Lifschitz (1986), the displacement field $\boldsymbol{u}$ due to changes in temperature in and around the cooling pluton requires knowledge of just two coefficients: the thermal expansion coefficient (α), and the Poisson ratio (ν) of the medium according to:

$$\frac{3(1-\nu)}{1+\nu}\,\mathrm{grad\,div}\,\boldsymbol{u} - \frac{(3(1-2\nu)}{2(1+\nu)}\,\mathrm{curl\,curl}\,\boldsymbol{u} = \alpha\,\mathrm{grad}\,T. \quad (2)$$

The strain field that follows depends on the evaluation of the integrals in equation (2). The solution to the problem requires a rather technical set

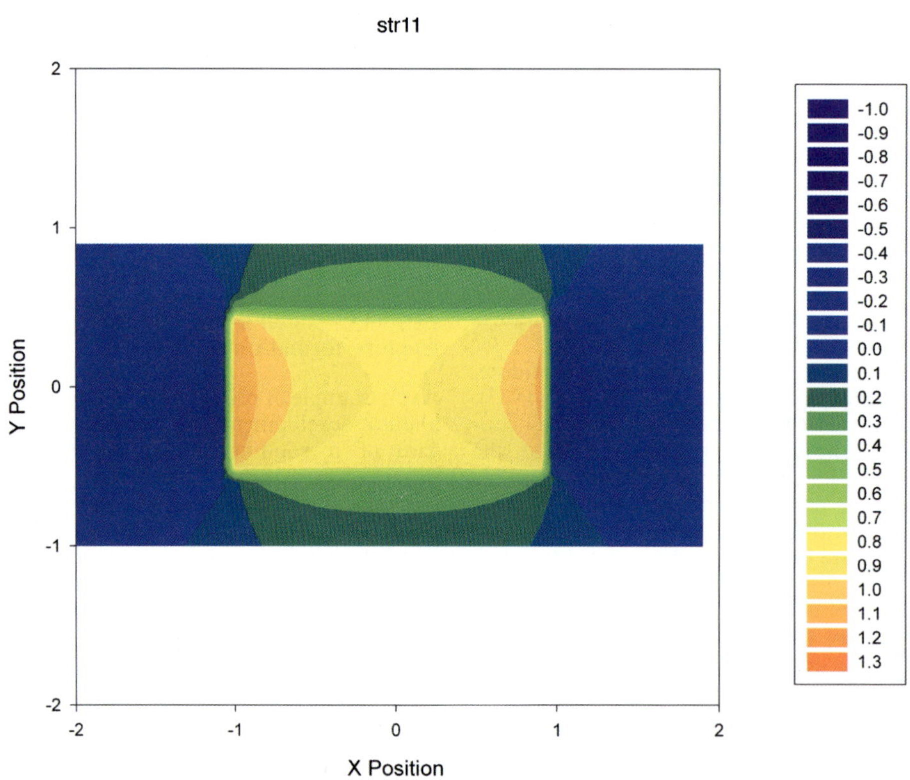

Fig. 1. $\partial u_1/\partial x_1$ strain for large times calculated using equation (3a). Note the negative strain ($\partial u_1/\partial x_1 < 0$) outside the L_2 pluton margin (blue region), and the high positive values inside the intrusion (yellow and orange), which increase towards the pluton margins. Intrusion dimensions are $L_1 = 1$ and $L_2 = 0.5$.

of operations that for brevity are not included in this short note. The final form of the equations that relate the temperature field (equation (1)) to the displacement field (equation (2)) are:

$$\frac{\partial u_1}{\partial x_1} = \frac{-\alpha(\nu+1)}{6(\nu-1)}(T_1 - T_0) \times \left(B_{L_1}(x)B_{L_2}(y) + \frac{1}{\pi}\left[\tan^{-1}\left(\frac{L_2-y}{L_1+x}\right) + \tan^{-1}\left(\frac{L_2+y}{L_1-x}\right) + \tan^{-1}\left(\frac{L_2-y}{L_1-x}\right) + \tan^{-1}\left(\frac{L_2+y}{L_1+x}\right)\right]\right) \quad (3a)$$

$$\frac{\partial u_2}{\partial x_2} = \frac{-\alpha(\nu+1)}{6(\nu-1)}(T_1 - T_0) \times \left(B_{L_1}(x)B_{L_2}(y) - \frac{1}{\pi}\left[\tan^{-1}\left(\frac{L_2-y}{L_1+x}\right) + \tan^{-1}\left(\frac{L_2+y}{L_1-x}\right) + \tan^{-1}\left(\frac{L_2-y}{L_1-x}\right) + \tan^{-1}\left(\frac{L_2+y}{L_1+x}\right)\right]\right) \quad (3b)$$

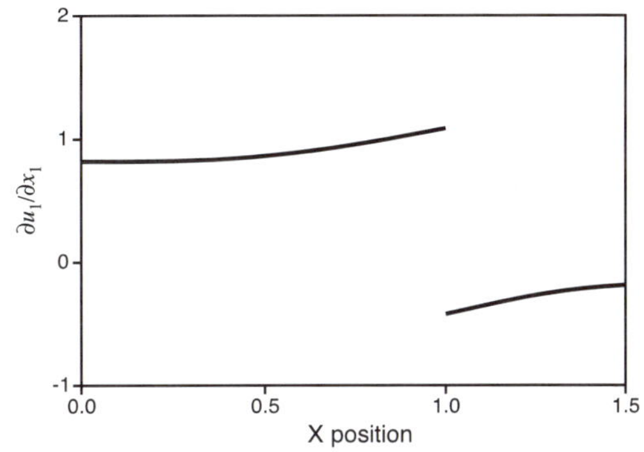

Fig. 2. $\partial u_1/\partial x_1$ plotted as a function of position from the pluton centre ($x = 0.0$) to its margin ($x = 1.0$) evaluated at $y = 0$. Note the large sudden drop in strain from positive to negative values across the pluton-country rock contact (country rock $= x > 1.0$). $L_1 = 1$, $L_2 = 0.5$.

$$\frac{\partial u_1}{\partial x_2} = \frac{-\alpha(\nu+1)(T_1 - T_0)}{12\pi(\nu-1)} \times \ln\left(\frac{(L_1^2 + 2L_1x + L_2^2 + 2L_2y + x^2 + y^2)(L_1^2 - 2L_1x + L_2^2 - 2L_2y + x^2 + y^2)}{(L_1^2 + 2L_1x + L_2^2 - 2L_2y + x^2 + y^2)(L_1^2 - 2L_1x + L_2^2 + 2L_2y + x^2 + y^2)}\right) \quad (3c)$$

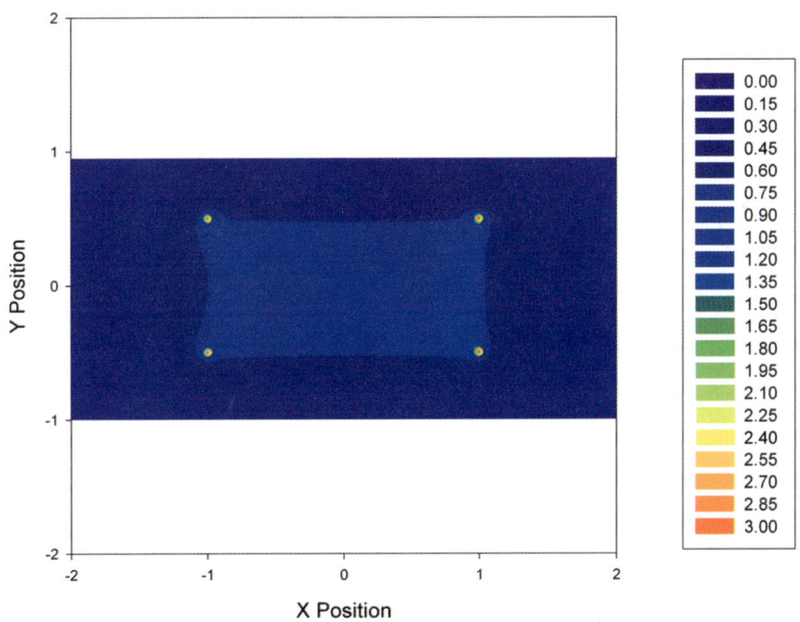

Fig. 3. Plot showing the strain magnitude e as a function of position for large times; $L_1 = 1$, $L_2 = 0.5$. Values are lower (dark blue) in the region outside the intrusion (generally 0.2–0.4) and higher (0.7–1.0) inside the intrusion, increasing towards the margins and corners. Maximum values (orange spots) occur at the four corners, indicating the most likely site of cracking.

where B is the boxcar function. Note that this set refers to the strain after a long period of cooling time. Solutions at shorter time scales, although easily calculated, are not directly relevant to this study and will be presented elsewhere.

Results

By plotting solutions to equations (3a–c) (Figs 1–5), some key properties of the strain field due to thermally induced stresses in and around the cooling intrusion can be assessed.

Strain 11

A plot of the spatial pattern of the maximum strain component in the x_1 direction after cooling (equation (3a)) is shown in Figure 1. The units are scaled to highlight the general pattern of the strain and colour-coded, with blue shades corresponding to negative values and yellow-orange regions depicting high strains. The edge of the intrusion is clearly defined, with values of strain outside the intrusion generally $-1.0 < \partial u_1/\partial x_1 < 0.5$. The maximum strains (*c*. 1.3) are generated just inside the pluton, parallel with the smallest diameter margin and corresponding to the direction L_2. The strain distribution is symmetrical around the pluton and controlled strongly by its geometry. Thus, strains in the country rock adjacent and parallel with the longest pluton dimension (L_1) have positive values of *c*. 0.2–0.7, which increase towards the contact.

The changing strain profile across the pluton is more easily seen when plotted in one dimension at a constant value of $y = 0$. This is shown in Figure 2. The contact between the edge of the intrusion ($x = 1$) and the country rock ($x > 1$) is marked by a large jump in strain, from a value of 1.3 just inside the pluton (orange region, Fig. 1), to <0 in the adjacent country rock. The physical interpretation of this is that while the interior of the intrusion is experiencing cooling and contraction, the surrounding country rock is heating up and undergoing compression.

Magnitude of strain

The principal strains follow from the strain tensor and can be found by evaluating the equation:

$$\det\begin{pmatrix} e_{11} - \lambda & e_{12} \\ e_{12} & e_{22} - \lambda \end{pmatrix} = 0. \qquad (4)$$

This gives rise to two solutions for the parameter λ which represent the principal strains e_1 and e_2. The magnitude of the strain e can now be defined as:

$$e = \sqrt{e_1^2 + e_2^2}. \qquad (5)$$

The magnitude of strain is shown in Figure 3 (colour-coded as in Fig. 1), with large strains ($2.5 > e > 3.0$), developing in the vicinity of the four corners of the intrusion. Values are generally low in the country rock outside the intrusion ($e < 0.5$), but greater inside ($e = 0.8$–1.0), and highest close to the shortest pluton margin L_2. It is perhaps worth cautioning that these high values are dependent largely on the sharpness of the corners, and that more rounded geometries will result in lower magnitudes of strain.

For fracture sensitivity it is advantageous to show that the *ratio* of the major and minor principal strain is both negative and large. To this end, a plot of

$$\frac{e_1}{e_2}\sqrt{e_1^2 + e_2^2} \qquad (6)$$

(where the choice of e_1 and e_2 is such that $|e_1| > |e_2|$) has been made (Fig. 4). Once again, the large values are always obtained near, but just inside, the corners of the intrusion (cf. Fig. 3), indicating the position where primary igneous fracturing due to cooling is most likely to occur.

Finally the directions of the minor principal strain can be plotted to gain an impression of the fracture directions as a function of position. To achieve this, the strain magnitude as a function of position has been integrated to form lines. The result is shown in Figure 5. Note the sharp change in direction along the side of the intrusion, crossing the interface with the country rock in the dimension L_2.

Discussion

Three-dimensional pluton shape and primary fracture distribution and direction

The above analysis rests heavily on the initial shape of the pluton, in particular the ratio L_1/L_2. This appears as a first-order effect, and will govern the subsequent development of primary fracture orientation in and around the intrusion, along with surface heat flow and style of fluid circulation (Norton & Knight 1977; Knapp & Norton 1981; Sams & Thomas-Betts 1988). It is thus crucial for well-constrained

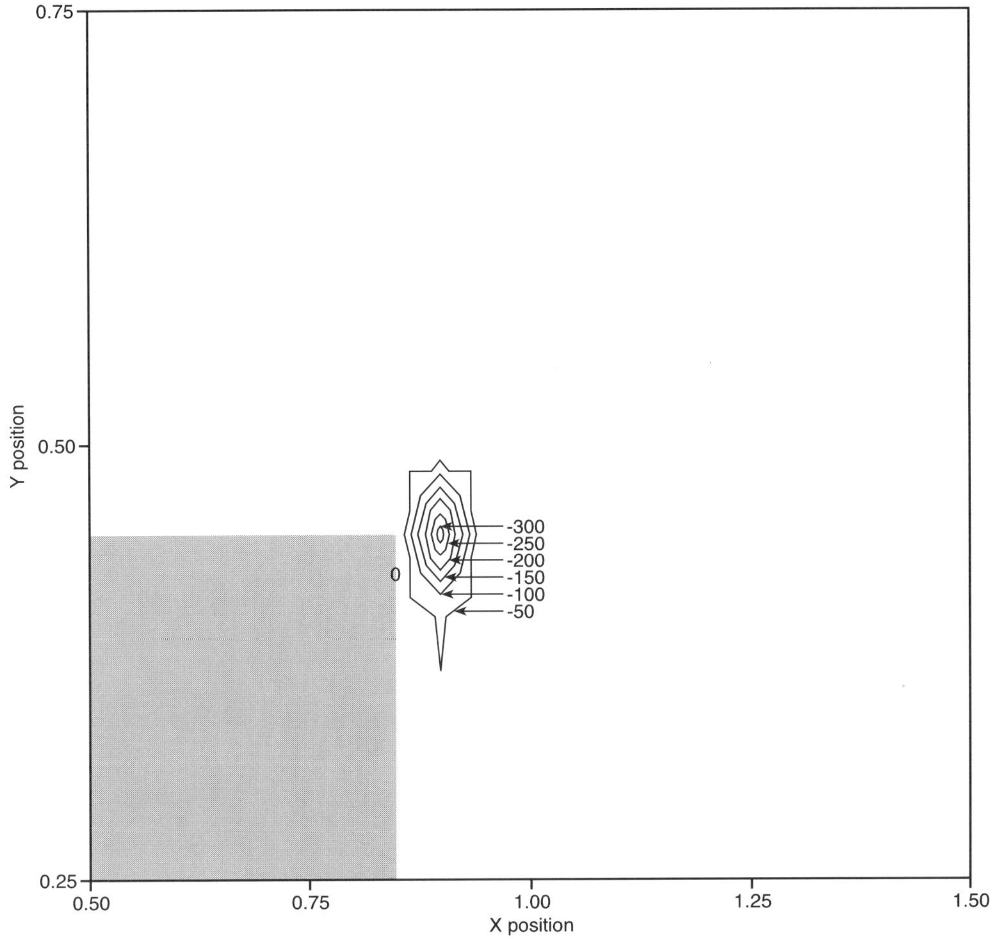

Fig. 4. Plot of the ratio of the major and minor principal strain ($(e_1/e_2)\sqrt{e_1^2 + e_2^2}$), as a function of position for large times; $L_1 = 1$, $L_2 = 0.5$. Fracturing is most likely where the strain ratio is negative (and large), in this instance the corners just within the pluton. Only the corner of the intrusion is shown for clarity.

estimates of pluton shape to be derived before a proper assessment of the important fracture orientations and associated permeability features can be calculated. In recent years, the traditional idea of a granite pluton as a diapir has been challenged (e.g. Clemens *et al.* 1997), while an analysis of pluton shapes has revealed an apparent preference for relatively thin, sheet-like geometries that follow a power law (self-affine) size distribution of the form:

$$L = kT^a, \qquad (7)$$

where L and T are pluton length and thickness, and a is the exponent (McCaffrey & Petford 1997; Cruden & McCaffrey 2001). A value of $a < 1$ indicates that plutons are longer than they are thick in the vertical direction, and maintain this geometry during emplacement (Fig. 6). This relationship appears to hold even for batholith-sized intrusions (Petford *et al.* 2000). Supporting evidence for these largely theoretical models comes from the relatively few numbers of seismic surveys across granitic plutons (e.g. Evans *et al.* 1994; Brown & Tryggvason 2001), and recent gravity modelling (Haederle & Atherton 2002), showing that the geometry of 'granitic' basement reservoirs may not be a collection of diapiric structures but rather a series of relatively thin (*c.* 1–5 km thick) sheets.

Given the sensitivity of primary fracture distribution and direction on pluton shape, it is clearly important to constrain this as accurately as possible prior to modelling thermally induced fracture orientations. For example, we would predict a pattern of largely radial to tangential primary fracture sets around the margins of a

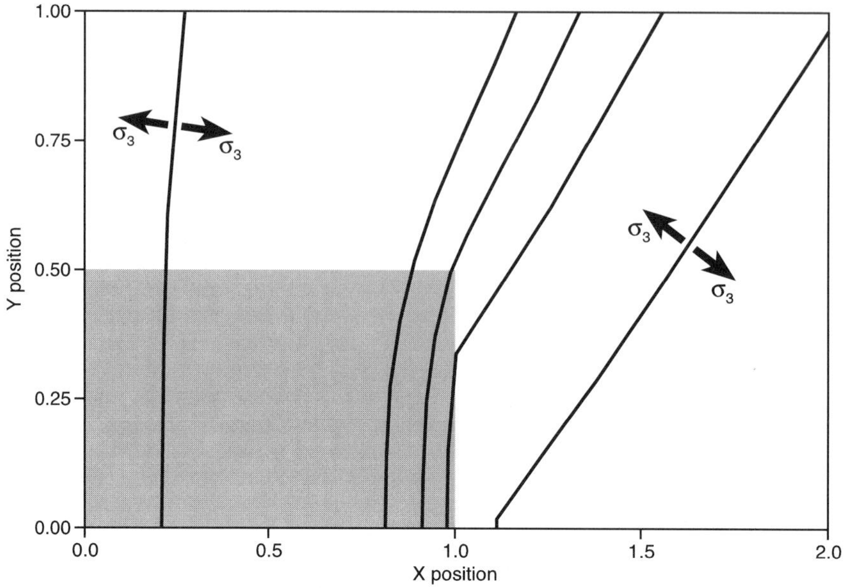

Fig. 5. Predicted primary fracture directions as a function of position for large times ($L_1 = 1$, $L_2 = 0.5$). Note that the direction is margin parallel inside the intrusion and the sharp deflection predicted at position $x = 1$ across the pluton–country rock contact. Fractures outside the pluton are always inclined to L_2.

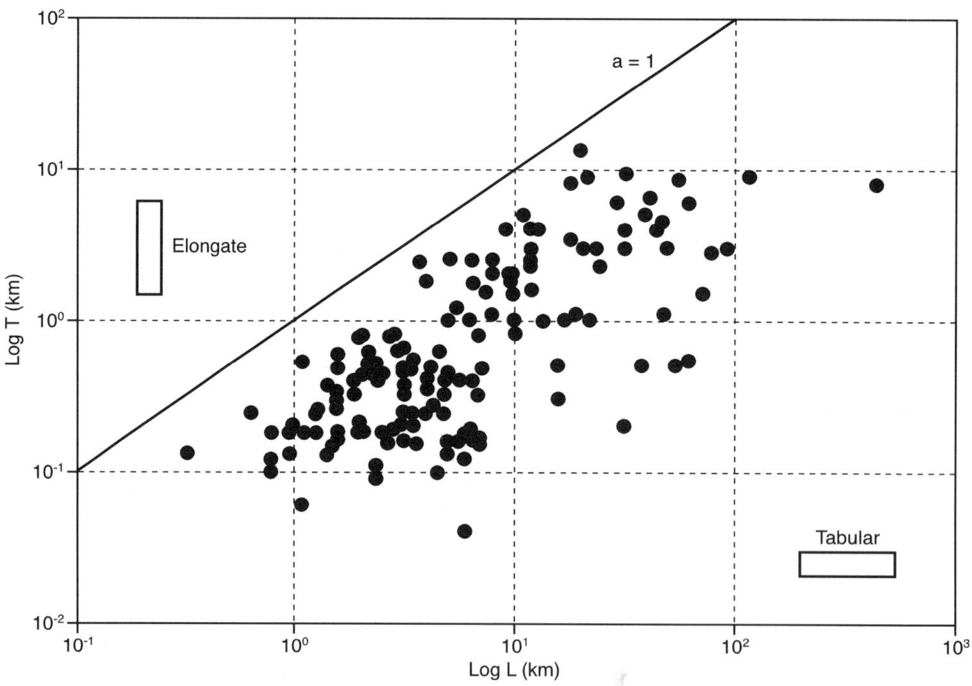

Fig. 6. Plot showing the measured thickness (T) versus length (L) for 159 plutons and laccoliths of different ages and from varying tectonic settings. The line $a = 1$ marks the divide between tabular and vertically elongated intrusion geometries. All intrusions plot below the line $a = 1$ in the tabular ($a < 1$) field (McCaffrey & Petford 1997; Petford *et al.* 2000).

strongly curved diapiric or highly ballooned pluton. Previous numerical studies (e.g. Knapp & Norton 1981) on plutons with high prescribed aspect ratios ($a > 1$ in Fig. 6), have stressed the potential role of magma and pore fluid pressure, in addition to thermal stresses, in generating cracks, and have shown a marked dependency in fracture orientation with intrusion depth. It is thus encouraging that our model, predicting relatively low primary fracture densities in the cores of tabular plutons, is consistent with the observation of Balk (1937) that many pluton interiors are notably free from joints unless they have been extensively deformed by subsequent tectonic processes. A strong test of the tabular geometry model would be field observations showing primary fracture orientations consistent with the directions indicated in Figure 5. Our initial results suggest that fracturing, and by implication, primary fracture permeability, will be highest along the margins and at the corners of tabular plutons. However, we acknowledge that the link between a theoretical description of likely sites of fracture development during monotonic cooling, and actual fracture permeability as measured in the field, is far from complete. Future refinements to our model will include a more complete consideration of the role of pluton shape in governing fracture patterns around cooling intrusions (e.g. Bergbauer *et al.* 1998).

Summary

A first-order estimate of the magnitude and direction of fractures that form during the cooling of a granitic intrusion from an initial intrusion temperature has been made. When scaled, the problem depends on geometry of the intrusion, expressed by the ratio L_1/L_2. Assuming an initial tabular geometry, we find that most of the strain due to thermal contraction is taken up at the edges of the pluton and the immediate contact with surrounding country rock, and primary (igneous) fracture porosity will be highest in these regions. In contrast, the central part of the pluton remains relatively undeformed and fracture porosity due solely to cooling will be lower. The primary direction of fractures in a monotonically cooling pluton is vertical, parallel with the intrusion margins, although fracture orientations diverge away from the vertical in the country rock immediately adjacent to the pluton walls.

We would like to acknowledge the Japan Vietnam Petroleum Company for funding part of this study. D. Sanderson and K. McCaffrey are thanked for helpful comments.

References

BALK, R. 1937. *Structural Behaviour of Igneous Rocks*. Geological Society of America, Memoirs, **5**.

BERGBAUER, S. & MARTEL, S. J. 1999. Formation of joints in cooling plutons. *Journal of Structural Geology*, **21**, 821–835.

BERGBAUER, S., MARTEL, S. J. & HIERONYMUS, C. F. 1998. Thermal stress evolution in cooling plutonic environments of different geometries. *Geophysical Research Letters*, **25**, 707–710.

BROWN, D. & TRYGGVASON, A. 2001. Ascent mechanism of the Dzhabyk batholith, southern Urals: constraints from URSEIS reflection seismic profiling. *Journal of the Geological Society, London*, **158**, 881–884.

CARSLAW, H. S. & JAEGER, J. C. 1959. *Conduction of Heat in Solids*, 2nd edn. Clarendon, Oxford.

CLEMENS, J. D., PETFORD, N. & MAWER, C. K. 1997. Ascent mechanisms of granitic magmas: causes and consequences. *In*: HOLNESS, M. B. (ed). *Deformation-Enhanced Fluid Transport in the Earth's Crust and Mantle*. Chapman & Hall, London, 145–172.

CLOOS, H. 1925. *Einfuhrung in die tektonische Behandlung magmatischer Erscheinungen, 1. Das Riesengebrige in Schlesien*. Gebrude Borntraeger, Berlin.

CRUDEN, A. R. & MCCAFFREY, K. J. W. 2001. Growth of plutons by floor subsidence: implications for rates of emplacement, intrusion spacing and melt extraction mechanisms. *Physics and Chemistry of the Earth*, **26**, 303–315.

DMITRIYEVSKIY, A. N., KIREYEV, F. A., BOCHKO, R. A. & FEDOROVA, T. A. 1993. Hydrothermal origin of oil and gas reservoirs in basement rock of the South Vietnam continental shelf. *International Geology Review*, **35**, 621–630.

EVANS, D. J., ROWLEY, W. J., CHADWICK, R. A., KIMBELL, G. S. & MILLWARD, D. 1994. Seismic reflection data and the internal structure of the Lake District batholith, Cumbria, northern England. *Proceedings of the Yorkshire Geological Society*, **50**, 11–24.

GERLA, J. P. 1988. Stress and fracture evolution in a cooling pluton: an example from Diamond Joe stock, western Arizona, USA. *Journal of Volcanology and Geothermal Research*, **34**, 267–282.

HAEDERLE, M. & ATHERTON, M. P. 2002. Shape and intrusion style of the Coastal Batholith, Peru. *Tectonophysics*, **345**, 17–28.

HEUZE, F. E. 1983. High temperature mechanical, physical and thermal properties of granitic rocks—a review. *International Journal of Rock Mechanics*, **20**, 3–10.

KNAPP, R. B. & NORTON, D. 1981. Preliminary numerical analysis of processes related to magma crystallisation and stress evolution in cooling plutonic environments. *American Journal of Science*, **281**, 35–68.

LANDAU, L. D. & LIFSCHITZ, E. M. 1986. *Theory of Elasticity*. Pergamon, Oxford.

LINDGREN, W. 1907. The relation of ore deposition to physical conditions. *Economic Geology*, **2**, 105–127.

MCCAFFREY, K. J. W. & PETFORD, N. 1997. Are granitic intrusions scale invariant? *Journal of the Geological Society, London*, **154**, 1–4.

NORTON, D. & KNIGHT, J. 1977. Transport phenomena in hydrothermal systems: cooling plutons. *American Journal of Science*, **277**, 937–981.

PETFORD, N., CRUDEN, A. R., MCCAFFREY, K. J. W. & VIGNERESSE, J. L. 2000. Granite magma formation, transport and emplacement in the Earth's crust. *Nature*, **408**, 669–673.

POLLARD, D. D. & AYDIN, A. 1988. Progress in understanding jointing over the last century. *Geological Society of America Bulletin*, **100**, 1181–1204.

PRICE, N. 1966. *Fault and Joint Development in Brittle and Semi-Brittle Rock*. Pergamon Press, Oxford.

SAMS, M. S. & THOMAS-BETTS, A. 1988. 3-D numerical modelling of the conductive heat flow of SW England. *Geophysical Journal*, **92**, 323–334.

SANDERS, C. A. E., FULLARTON, L. & CLAVET, S. 2003. Modelling fracture systems in extensional crystalline basement. *In*: PETFORD, N. & MCCAFFREY, K. (eds) *Hydrocarbon in Crystalline Rocks*. Geological Society, London, Special Publications, **214**, 221–236.

SCHUTTER, S. R. 2003. Occurrences of hydrocarbons in and around igneous rocks. *In*: PETFORD, N. & MCCAFFREY, K. (eds) *Hydrocarbon in Crystalline Rocks*. Geological Society, London, Special Publications, **214**, 35–68.

SEGALL, P. & POLLARD, D. D. 1983. Joint formation in granitic rock of the Sierra Nevada. *Geological Society of America, Bulletin*, **94**, 563–575.

VELDE, B., DUBOIS, J., MOORE, D. & TOUCHARD, G. 1991. Fractal patterns of fractures in granites. *Earth and Planetary Science Letters*, **104**, 25–35.

A review of the occurrence and origin of abiogenic hydrocarbons in igneous rocks

J. POTTER[1,2] & J. KONNERUP-MADSEN[3]

[1]*Institut für Mineralogie und Mineralogische Rohstoffe, Technische Universität Clausthal, Adolph-Roemer-Str. 2a, 38678 Clausthal-Zellerfeld, Germany*
[2]*School of Earth Sciences and Geography, Kingston University, Penrhyn Road, Kingston-upon-Thames, Surrey, KT1 2EE, UK*
(e-mail: joanna.potter@tu-clausthal.de)
[3]*Geological Institute, Copenhagen University, Øster Volgade 10, DK-1350, Copenhagen, Denmark*

Abstract: Reports on the presence of hydrocarbons in igneous rocks have been on the increase and generating greater interest in the scientific community over the last 20 years. Most of the occurrences are due to the incorporation of organic material into the magmatic systems. However, reports on the presence of hydrocarbons formed by abiogenic processes have also increased in recent years, suggesting that these hydrocarbons may not be as rare as previously thought and may have implications for natural gas resources in the future. This paper reviews these occurrences and the models proposed for the generation of these hydrocarbons, in particular the nature of the hydrocarbon-bearing fluids in the alkaline complexes Khibina, Lovozero and Ilímaussaq. The origin of these hydrocarbons remains controversial, whether they are (1) derived directly from the mantle, (2) formed during late crystallization stages by respeciation of a C–O–H fluid below 500 °C, or (3) formed during postmagmatic alteration processes involving Fisher–Tropsch type reactions catalysed in the presence of Fe-oxides and silicates. The reports suggest that a direct mantle origin for the hydrocarbon fluid is unlikely. A model involving near-solidus reequilibration of a C–O–H fluid to a CH_4-rich composition is possible, although only for extreme melt compositions that have large crystallization temperature ranges (i.e. hyperagpaitic melts). The Fischer–Tropsch synthesis of hydrocarbons in igneous rocks seems to be a more applicable model for a wide variety of igneous rocks.

The origin of abiogenic hydrocarbons in rocks has recently attracted considerable research interest in geology and applied geology. Most of the Earth's hydrocarbons occur in sedimentary rocks and have been produced from biogenic material during burial and diagenesis (e.g. Schidlowski 1982; Belokon *et al.* 1995). Hydrocarbons in fluid inclusions from metamorphic rocks are also well documented and are thought to have been produced abiogenically through reactions between graphite or bitumen, present in the initial sedimentary rock, and a H_2O-bearing fluid during metamorphism (Holloway 1984; Samson & Williams-Jones 1991; Andersen & Burke 1996). In contrast, fluids associated with igneous rocks are generally CO_2 and H_2O rich (e.g. Roedder 1984; Andersen 1986; Vard & Williams-Jones 1993; Samson *et al.* 1995; Morogan & Lindblom 1995). However, the discovery of large volumes of hydrocarbons in the alkaline intrusions of Khibina and Lovozero of the Kola Peninsula, NW Russia, and Ilímaussaq in Greenland in the late 1950s (Petersilie *et al.* 1961; Petersilie 1962; Petersilie & Sørensen 1970) produced much interest in possible natural gas resources in igneous rocks. These findings led to the proposition of the existence of potentially large natural gas reservoirs produced abiogenically by streaming of hydrocarbons directly from the mantle (Porfir'ev 1974; Gold 1979; Giardini *et al.* 1982) that ended in the ill-fated Graveberg-1 well drilled into the Siljan Ring Complex, Sweden (Jeffrey & Kaplan 1988; Kerr 1990). The recognition in recent years of the potential for the presence of large hydrocarbon reservoirs in igneous rocks has led to the discovery of many oil–gas fields in igneous and metamorphic basement rocks as well as hydrocarbon occurrences on a smaller scale. The majority of these hydrocarbons are biogenic in origin, having infiltrated through fractures from the surrounding sedimentary source rocks into the igneous assemblage or incorporated into hydrothermal springs associated with igneous terrains (Des Marais *et al.* 1981; Welhan & Lupton 1987; Simoneit 1988; Gize & McDonald 1993; Darling *et al.* 1995; Darling 1998). However, over the last 20 years there have also been many reports emerging on abiogenic hydrocarbons discovered in a variety of igneous rock

From: PETFORD, N. & MCCAFFREY, K. J. W. (eds) 2003. *Hydrocarbons in Crystalline Rocks*. Geological Society, London, Special Publications, **214**, 151–173. 0305-8719/03/$15

types, from ultrabasic and basic rocks (Jeffrey & Kaplan 1988; Abrajano *et al.* 1988, 1990; Larsen *et al.* 1992; Sherwood-Lollar *et al.* 1993; Sugisaki & Mimura 1994; Kelley 1996; Kelley & Früh-Green 2001), to alkaline rocks (Gerlach 1980; Konnerup-Madsen *et al.* 1985; Jeffrey & Kaplan 1988; Ting *et al.* 1994; Salvi & William-Jones 1997; Potter *et al.* 1998), mantle xenoliths (Mathez 1987; Krot *et al.* 1994) and hydrothermal gas plumes (Welhan & Craig 1983; Botz *et al.* 1996; Charlou *et al.* 1998).

The proliferation of reports on hydrocarbons in igneous rocks indicates that they may be more widespread than previously thought. However, the origin of these hydrocarbons is still somewhat controversial, as is the mechanism which generated them. This paper reviews the occurrence and composition of abiogenic hydrocarbons in igneous rocks and will summarize and discuss the models that have been proposed for the origin of these hydrocarbons. Critical evaluation of these models could lead to the possibility of predicting the presence of abiogenic hydrocarbons in igneous rocks and their potential as a natural gas resource in the future.

Abiogenic hydrocarbon occurrences in igneous rocks

The origin of abiogenic hydrocarbons in igneous rocks has been of interest to Russian scientists since the beginning of the 20th century. However, little information on this phenomenon has

Table 1. *A summary of reports on abiogenic hydrocarbon occurrences in a variety of igneous rocks*

Reference	Rock type	Locality	Interpretation of source
Petersilie *et al.* 1961	Alkaline ne-syenite	Kola Peninsula	magmatic
Petersilie 1962	Alkaline ne-syenite	Kola Peninsula	magmatic
Zakrzhevskaya 1964	Alkaline ne-syenite	Kola Peninsula	magmatic
Karzhavin & Vendillo 1970	Alkaline ne-syenite	Kola & Siberia	magmatic
Gerlach 1980	Alkaline ne-syenite	Kola Peninsula	late magmatic
Kogarko *et al.* 1987	Alkaline ne-syenite	Kola Peninsula	late magmatic
Ikorski 1991	Alkaline ne-syenite	Kola & Siberia	abiogenic
Voytov 1992	Alkaline ne-syenite	Kola Peninsula	abiogenic
Ikorski *et al.* 1993	Alkaline ne-syenite	Kola Peninsula	late magmatic
Nivin *et al.* 1995	Alkaline ne-syenite	Kola Peninsula	post magmatic
Potter *et al.* 1998	Alkaline ne-syenite	Kola Peninsula	post magmatic
Petersilie & Sørensen 1970	Alkaline ne-syenite	Kola & Greenland	magmatic
Konnerup-Madsen & Rose-Hansen 1982	Alkaline ne-syenite	Ilímaussaq, Greenland	late magmatic
Konnerup-Madsen *et al.* 1985	Alkaline ne-syenite	Ilímaussaq, Greenland	late magmatic
Larsen *et al.* 1992	Gabbro	Skaergaard, Greenland	magmatic
Sherwood-Lollar *et al.* 1993	Basic shield rocks	Canada & Finland	post magmatic
Sherwood-Lollar *et al.* 2002	Basic shield rocks	Kidd Creek, Canada	post magmatic
Jeffrey & Kaplan 1988	Granite & dolerite	Siljan, Finland	post magmatic
Salvi & Williams-Jones 1992	Alkaline granite	Strange Lake, Quebec	post magmatic
Salvi & Williams-Jones 1997	Alkaline granite	Strange Lake, Quebec	post magmatic
Welhan & Craig 1983	Hydrothermal	21°N EPR	post magmatic
Neal & Stanger 1983	Ophiolite	Oman	post magmatic
Mathez 1987	Mantle xenolith	Hawaii	abiogenic
Abrajano *et al.* 1988	Ophiolite	Zambales, Philippines	post magmatic
Abrajano *et al.* 1990	Ophiolite	Zambales, Philippines	post magmatic
Sugisaki & Mimura 1994	Basic	50 localities	abiogenic
Krot *et al.* 1994	Mantle garnet	Mir, Siberia	magmatic
Kharmalov *et al.* 1981	Carbonatite	Kovdor, Kola	magmatic
Ting *et al.* 1994	Carbonatite	Sukulu, Uganda	abiogenic
Gerlach 1980	Carbonatite	Nyiragongo	late magmatic
Botz *et al.* 1996	Hydrothermal	Milos, Greece	post magmatic
Kelley 1996	Basalt	SWIR	post magmatic
Kelley & Früh-Green 2001	Basalt	SWIR	late magmatic
Charlou *et al.* 1998	Hydrothermal	MAR	post magmatic

EPR, East Pacific Ridge; SWIR, South-West Indian Ridge; MAR, Mid-Atlantic Ridge. *Note*: Where 'abiogenic' is listed no further interpretation of the origin was reported. Reports of hydrocarbon occurrences in volcanic rocks, where the hydrocarbons have migrated in from external reservoirs, are not included here.

been accessible outside Russia. Therefore, the presence of abiogenically formed hydrocarbons in crystalline rocks has generally not been noted by western scientists.

A summary of recent reports on hydrocarbons in igneous rocks is shown in Table 1, with their interpreted origin. The review by Porfir'ev (1974) listed many early reports of hydrocarbons found in crystalline rocks including many intrusive rocks in Siberia (i.e. Anabar, Timan, Volga, Kamchatka) as well as large hydrocarbon accumulations in the shield rocks of California, the Urals, Ukraine, southern Norway, Arizona and Nevada. He argued that most of these hydrocarbons could be inorganic in origin and that they could come directly from the upper mantle through drainage into deep-seated faults. Petersilie *et al.* (1961), Petersilie (1962), Zakrzhevskaya (1964) and Petersilie & Sørensen (1970) described large volumes of hydrocarbons, (up to 168 cm^3 of CH_4/kg of rock), found in the alkaline intrusions of Khibina and Lovozero on the Kola Peninsula and the Ilímaussaq intrusion, Greenland. They interpreted them as magmatic in origin (Table 1). Thermodynamic calculations demonstrated that a CH_4-rich fluid could be stable at magmatic conditions (Karzhavin & Vendillo 1970). Further hydrocarbon occurrences were noted in the alkaline intrusions, Kiyar-Shaltyr and Sredniy-Tatar (Transangaara) in Siberia (Karzhavin & Vendillo 1970; Ikorski 1991). Later investigations of the hydrocarbon inclusions in Khibina, Lovozero and Ilímaussaq linked the trapping of hydrocarbon-bearing inclusions to late- or post-magmatic stages below 600 °C, between 0.5–1.5 kbar (Gerlach 1980; Konnerup-Madsen *et al.* 1985; Kogarko *et al.* 1987; Konnerup-Madsen 1988; Ikorski *et al.* 1993; Nivin *et al.* 1995; Potter *et al.* 1998) (Table 1). Hydrocarbon occurrences in alkaline igneous rocks have been extended to include peralkaline granites and carbonatites with reports of hydrocarbons in the Strange Lake granitic complex, Quebec (Salvi & Williams-Jones 1992; Salvi & Williams-Jones 1997), the Sukulu carbonatite, Uganda (Ting *et al.* 1994) and Nyiragongo crater, Tanzania (Gerlach 1980) (Table 1). The Strange Lake granite contains mixed aqueous–carbonic fluid inclusions that are composed mainly of CH_4, H_2, C_2H_6, CO_2, H_2O and NaCl with smaller amounts of hydrocarbons up to C_6 (Salvi & Williams-Jones 1997). The Sukulu carbonatite contains CO_2, H_2O and CH_4-bearing inclusions. These CH_4-bearing inclusions were also been interpreted as abiogenic in origin and linked to late and post-magmatic processes within the intrusions (Ting *et al.* 1994).

Mathez (1987) described the presence of carbonaceous matter in mantle xenoliths and ultra-basic magmatic cumulates, occurring as films on crack surfaces, and on walls of fluid inclusions. These films were thought to be condensates from volcanic gases during cooling of the host rock. Complex hydrocarbon inclusions were also found in garnet phenocrysts in the Mir kimberlite, Siberia (Krot *et al.* 1994) and are thought to be magmatic in origin, stable at the high pressures where the kimberlite magma formed (~10 GPa) and preserved due to the rapid transportation of the kimberlite to the surface.

The presence of hydrocarbons in basic and ultrabasic rocks was noted as early as 1902 by Mendelyev (see review by Porfir'ev 1974). Neal & Stanger (1983) noted the presence of up to 4.1 vol% CH_4 and 99 vol% H_2 in the gas phase from spring water emerging from the Oman ophiolitic suite. The generation of CH_4 and H_2 was linked with the serpentinization of these basic rocks. This process was also noted in the Zambales ophiolite in the Philippines by Abrajano *et al.* (1990) where CH_4 and H_2 were the dominant gases in seeps coming from the partially serpentinized body. Other authors have reported the presence of hydrocarbons and H_2 in serpentinized basic rocks from carbonic fluid inclusions in mid-ocean ridge basalt (Kelley 1996), in the basic rocks of the Canadian and Fennoscandian shields (Sherwood-Lollar *et al.* 1993, 2002), in dolerite dykes in the Siljan ring complex (Jeffrey & Kaplan 1988), in hydrothermal vents releasing up to 1.45 cm^3/kg CH_4 and 38 cm^3/kg H_2 at the East Pacific Ridge (Welhan & Craig 1983), and intense CH_4 plumes at the Mid-Atlantic ridge (Charlou *et al.* 1991, 1998). Abundant H_2 was observed in gases in ventilation shafts in the highly serpentinized dunites and peridotites of the Kempirsay intrusion, Siberia (Devirts *et al.* 1993). Larsen *et al.* (1992) described the presence of primary CH_4–H_2O–NaCl fluid inclusions in gabbroic pegmatites in the Skaergaard intrusion, Greenland, and interpreted the CH_4 to have evolved from a magmatic C–O–H fluid during the last stages of crystallization between 655–770 °C at low oxygen fugacities. Similar early CO_2–CH_4–NaCl fluids were found in basalts from the Mid-Atlantic Ridge, implying a similar magmatic origin (Kelley & Früh-Green 2001). Sugisaki & Mimura (1994) analysed hydrocarbons in a range of basic rocks from 50 localities. They proposed three possible origins for these hydrocarbons: (1) they were synthesized by Fischer–Tropsch reactions in the mantle from CO_2 and CO, or (2) they are primeval in origin, delivered by meteorites during the accretion of the Earth and preserved in the mantle, or (3) they are derived from recycled, subducted biogenic hydrocarbon material.

The nature of hydrocarbon-bearing fluid inclusions in silica-undersaturated alkaline igneous complexes

Khibina and Lovozero complexes, Kola

A more detailed description of the fluid inclusions found in these complexes can be found in Potter *et al.* (1998) or Potter (2000). In summary, the most abundant fluid inclusions in both complexes are hydrocarbon-rich inclusions consisting predominantly of a low-density vapour ($0.16\,gcm^{-3}$). These occur in healed fractures and cleavages within the host minerals (nepheline, apatite, eudialyte, sodalite). Other, less common, associated fluid inclusions consist of a low-salinity, aqueous fluid and rare mixed hydrocarbon–aqueous inclusions. These occur within the same fracture planes as the hydrocarbon-rich inclusions. This implies that the fluids were coeval, but immiscible. The inclusions are secondary in nature and are commonly observed to be present in rock samples that (1) contain titano-magnetite grains showing reaction rims of biotite, aegirine and pure magnetite; (2) contain arfvedsonite grains partially replaced by aegirine; (3) contain hydrated Na/K silicates; or (4) occur as inclusions attached to aegirine microlites (Potter *et al.* 1999).

Microthermometric data from the hydrocarbon-rich inclusions show that the majority of the inclusions are composed of pure CH_4 (critical homogenization temperature of $CH_4 \sim -82\,^\circ C$). Anomalously low CH_4 homogenization temperatures (down to $-119\,^\circ C$) indicate the presence of H_2, confirmed by laser-Raman analysis. Higher CH_4 homogenization temperatures (up to $-25\,^\circ C$) indicate the presence of higher hydrocarbons (up to 40 mol% C_2H_6), confirmed by laser-Raman analysis (Potter 2000). The total homogenization temperatures and decrepitation temperatures of the aqueous, hydrocarbon-rich and mixed hydrocarbon–aqueous inclusions occur near the CH_4–H_2O solvus at 350 °C (Zhang & Frantz 1992). PVTX modelling of these fluids (i.e. calculating the volumetric and compositional properties of these fluids) infers trapping pressures and temperatures of around 0.5–1.5 kbars (equivalent to 1.5–5 km depth assuming lithostatic pressure) and 350 °C (Potter *et al.* 1998).

The Ilímaussaq complex, Greenland

A more detailed description of the fluid inclusions in the Ilímaussaq complex can be found in Konnerup-Madsen *et al.* (1981), Konnerup-Madsen & Rose-Hansen (1982) and Konnerup-Madsen *et al.* (1985). In summary, hydrocarbon-rich gaseous inclusions predominate in all examined minerals from the Ilímaussaq nepheline syenites and in hydrothermal veins considered to have formed from fluids expelled from the late nepheline syenites of the intrusion. Aqueous inclusions are only present in very limited numbers. Mixed hydrocarbon–aqueous inclusions are only very rarely observed in minerals from the nepheline syenites and, in general, there does not appear to be any connection between the hydrocarbon-rich and the aqueous fluids. The few aqueous inclusions observed in minerals from the nepheline syenites are isolated high-salinity inclusions, suggesting that they were entrapped prior to the hydrocarbon inclusions that are largely confined to more or less effectively healed fractures. A minor number of the hydrocarbon gaseous inclusions, however, occur in isolation, are occasionally associated with aegirine microlites in nepheline, and may be of a more primary nature. In the hydrothermal vein minerals, the commonly observed association of highly saline aqueous inclusions and hydrocarbon-rich inclusions indicates the simultaneous entrapment of non-miscible fluids at this stage.

Microthermometric data on the hydrocarbon-rich inclusions show the earliest entrapped fluids to be composed of pure CH_4 in the nepheline syenites, whereas in the late hydrothermal veins higher contents of e.g. ethane (Th_{CH_4} at -60° to $-55\,^\circ C$) are indicated, confirmed by laser-Raman analyses. Homogenization to liquid of the earliest entrapped hydrocarbon-rich inclusions range from -90° to $-82\,^\circ C$ in minerals from the nepheline syenites, and from -110° to $-90\,^\circ C$ in the late hydrothermal veins. PVTX modelling infers trapping pressures around 1–2 kbars, equivalent to 3–6 km depth, assuming lithostatic pressure and temperatures of 400° to 600 °C for these fluids (Konnerup-Madsen 2001).

Chemistry of abiogenic hydrocarbons in igneous rocks

Bulk gas compositions

Bulk gas compositions of hydrocarbon-bearing fluids in the alkaline complexes Khibina,

Lovozero, Ilímaussaq and Strange Lake have been obtained by mechanical crushing of the rock samples in a vacuum ball mill with the subsequent released gases transported into a chromatograph. Tables 2 and 3 summarize the compositions of the gases in these alkaline rocks along with some borehole gases collected from pressurized pockets within the Canadian and Fennoscandian shield rocks. The majority of the analyses are dominated by CH_4 (80–90 vol%), with higher hydrocarbons decreasing exponentially with increasing carbon number. Hydrocarbons up to C_5 have been detected (Petersilie 1962; Petersilie & Sørensen 1970; Konnerup-Madsen & Rose-Hansen 1982; Voytov 1992; Salvi & Williams-Jones 1997). Methane concentrations can reach as high as 168.7 cm^3/kg (Petersilie 1962). Carbon dioxide is either absent or present in small concentrations (up to 4 vol%) (Konnerup-Madsen *et al.* 1979). In the Lovozero samples, H_2 is relatively abundant as the second dominant species after CH_4. This H_2 is present as a free gas in fluid inclusions. Laser-Raman investigations determined that the hydrocarbon-bearing inclusions can have compositions up to 65 mol% CH_4 and 35 mol% H_2 (Potter 2000).

Notable exceptions are analyses 11 and 60 of altered samples that contain distinctly lower CH_4 concentrations and a significant increase in CO_2 in comparison to their unaltered equivalents. This has been attributed to the oxidation of the CH_4 fluid to CO_2 during low-temperature weathering (<150 °C) of the minerals to clays (Petersilie 1962). Other exceptions are analyses 3, 8, 31, 39, 40 and 50. These monchiquite dykes, alkali effusives and early olivine gabbros are dominated by H_2O fluids with low concentrations of gaseous volatiles consisting of CO_2 and CO very different to the fluid inclusion assemblages observed in the other samples. The high H_2 concentrations detected probably came from dissociation of H_2O during crushing, as no evidence from the fluid inclusions suggests that free H_2 gas is present in these samples. Characterization of fluid inclusions in samples 39 and 40 revealed that they contained H_2O-dominant inclusions (Potter 2000). The analyses revealed low gaseous concentrations (<7 cm^3/kg) dominated by H_2 and CO_2 that may have come from dissociation of H_2O and decarbonation of carbonate minerals in these samples (i.e. cancrinite and calcite). Analyses from gases from the Canadian and Fennoscandian shield rocks also show high percentage volumes of N_2 and He in contrast to the gas analyses from the alkaline igneous rocks (Table 2).

Stable isotope characteristics

Table 4 shows a summary of the stable isotope results for a number of reports on hydrocarbon-bearing fluids in a variety of igneous terrains. Figure 1 shows the collated $\delta^{13}C_{CH_4}$ and δD_{CH_4} data for a variety of hydrocarbon-bearing fluids in igneous rocks, plotted alongside well-established source fields. The $\delta^{13}C_{CH_4}$ results show a large range in values from −3.2‰ to −44.9‰. The majority of examples, however, are in the range of −20‰ to −28‰ (Fig. 1; Table 4). This range of $\delta^{13}C_{CH_4}$ results would indicate that the hydrocarbon-bearing fluid is abiogenic in origin (biogenic signatures tend to be more depleted for CH_4). However, the results do not fall within the mantle field (−3‰ to −9‰), so a direct mantle origin is unlikely. The $\delta^{13}C_{CO_2}$ results for associated CO_2 fluids range from 0.6‰ to 7.1‰. The CO_2 analysed by Botz *et al.* (1996) was interpreted as originating from marine carbonates ($\delta^{13}C_{CO_2} \sim 0$‰). The $\delta^{13}C_{CO_2}$ results of around −7‰ for the CO_2 fluids in the Khibina complex (Potter 2000) and the East Pacific Rise (Welhan & Craig 1983) fall in the mantle field and are interpreted as magmatic in origin. In the Ilímaussaq intrusion, the $\delta^{13}C_{CH_4}$ results show a more restricted range in values from −1.0‰ to −7.0‰. The calculated $\delta^{13}C_{CH_4}$ of the bulk carbon of the hydrocarbon fluids is −4.5‰ ± 1.5‰, interpreted to be a clear magmatic/juvenile signature (Konnerup-Madsen 2001).

The δD_{CH_4} results tend to fall around −110‰ to −135‰ with the exception of gases from the Canadian Shield showing depleted values down to below −400‰. These values plot in the fumaroles and hot springs or juvenile fields on the $\delta^{13}C_{CH_4}$–δD_{CH_4} plot (Fig. 1). The majority of δD_{H_2} signatures are extremely depleted with values below −600‰. Such δD_{H_2} values have been noted in H_2 produced by serpentinization processes in basic rocks (Devirts *et al.* 1993).

The origin of hydrocarbons in igneous rocks

Two possible sources for hydrocarbons in igneous rocks can be envisaged: an organic (biogenic) origin, or an abiogenic origin. Three processes can generally be distinguished for organically derived material: bacterially generated hydrocarbons by either (1) CO_2 reduction or (2) fermentation or as (3) thermogenically derived hydrocarbons. In igneous rocks, biogenic hydrocarbons are generally all thermogenic in nature as the organic matter has been subjected

Table 2. *Bulk gas analyses of abiogenic hydrocarbons in igneous rocks, in vol%, from the literature*

Analysis #	Sample	CH_4 vol%	C_2H_6 vol%	C_3H_8 vol%	C_4H_{10}-I vol%	C_4H_{10}-*n* vol%	C_5H_{10}-I vol%	C_5H_{10}-*n* vol%	HHC vol%	H_2 vol%	CO vol%	CO_2 vol%	N_2 vol%	He vol%	O_2 vol%
Khibina															
1	Khibinite	90.00	—	—	—	—	—	—	2.15	1.77	1.32	0.19	4.59	—	—
2	Ijolite	89.70	—	—	—	—	—	—	2.53	3.45	4.13	0.19	—	—	—
5	Ne syenite	93.42	1.08	0.24	0.0500		—	—	—	0.91	2.30	0.00	2.20	—	—
7	Rischorrite	60.10	4.86	1.19	0.1300	0.3600	—	—	—	5.96	9.34	1.85	16.21	—	—
8	Monchiquite	17.45	0.00	0.00	0.0000	0.0000	—	—	—	30.95	51.60	0.00	—	—	—
9	Malignite	70.30	1.18	0.51	—	—	—	—	—	10.34	15.10	2.57	—	—	—
10	Ijolite	71.96	2.78	0.73	0.0500	0.1500	0.07	0.04	—	1.99	6.77	0.67	14.52	—	—
11	Altered Ijolite	2.06	0.00	0.00	0.0000	0.0000	—	—	—	10.54	16.40	71.00	—	—	—
12	Yukspor mine	79.50	—	—	—	—	—	—	1.90	13.90	—	0.00	4.30	0.00	—
13	Yukspor mine	24.06	—	—	—	—	—	—	0.65	2.40	—	1.90	70.30	0.69	—
14	Yukspor mine	56.00	—	—	—	—	—	—	2.17	2.88	—	2.88	35.98	0.09	—
15	Kukisvumchorr mine	42.09	—	—	—	—	—	—	3.78	1.68	—	0.00	52.31	0.15	—
16	Rasvumchorr mine	91.59	—	—	—	—	—	—	4.68	1.40	—	0.00	1.58	0.54	—
17	Rasvumchorr mine	57.94	—	—	—	—	—	—	4.61	10.04	—	0.00	27.26	0.15	—
12a	Yukspor mine	97.64	0.76	1.52	—	0.0400	0.00	—	—	—	—	—	—	—	—
13a	Yukspor mine	96.10	3.19	0.43	—	0.2800	0.00	—	—	—	—	—	—	—	—
14a	Yukspor mine	96.00	3.32	0.48	—	0.2000	0.00	—	—	—	—	—	—	—	—
15a	Kukisvumchorr mine	91.80	6.28	1.55	0.1100	0.1700	0.08	—	—	—	—	—	—	—	—
16a	Rasvumchorr mine	94.88	4.36	0.58	0.0300	0.1100	0.03	—	—	—	—	—	—	—	—
17a	Rasvumchorr mine	92.58	6.56	0.52	0.0250	0.0270	0.00	—	—	—	—	—	—	—	—
26	Khibinite	95.69	2.93	0.15	0.0005	0.0020	—	—	—	1.16	0.00	0.03	—	0.04	—
27	Khibinite	95.55	3.15	0.16	0.0002	0.0010	—	—	—	1.08	0.00	0.02	—	0.03	—
28	Urtite	96.75	1.93	0.05	0.0060	0.0200	—	—	—	1.16	0.00	0.00	—	0.08	—
29	Rischorrite	90.04	3.59	0.17	0.0040	0.0080	—	—	—	4.02	1.97	0.17	—	0.03	—
30	Foyaite	92.84	1.63	0.13	0.0000	0.0030	—	—	—	5.37	0.00	0.34	—	0.00	—
31	Monchiquite	38.09	0.00	0.00	0.0000	0.0000	—	—	—	61.91	0.00	0.00	—	0.00	—
35	Ap-ne ore	83.79	2.08	—	—	—	—	—	—	7.95	—	0.00	6.16	0.00	—
36	Urtite	95.27	0.00	—	—	—	—	—	—	1.48	—	0.00	3.08	0.18	—
37	Ijolite	89.70	3.28	—	—	—	—	—	—	4.56	—	0.00	2.26	0.19	—
38	Rischorrite	81.17	4.04	—	—	—	—	—	—	8.97	—	0.00	5.83	0.01	—
39	Foyaite	1.69	0.03	—	—	—	—	—	—	27.36	—	28.38	42.57	0.10	—
40	Carbonatite	5.16	3.69	—	—	—	—	—	—	63.27	—	18.43	9.44	0.02	—

Lovozero															
3	Alkali effusive	8.52	—	—	—	—	—	—	0.12	65.32	25.71	0.30	—	—	—
4	Ne syenite	78.55	—	—	—	—	—	—	6.25	10.84	4.20	0.15	—	—	—
6	Juvite	81.20	5.41	0.44	0.26	—	—	—	—	8.34	4.35	0.00	0.00	—	—
18	Urtite	59.90	4.40	—	—	—	—	—	0.30	20.10	—	—	13.90	0.10	1.30
19	Foyaite	76.10	4.20	—	—	—	—	—	0.20	4.80	—	—	12.30	0.03	2.20
20	Foyaite	71.10	2.90	—	—	—	—	—	0.20	8.60	—	—	14.40	0.02	2.80
21	Urtite	64.70	5.80	—	—	—	—	—	0.00	16.00	—	—	9.60	1.90	2.00
22	Urtite	53.80	2.30	—	—	—	—	—	0.00	39.20	—	—	3.00	0.90	0.80
23	Urtite	15.70	0.50	—	—	—	—	—	0.00	34.40	—	—	37.50	0.60	11.30
24	Foyaite	55.70	2.60	—	—	—	—	—	0.00	35.00	—	—	4.80	0.70	1.20
25	Foyaite	60.70	2.50	—	—	—	—	—	0.00	29.00	—	—	5.90	0.80	1.10
32	Syenite	79.47	4.10	0.85	0.01	0.05	—	—	—	—	—	1.50	0.33	0.04	—
33	Urtite	84.36	5.45	3.03	0.00	0.00	—	—	—	—	—	0.00	0.41	0.00	—
34	Foyaite	91.36	2.77	1.29	0.00	0.24	—	—	—	—	—	0.00	0.12	0.00	—
41	Eud lujavrite	87.17	2.86	—	—	—	—	—	—	5.13	—	0.00	4.84	0.00	—
42	Lop juvite	62.68	4.40	—	—	—	—	—	—	25.90	—	0.00	6.48	0.05	—
43	Lujavrite	65.30	6.91	—	—	—	—	—	—	24.66	—	0.00	3.12	0.02	—
44	Foyaite	89.43	1.92	—	—	—	—	—	—	0.04	—	0.00	4.51	0.00	—
45	Urtite	62.72	6.80	—	—	—	—	—	—	27.76	—	0.00	2.69	0.02	—
Ilímaussaq															
46	Augite syenite	83.72	2.50	0.23	0.04	0.04	—	—	—	12.59	0.00	0.01	—	0.81	—
47	Naujaite	80.82	9.90	1.74	0.12	0.31	—	—	—	5.04	0.00	0.04	—	1.97	—
48	Sod foyaite	70.95	7.37	1.34	0.11	0.37	—	—	—	12.64	6.93	0.06	—	0.16	—
49	Arf lujavrite	24.44	3.15	0.21	0.01	0.00	—	—	—	68.44	0.00	0.02	—	3.66	—
50	Olivine gabbro	5.30	0.05	0.01	0.00	0.00	—	—	—	93.11	0.00	0.00	—	1.53	—
51	Sodalite and nepheline	73.40	8.60	1.30	0.20	0.40	0.10	0.05	—	8.30	—	1.20	6.00	0.20	—
52	Arfvedsonite	66.30	6.20	0.80	0.10	0.30	0.03	0.01	—	18.80	—	0.70	5.10	—	—
53	Chlakovite	92.10	2.10	—	—	—	—	—	—	5.80	—	—	—	—	—
54	Sodalite	75.40	9.10	1.90	—	0.60	—	—	—	4.20	—	0.40	8.10	—	—
55	Sodalite	70.00	7.30	1.10	—	0.28	—	0.07	—	3.50	—	4.00	13.00	—	—
56	Arfvedsonite	57.00	5.60	0.71	—	0.17	—	0.02	—	3.00	—	0.50	10.00	—	—
57	Nepheline	80.00	8.50	1.10	—	0.26	—	0.03	—	6.20	—	0.40	3.60	—	—
58	Eudialyte	43.00	13.00	2.00	—	0.49	—	0.06	—	34.00	—	0.40	4.00	—	—
Strange Lake															
59	Fresh pegmatite	68.66	5.78	1.04	—	0.35	—	0.12	—	19.09	—	3.76	1.92	—	—
60	Altered pegmatite	36.70	3.58	0.49	—	0.20	—	0.07	—	19.23	—	48.35	6.27	—	—

Table 2. *(continued)*

Analysis #	Sample	CH_4 vol%	C_2H_6 vol%	C_3H_8 vol%	C_4H_{10}-I vol%	C_4H_{10}-*n* vol%	C_5H_{10}-I vol%	C_5H_{10}-*n* vol%	HHC vol%	H_2 vol%	CO vol%	CO_2 vol%	N_2 vol%	He vol%	O_2 vol%
Canadian Shield															
61	Sudbury N	12.55	0.07	0.02	—	<0.02	—	—	—	—	—	1.44	79.78	6.12	—
62	Sudbury CCS	74.00	4.21	0.31	—	0.13	—	—	—	7.13	—	0.21	8.77	5.30	—
63	Elliot Lake	54.20	0.11	0.02	—	0.03	—	—	—	—	—	0.11	26.40	19.10	—
64	Timmins	78.55	4.67	0.71	—	0.23	—	—	—	—	—	0.10	14.15	1.59	—
65	Red Lake – Dickenson	41.37	1.30	0.73	—	0.06	—	—	—	—	—	1.11	51.97	3.44	—
66	Red Lake – Campbell	56.35	2.20	0.99	—	<0.01	—	—	—	0.28	—	0.32	31.98	7.87	—
Fennoscandian Shield															
67	Juuka	78.90	0.79		<0.01	—	<0.01	—	—	—	12.80	—	0.04	6.32	1.09
—															

1–4 are analyses from Kogarko *et al.* (1987). HHC = C_2–C_5 or C_{3-5} where individual C*n* were not reported. 5–6 are from Petersilie (1962). 7–11 from Petersilie *et al.* (1961). 12–17 are free gas analyses from gas jets in the Khibina mines taken from Voytov (1992). The vol% of the individual hydrocarbons present are displayed in 12a–17a. 18–20 are occluded gas analyses, 21–25 are free gas analyses from Nivin *et al.* (1995). 26–34, 46–50 are from Petersilie & Sørensen (1970). 35–45 are from Potter (2000). 51–53 are from Konnerup-Madsen & Rose-Hansen (1982). 54–58 are from Konnerup-Madsen *et al.* (1979). 59–60 are from Salvi & Williams-Jones (1997). 61–67 are average compositions from Sherwood-Lollar *et al.* (1993). *Note*: the majority of reports of hydrocarbons do not report bulk gas contents.

Table 3. *Bulk gas analyses of abiogenic hydrocarbons in igneous rocks, in cm^3/kg, from the literature*

Analysis #	Sample	Total gas cm^3/kg	CH_4 cm^3/kg	C_2H_6 cm^3/kg	C_3H_8 cm^3/kg	C_4H_{10}-I cm^3/kg	C_4H_{10}-n cm^3/kg	C_5H_{10}-I cm^3/kg	C_5H_{10}-n cm^3/kg	HHC cm^3/kg	H_2 cm^3/kg	CO cm^3/kg	CO_2 cm^3/kg	N_2 + inert cm^3/kg
Khibina														
1	Khibinite	41.78	37.60	—	—	—	—	—	—	0.89	0.74	0.55	0.08	1.92
2	Ijolite	33.90	30.41	—	—	—	—	—	—	0.86	1.17	1.40	0.06	—
5	Ne syenite	176.66	168.70	1.910	0.400	0.1000	—	—	—	—	1.50	4.05	0.00	—
7	Rischorrite	20.97	12.59	1.020	0.250	0.0380	0.0760	—	—	—	1.25	1.96	0.39	3.40
8	Monchiquite	25.00	3.49	0.000	0.000	0.0000	0.0000	—	—	—	6.19	15.32	0.00	—
9	Malignite	19.33	13.57	0.230	0.010	—	—	—	—	—	2.00	2.93	0.59	—
10	Ijolite	33.78	24.50	0.950	0.019	0.0520	0.0270	0.0130	—	—	0.68	2.31	0.23	5.00
11	Altered Ijolite	21.34	0.44	0.000	0.000	0.0000	0.0000	—	—	—	2.25	3.50	15.15	—
26	Khibinite	54.20	51.86	1.590	0.080	0.0003	0.0010	—	—	—	0.63	0.00	0.02	0.02
27	Khibinite	44.36	42.38	1.400	0.070	0.0001	0.0005	—	—	—	0.48	0.00	0.01	0.02
28	Urtite	50.85	49.20	0.980	0.020	0.0030	0.0110	—	—	—	0.59	0.00	0.00	0.04
29	Rischorrite	11.71	10.54	0.420	0.020	0.0004	0.0009	—	—	—	0.47	0.23	0.02	—
30	Foyaite	11.46	10.64	0.190	0.020	—	0.0004	—	—	—	0.62	0.00	0.04	—
31	Monchiquite	0.84	0.32	—	—	—	0.0000	—	—	—	0.52	0.00	0.00	—
35	Ap-ne ore	139.96	11.70	0.290	—	—	—	—	—	—	1.11	—	0.00	0.86
36	Urtite	113.11	107.76	0.000	—	—	—	—	—	—	1.67	—	0.00	3.45
37	Ijolite	45.15	40.50	1.480	—	—	—	—	—	—	2.06	—	0.00	1.11
38	Rischorrite	22.30	18.10	0.900	—	—	—	—	—	—	2.00	—	0.00	1.30
39	Foyaite	2.96	0.05	0.000	—	—	—	—	—	—	0.81	—	0.84	1.26
40	Carbonatite	6.78	0.35	0.250	—	—	—	—	—	—	4.29	—	1.25	0.64
Lovozero														
3	Alkali effusive	6.69	0.57	—	—	—	—	—	—	0.01	—	4.37	0.02	—
4	Ne syenite	18.55	14.57	—	—	—	—	—	—	1.16	—	2.01	0.03	—
6	Juvite	57.80	46.70	3.400	0.250	0.1500	—	—	—	—	4.80	2.50	0.00	—
32	Syenite	15.37	12.21	0.630	0.130	0.0016	0.0075	—	—	—	2.10	0.23	0.05	0.01
33	Urtite	19.44	16.40	1.060	0.590	—	—	—	—	—	1.31	—	0.08	—
34	Foyaite	33.22	30.35	0.920	0.430	—	0.0800	—	—	—	1.40	—	0.04	—
41	Eud Lujavrite	23.77	20.72	0.680	—	—	—	—	—	—	1.22	—	0.00	1.15
42	Lop Juvite	8.18	5.13	0.360	—	—	—	—	—	—	2.12	—	0.00	0.57
43	Lujavrite	7.38	4.82	0.360	—	—	—	—	—	—	1.82	—	0.00	0.23
44	Foyaite	26.59	23.78	0.510	—	—	—	—	—	—	1.10	—	0.00	1.21
45	Urtite	28.24	17.70	1.920	—	—	—	—	—	—	7.84	—	0.00	0.77
Ilímaussaq														
46	Augite syenite	15.24	12.76	0.382	0.036	0.0061	0.0064	—	—	—	1.92	0.00	0.01	0.12
47	Naujaite	60.10	48.59	5.954	1.049	0.0750	0.1860	—	—	—	3.03	0.00	0.02	1.18
48	Sod Foyaite	5.75	4.08	0.424	0.077	0.0064	0.0210	—	—	—	0.73	0.40	0.00	0.01
49	Arf lujavrite	4.06	0.99	0.128	0.008	0.0004	0.0009	—	—	—	2.78	0.00	0.01	0.15
50	Olivine gabbro	2.72	0.15	0.001	0.002	0.0000	0.0000	—	—	—	2.54	0.00	0.00	0.04

1–4 are total gas contents for occluded gases from Kogarko *et al.* (1987). 5–6 are analyses from Petersilie (1962). 7–11 are analyses from Petersilie *et al.* (1961). Analyses 26–34, 46–50 are from Petersilie & Sørensen (1970). HHC = C_2–C_5. 35–45 are from Potter (2000). *Note*: Bulk gas analyses from other igneous terranes are not reported.

Table 4. *A summary of isotopic data reported for abiogenic hydrocarbon-bearing fluids from various igneous terranes*

Locality	Type	Origin	$\delta^{13}C_{CH_4}$	δD_{CH_4}	δD_{H_2}	$\delta^{13}C_{CO_2}$	Reference
Oman	Ophiolite	F–T synthesis~	—	—	−697 to −714	—	Neal & Stanger (1983)
21°N EPR	Hydrothermal	F–T synthesis~	−15 to −17.6	−102 to −126	−388 ± 15	−7 ± 0.1	Welhan & Craig (1983)
Sweden	Siljan dolerite	F–T synthesis	−16.5 to −26.3	—	—	—	Jeffrey & Kaplan (1988)
Sweden	Siljan granite	F–T synthesis	−19.4 to −36.7	—	—	—	Jeffrey & Kaplan (1988)
Philippines	Zambales ophiolite	F–T synthesis~	−6.1 to −7.5	−118 to −137	−581 to −599	—	Abrajano *et al.* (1990)
Kola Peninsula	Khibina	F–T synthesis	−3.2 to −12.8	—	—	—	Voytov (1992)
Canada & Scandinavia	Shield rocks	F–T synthesis~	−22.4 to −44.9	−133 to −372	−619 to −659	—	Sherwood-Lollar *et al.* (1993)
Canada & Scandinavia	Shield rocks	F–T synthesis~	−33.0 to −40.7	−390 to −419			Sherwood-Lollar *et al.* (2002)
Siberia	Kempirsay Ultramafite	F–T synthesis~	—	—	−744 to −766	—	Devirts *et al.* (1993)
Kola Peninsula	Lovozero	Post-magmatic	—	−132 to −164	−359 to −629	—	Nivin *et al.* (1995)
Various	Igneous Rocks	F–T synthesis	−26 to −29	—	—	—	Sugisaki & Mimura (1994)
Milos, Greece	Geothermal	F–T synthesis	−9.4 to−17.8	−104 to −377	—	−0.6 to −1.1	Botz *et al.* (1996)
Kola Peninsula	Khibina & Lovozero	F–T synthesis	−12.8 to −28.6	—	—	−7.1	Potter (2000)
Greenland	Ilímaussaq	Late-magmatic	−1 to −7	−132 to −145	—	—	Konnerup-Madsen (2001)

Abiogenic mechanisms proposed for the generation of the hydrocarbons: F–T synthesis—Fischer–Tropsch type reactions, ~ related to serpentinization of the assemblage, late-magmatic-closed system fluid respeciation. — not reported.

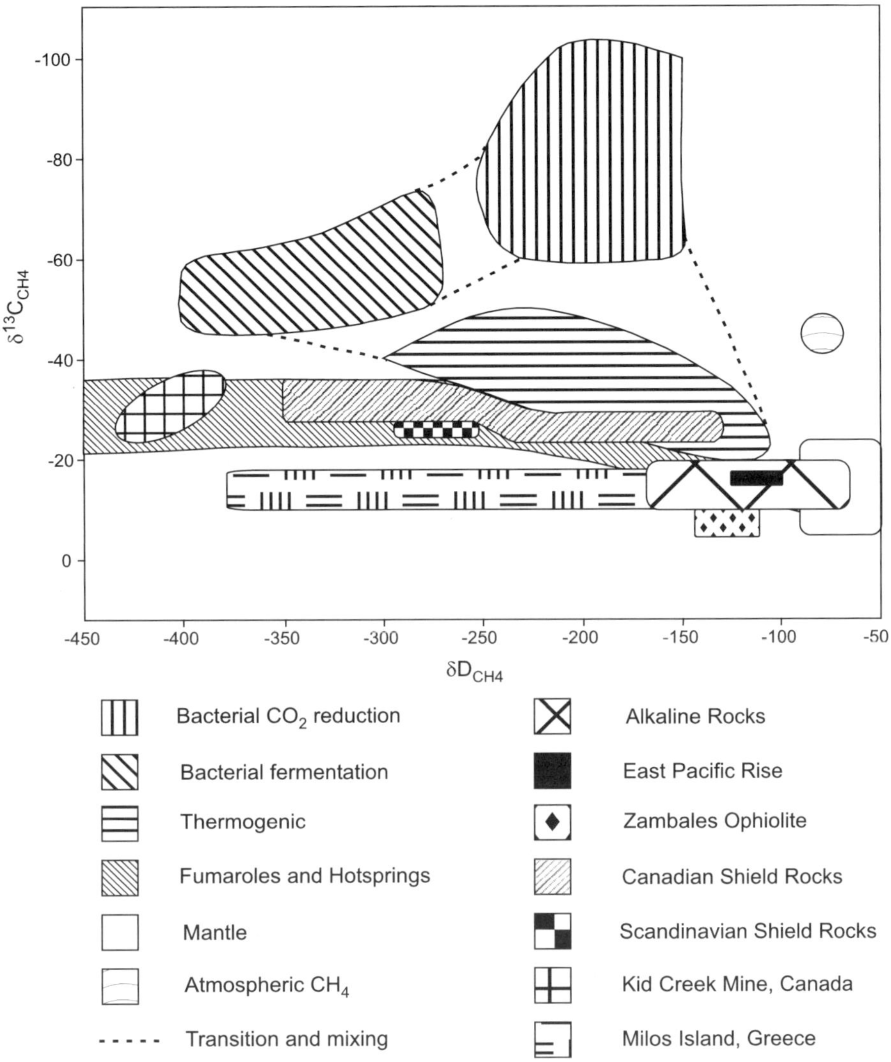

Fig. 1. A $\delta^{13}C_{CH_4}$–δD_{CH_4} diagram showing where reported abiogenically generated hydrocarbons fall in respect to a number of well-established fields for biogenically generated hydrocarbons and those directly from the mantle. These include the alkaline igneous complexes of Khibina, Lovozero and Ilímaussaq. Data have been taken from Welhan & Craig (1983), Schoell (1988), Abrajano *et al.* (1990), Voytov (1992), Sherwood-Lollar *et al.* (1993, 2002), Nivin *et al.* (1995), Botz *et al.* (1996), Potter (2000) and Konnerup-Madsen (2001).

to thermal alteration by the igneous body. The organic material may be leached out of the country rocks by late-magmatic or hydrothermal fluids and found in association with the igneous body (e.g. Welhan & Lupton 1987; Simoneit 1988; Gize & McDonald 1993) or entirely incorporated into and recycled in the magmatic hydrothermal system (Gunter & Musgrave 1971; Des Marais *et al.* 1981; Darling *et al.* 1995; Sakata *et al.* 1997; Darling 1998). Biogenic sources can be distinguished from abiogenic sources by the $\delta^{13}C$ values of the hydrocarbons. The pattern of $\delta^{13}C$ values between the individual hydrocarbon species for hydrocarbons generated by thermal breakdown of organic matter and hydrocarbons generated by abiogenic polymerization processes can be distinctly different. The resulting $\delta^{13}C_{CH_4}$ values are generally used to differentiate between the two sources (Sherwood-Lollar *et al.* 2002; Horita & Berndt 1999). Biogenic sources tend to have more depleted $\delta^{13}C_{CH_4}$ values (below −30‰), whereas more enriched $\delta^{13}C_{CH_4}$ values indicate an abiogenic origin (Schoell 1983, 1988).

There are three main theories on the abiogenic generation of hydrocarbons, which are not mutually exclusive:

(a) a direct mantle origin (e.g. Petersilie *et al.* 1961; Petersilie 1962; Petersilie & Sørensen 1970; Gold 1979) where hydrocarbons are either synthesized in the mantle via reactions with CO and CO_2, or are present as primeval hydrocarbons delivered by meteorites and preserved in the mantle since the accretion of the Earth (Sugisaki & Mimura 1994);

(b) closed system respeciation of a primary CO_2-bearing fluid during late to post-magmatic stages below 600 °C (Gerlach 1980; Konnerup-Madsen *et al.* 1985; Kogarko *et al.* 1987);

(c) generation of hydrocarbons during post-magmatic mineral–fluid reactions involving Fischer–Tropsch reactions (i.e. serpentinization) (Abrajano *et al.* 1990; Sherwood-Lollar *et al.* 1993; Kelley 1996; Salvi & Williams-Jones 1997; Charlou *et al.* 1998; Potter *et al.* 1998).

Mantle origin

The potential for exploitable abiogenic hydrocarbon reservoirs outgassing from the primeval mantle was first discussed by Gold (1979). He noted that most hydrocarbons found in igneous rocks have a uniform $\delta^{13}C$ isotopic signature between −25‰ and −28‰. Although this signature is within the geothermal field, the uniformity of these $\delta^{13}C$ signatures suggested a common large homogenous source such as the mantle. Petersilie (1962) and Zakrzhevskaya (1964) interpreted the hydrocarbons found in the alkaline intrusions of Khibina and Lovozero as having a primary magmatic origin, and suggested there may be economic accumulations of hydrocarbon gases in deep zones within these massifs. The Graveberg-1 Siljan Ring drilling project was the first well drilled in crystalline rocks. The well was set down in a meteorite impact crater where numerous oil seeps had been observed. Although one hypothesis was that the hydrocarbons observed were generated by the abrupt baking of the surrounding source sediments at the time of impact (Vlierboom *et al.* 1986), the project was funded on the basis of Gold's hypothesis that large reserves of mantle gas might be stored within the crystalline basement (Gold 1979; Kerr 1990). However, only very small concentrations of CH_4 were found in the igneous rocks and these were interpreted as postmagmatic in origin (Jeffrey & Kaplan 1988). More recent fluid inclusion studies on the hydrocarbons found in the alkaline igneous intrusions Khibina and Lovozero have suggested these fluids are secondary and were trapped at temperatures and pressures around 350 °C and 0.5–1.2 kbars, not directly from the mantle (Kogarko *et al.* 1987; Nivin *et al.* 1995; Potter *et al.* 1998). Similarly, a late- to post-magmatic origin for the hydrocarbons in the Ilímaussaq intrusion has been proposed. Therefore, it can be concluded that a direct mantle origin is unlikely.

A late magmatic origin

The majority of abiogenic hydrocarbon occurrences in crystalline rocks have been suggested to be formed by the respeciation of a carbonic fluid in the C–O–H system (Karzhavin & Vendillo 1970; Gerlach 1980; Konnerup-Madsen & Rose-Hansen 1982; Holloway 1984; Kogarko *et al.* 1987; Samson & Williams-Jones 1991; Cesare 1995; Andersen & Burke 1996). In the presence of graphite ($a_{GRAPHITE} = 1$), the relative proportions of the main volatile species CO_2, CH_4 and H_2O are controlled by the four independent reaction equilibria:

$$C + O_2 \leftrightarrow CO_2 \quad \text{(R1)}$$

$$CO + \tfrac{1}{2}O_2 \leftrightarrow CO_2 \quad \text{(R2)}$$

$$H_2 + \tfrac{1}{2}O_2 \leftrightarrow H_2O \quad \text{(R3)}$$

$$CH_4 + 2O_2 \leftrightarrow CO_2 + 2H_2O, \quad \text{(R4)}$$

plus

$$CH_4 + 2H_2O \leftrightarrow CO_2 + 4H_2. \quad \text{(R5)}$$

Studies by e.g. Holloway (1984), Huizenga (1995) and Cesare (1995) clearly demonstrated the effects of varying pressure, temperature, oxygen fugacity and activity of graphite for the speciation of a variety of fluids in the C–O–H system. Figure 2 shows schematically some important aspects of the C–O–H system related to the actual concentration of CH_4 to be observed in fluid inclusions.

Figure 2a schematically illustrates the phase relationships for the C–O–H system at some specified pressure and temperature conditions and low oxygen fugacities. Two initial fluid compositions (**a** and **b**; Fig. 2a) differ in their initial H_2O/CO_2 ratio. Re-equilibration of composition **a** (70% H_2O, 30% CO_2) to an oxygen fugacity of QFM-2 will change the initial fluid composition first away from the O apex onto the graphite saturation curve, and then along the graphite saturation curve, with associated precipitation

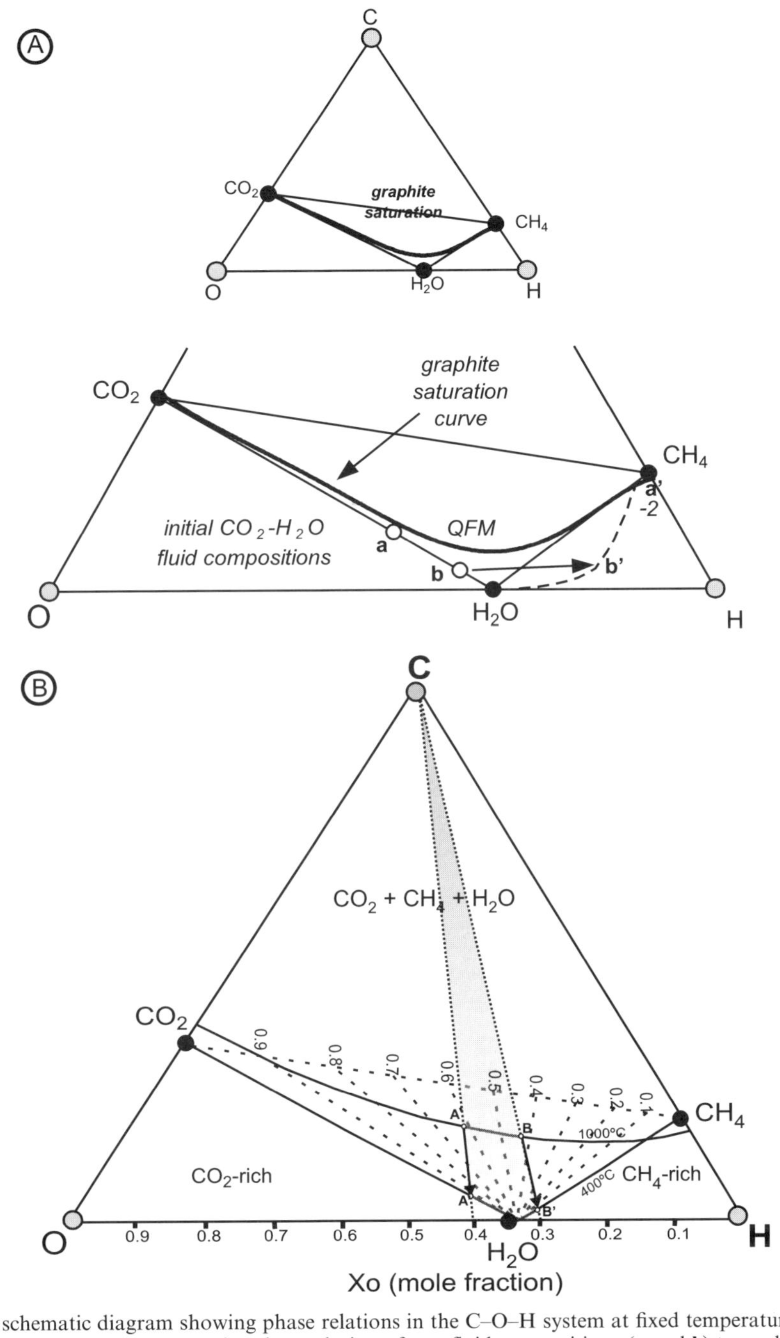

Fig. 2. (**a**) A schematic diagram showing phase relations in the C–O–H system at fixed temperatures and pressures (500 °C, 2 kbar) demonstrating the evolution of two fluid compositions (**a** and **b**) towards CH_4-rich compositions (**a**′ and **b**′) with decreasing fO_2 conditions (QFM to QFM-2). (**b**) Another schematic C–O–H diagram at fixed pressure (1 kbar) showing the down temperature evolution of C–O–H fluids at constant Xo (Xo = O/O + H) towards CO_2-rich, CH_4-rich or H_2O-rich compositions depending on initial Xo. The shaded area represents the H_2O-rich fluid region. Paths **A-A**′ and **B-B**′ represent temperature evolution of an H_2O-rich fluid with a minor CO_2-rich or CH_4-rich minor gas phase, respectively. Dashed lines represent $CO_2/CO_2 + CH_4$ ratio isopleths. See text for further discussion. Based on Konnerup-Madsen (2001), Huizenga (2001) and Cesare (1995). QFM = quartz–fayalite–magnetite buffer.

of graphite, to point **B′** (QFM-2 on the graphite saturation curve). Composition **b**, which is richer in H_2O (90% H_2O, 10% CO_2), will evolve away from the O apex along the arrow until point **b′** without the precipitation of graphite. Thus both compositions will re-equilibrate to CH_4-rich fluid compositions, composition **a** to an essentially H_2O–CH_4 fluid, and composition **b** to a H_2O–CH_4–H_2 fluid. Therefore, a CH_4-rich fluid could evolve in a closed system by re-equilibration from a primary more oxidized CO_2–CH_4–H_2O fluid under geologically reasonable fO_2 conditions below about 500 °C, between QFM and QFM-3. The CO_2/H_2O ratio of the initial fluid will determine whether or not graphite precipitates during re-equilibration.

Figure 2b also illustrates schematically how a CO_2–CH_4–H_2O mixed fluid might evolve upon cooling. Depending on the initial Xo ratio ($Xo = O/O+H$), the C–O–H fluid could evolve to a CO_2-rich fluid ($Xo > 0.4$), a CH_4-rich fluid ($Xo < 0.3$) or an H_2O-rich fluid with a minor gas phase ($Xo = 0.3$–0.4). The Xo ratio remains constant throughout. Upon cooling, the fluid will pass through the graphite saturation curve, resulting in precipitation of graphite and a change in the CO_2/CO_2+CH_4 ratio. Two initial compositions, **A** and **B**, with Xo of 0.3 and 0.4 are shown at 1000 °C. Upon cooling to 400 °C these compositions will evolve to an H_2O-rich fluid. However, the minor gas phase in **A**, at 400 °C, is represented by the intersection of the CO_2/CO_2+CH_4 isopleth with the graphite saturation curve (point **A′**). The minor gas phase would consist of 90% CO_2, whereas, at point **B′**, the final gas phase would consist of 90% CH_4.

Therefore, Figure 2 clearly demonstrates how an early C–O–H fluid could evolve to a CH_4-rich composition below 500 °C. This model can be readily applied to CH_4 generation in metamorphic terranes (Holloway 1984), but for magmatic systems unusually low solidus temperatures are required. More recent studies on the hydrocarbons present in the alkaline intrusions Khibina, Lovozero, Ilímaussaq and Strange Lake have indicated that these hydrocarbon-bearing fluids could have been trapped during the final stages of crystallization of these intrusions (Gerlach 1980; Konnerup-Madsen *et al.* 1985; Kogarko *et al.* 1987; Salvi & Williams-Jones 1992), and in particular the Ilímaussaq intrusion, which during final solidification stages evolved into a hyper-agpaitic melt. These melts have the necessary low solidus temperatures (down to 450 °C) required to evolve a CH_4-rich gas phase due to high concentrations of volatiles (i.e. F, Cl) (Konnerup-Madsen 1988).

Post-magmatic origin

Finally, abiogenic hydrocarbons could be generated during post-magmatic hydrothermal alteration from Fischer–Tropsch types of reactions between an exsolved magmatic CO_2-dominant fluid and H_2 produced from hydrothermal reactions (Abrajano *et al.* 1988; Szatmari 1989; Abrajano *et al.* 1990; Sherwood-Lollar *et al.* 1993, 2002; Sugisaki & Mimura 1994; Kelley 1996; Salvi & Williams-Jones 1997; Potter *et al.* 1998, 1999).

The disequilibrium Fischer–Tropsch reaction has been well known as a way of producing hydrocarbons from combustion of coal in industry (Anderson 1984; Szatmari 1989). Fischer–Tropsch synthesis involves a series of step reactions, which can be represented by equations such as:

$$nCO + (2n+1)H_2 \rightarrow CnH_{2n+2} + nH_2O, \quad (R6)$$

$$nCO_2 + (3n+1)H_2 \rightarrow CnH_{2n+2} + 2nH_2O \quad (R7)$$

For CO_2-dominant initial fluid compositions, these reactions can be simplified to:

$$CO_2 + 4H_2 \rightarrow CH_4 + 2H_2O \quad (Cn = 1), \quad (R8)$$

$$2CO_2 + 7H_2 \rightarrow C_2H_6 + 4H_2O \quad (Cn = 2),$$

etc. (R9)

The Fischer–Tropsch reaction is catalysed in the presence of group VIII metals, Fe-oxides, Fe-silicates and hydrated silicates (Lancet & Anders 1970; Porfir'ev 1974; Pommier *et al.* 1985; Szatmari 1989) and may therefore be a mechanism for the production of CH_4 and higher hydrocarbons in natural rocks. The production of CH_4 in ultrabasic and basic rocks during serpentinization processes has previously been attributed to this mechanism (Porfir'ev 1974; Abrajano *et al.* 1988, 1990; Szatmari 1989; Sherwood-Lollar *et al.* 1993; Sugisaki & Mimura 1994; Kelley 1996), the H_2 needed being provided during serpentinization of the primary mineral assemblage. Serpentine and brucite tend to exclude Fe^{2+} ions from their lattices and the Fe^{2+} is partially oxidised to Fe^{3+} and incorporated into magnetite. H_2 is generated in this reaction, due to the dissociation of H_2O and oxidation of the ferrous ions through reactions such as:

$$\underset{\text{Forsterite}}{5Mg_2SiO_4} + \underset{\text{Fayalite}}{Fe_2SiO_4} + 9H_2O \rightarrow$$

$$\underset{\text{Serpentine}}{3Mg_3Si_2O_5(OH)_4} + \underset{\text{Brucite}}{Mg(OH)_2} + 2Fe(OH)_2, \quad (R10a)$$

$$3Fe(OH)_2 \rightarrow \underset{\text{Magnetite}}{Fe_2^{2+}Fe^{3+}O_4} + H_2 + 2H_2O, \quad \text{(R10b)}$$

(Sherwood-Lollar *et al.* 1993).

Reports on abiogenic CH_4 generation at mid-ocean ridges, hydrothermal vents, seamounts and in ophiolites, plausible due to Fischer–Tropsch synthesis during serpentinization, are becoming increasingly more common (Welhan & Craig 1983; Abrajano *et al.* 1990; Charlou *et al.* 1991, 1998; Kiyosu *et al.* 1992; Kelley 1996; Sakata *et al.* 1997).

The association between post-magmatic alteration involving Fe oxidation, Fischer–Tropsch synthesis and hydrocarbon generation in igneous rocks can also be extended to include alkaline rocks (Salvi & Williams-Jones 1997; Potter *et al.* 1998, 1999). Hydrocarbon generation in the Khibina, Lovozero and Strange Lake complexes has been related to the alteration of the primary mineral assemblages (nepheline, augite, arfvedsonite, Ti-magnetite) to late biotite, cancrinite, aegerine, magnetite and natrolite via reactions such as:

$$3\underset{\text{Arfvedsonite}}{Na_3Fe_4^{2+}Fe^{3+}Si_8O_{22}(OH)_2} + 2H_2O \rightarrow$$

$$9\underset{\text{Aegirine}}{NaFe^{3+}Si_2O_6} + 2\underset{\text{Magnetite}}{Fe_2^{3+}Fe^{2+}O_4} + 6\underset{\text{Quartz}}{SiO_2} + 5H_2, \quad \text{(R11)}$$

(Salvi & Williams-Jones 1997)

$$\underset{\text{Arfvedsonite}}{Na_3Fe_4^{2+}Fe^{3+}Si_8O_{22}(OH)_2} + 2\underset{\text{Nepheline}}{NaAlSiO_4}$$

$$+ 4H_2O \rightarrow 5\underset{\text{Aegirine}}{NaFe^{3+}Si_2O_6} + 2Al(OH)_3 + 2H_2, \quad \text{(R12)}$$

$$4\underset{\text{Augite}}{(Ca_{0.5}Na_{0.5})(Fe_{0.5}^{2+}Fe_{0.5}^{3+})Si_2O_6} + 3\underset{\text{Ti-magnetite}}{(Fe_2^{2+}Ti)O_4}$$

$$+ 3\underset{\text{Nepheline}}{NaAlSiO_4} + 2\underset{\text{Orthoclase}}{KAlSi_3O_8} + Si^{4+} + 4H_2O \rightarrow$$

$$\underset{\text{Biotite}}{K_2Fe_5^{2+}Fe^{3+}(AlTi)_2Si_6O_{20}(OH)_4}$$

$$+ \underset{\text{Magnetite}}{Fe_2^{3+}Fe^{2+}O_4} + \underset{\text{Aegirine}}{NaFe^{3+}Si_2O_6}$$

$$+ 2\underset{\text{Titanite}}{Ca_2TiSiO_5} + 2\underset{\text{Albite}}{NaAl_3Si_3O_8} + 4H_2 \quad \text{(R13)}$$

and

$$3\underset{\text{Nepheline}}{NaAlSiO_4} + Na^+ + 2H_2O \rightarrow$$

$$\underset{\text{Cancrinite}}{Na_4Al_3Si_3O_{12}(OH)\cdot H_2O} + 0.5H_2. \quad \text{(R14)}$$

Therefore, this model can be readily applied to abiogenic hydrocarbon generation in a wide variety of igneous rock types and has become increasingly popular in recent years.

Overview and discussion of the models for the origin of hydrocarbons in igneous rocks

The generally secondary nature of the hydrocarbon-dominated inclusions in hitherto studied igneous rocks and the estimated low pressures and temperatures of their entrapment preclude a direct mantle origin as suggested by Gold (1979). Whether the hydrocarbon fluids in the silica-undersaturated alkaline rocks evolved in a late-magmatic setting through re-equilibration of a C–O–H fluid, or formed during post-magmatic mineral–fluid reactions, is still questionable (Gerlach 1980; Konnerup-Madsen 1988; Potter *et al.* 1998). However, it is generally accepted that the presence of hydrocarbons (and H_2) in basic rocks is due to the reduction of an early CO_2-rich fluid during post-magmatic serpentinization processes (e.g. Neal & Stanger 1983; Abrajano *et al.* 1990; Sherwood-Lollar *et al.* 1993).

The magmatic re-equilibration model suggested by Gerlach (1980) and Konnerup-Madsen *et al.* (1981) involving the evolution of a CH_4-rich fluid at low temperatures in a closed magmatic system is theoretically feasible for some particular alkaline melt compositions. Previous studies have indicated that the fulfilment of the most critical melt properties for the formation of a hydrocarbon-rich volatile phase is to be found in hyper-agpaitic nepheline syenites such as those at Ilímaussaq, Khibina and Lovozero. The formation of a hydrocarbon-rich volatile phase at the late-magmatic stage in this setting appears related to: (1) a wide temperature range of crystallization to very low solidus temperatures, enabling the buffering of any exsolved volatiles by the magma and minerals to temperatures as low as 450 °C; (2) low oxygen fugacities, with values roughly corresponding to those of the synthetic graphite–CH_4 curve during final solidification. Initial H_2O and C contents in the Ilímaussaq melt of about 4 wt% and 250 ppm, respectively, would at 2 kbars result in vapour saturation at

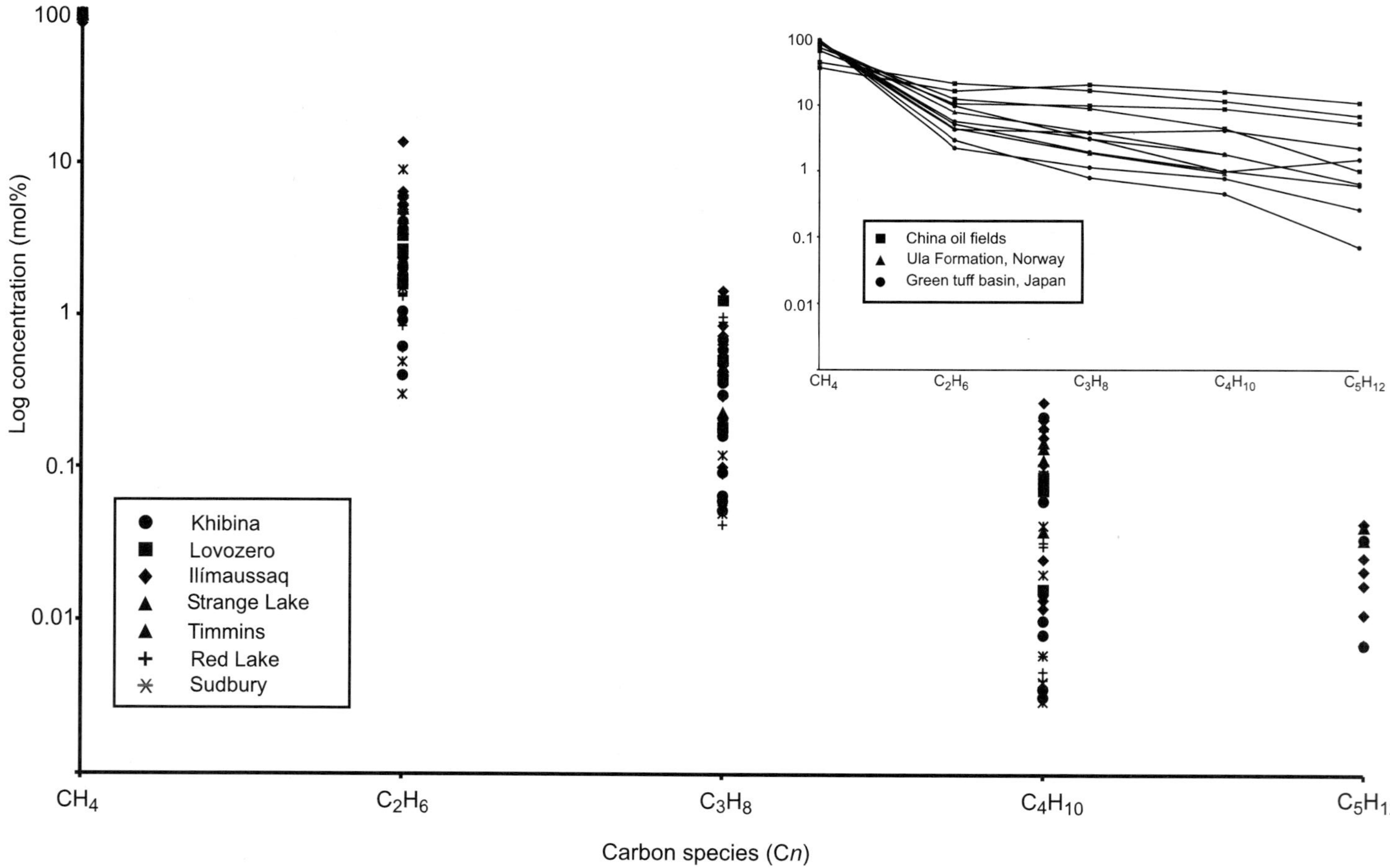

Fig. 3. A plot of log normalized hydrocarbon abundances (mol%) against increasing carbon number (C*n*) for abiogenic hydrocarbon gases at Khibina and Lovozero, Russia (Petersilie *et al.* 1961; Petersilie 1962; Petersilie & Sørensen 1970; Voytov 1992), Ilímaussaq, Greenland (Petersilie & Sørensen 1970; Konnerup-Madsen *et al.* 1979; Konnerup-Madsen & Rose-Hansen 1982), Strange Lake, Canada (Salvi & Williams-Jones 1997) and Sudbury, Timmins and Red Lake, Canada (Sherwood-Lollar *et al.* 1993). For comparison, hydrocarbon abundances are also shown for biogenic hydrocarbon gases in the China Oil fields (average mol%) (Chen *et al.*

close to 700 °C and the exsolution of a CO_2–H_2O rich fluid with X_{H_2O} about 0.7. As oxygen fugacities about 2–3 log units below the synthetic QFM buffer curve are indicated from mineral equilibria studies, respeciation of the exsolved fluid to very CH_4-rich compositions would result (Konnerup-Madsen 2001). The additional presence of about 0.5 wt% Cl in the melt furthermore would result in high salinities of the exsolved aqueous fluid and the subsequent formation of two non-miscible fluid phases, one rich in CH_4, the other rich in H_2O plus chlorides. This evolution is in agreement with the fluid inclusion arrays observed in the Ilímaussaq intrusion (Konnerup-Madsen *et al.* 1985).

However, for a large range of H_2O/CO_2 ratios of any exsolved fluid, re-equilibration to low oxygen fugacities would result in the precipitation of graphite (Fig. 2a), albeit in very small amounts. No graphite has been found in the fluid inclusions in the Khibina and Lovozero complexes and little has been found in Ilímaussaq (Konnerup-Madsen *et al.* 1981). Therefore this simple magmatic re-equilibration model is not considered to explain adequately all fluid inclusion features observed in alkaline (and other) igneous intrusions in general; also the model does not explain the significant concentrations of higher hydrocarbons.

The generation of hydrocarbons through Fischer–Tropsch synthesis during post-magmatic alteration may provide an additional mechanism for the formation of CH_4 and higher hydrocarbons in these silica-undersaturated alkaline complexes and also other igneous rocks. Evidence for this type of process may be found in the bulk gas, stable isotope data and from textural observations.

Using the published bulk gas data shown in Table 2, the distribution of hydrocarbons in the alkaline intrusions of Khibina, Lovozero, Ilímaussaq and Strange Lake, plus those found in the Canadian Shield, show a clear log-linear trend from C_2 to C_5 decreasing in concentration from 10^1 to 10^{-2} (Fig. 3). It is apparent that Khibina, Lovozero and Ilímaussaq have very low ethane/methane ratios (0.01 to 0.08) (Table 5). The ratios for the higher hydrocarbons either remain at the same low figures, indicating the dominance of lighter hydrocarbons, or increase, then level out at around 0.18 to 0.25 (Table 5). This is consistent with a Schulz–Flory distribution (Anderson 1984), with production of a hydrocarbon mixture dominated by light hydrocarbons with a C_{n+1}/C_n ratio of less than 0.6 and a clear indication that the hydrocarbons are being generated by Fischer–Tropsch reactions (Anderson 1984; Szatmari 1989). The hydrocarbon-bearing fluids at Strange Lake have ratios that level out at slightly higher values (~0.34), indicating a slightly higher proportion of higher hydrocarbons than the fluids in the Khibina, Lovozero and Ilímaussaq complexes. The ratios of the hydrocarbon fluids found in the Canadian and Fennoscandian shield rocks are also below 0.6 (with the exception of Elliot Lake). They have generally very low ratios, but tend to be more variable between sites. Comparing the Schulz–Flory distribution of these abiogenic hydrocarbons with the distribution of abundances of biogenic hydrocarbons (Fig. 3) in oil fields and geothermal (thermogenic) springs, the two are clearly different. The biogenic hydrocarbons have a flatter distribution (particularly for the oil fields). This clearly indicates that these hydrocarbons are of an abiogenic nature probably generated by Fischer–Tropsch reactions.

The fractionation of $\delta^{13}C$ isotope signatures during Fischer–Tropsch synthesis in a relatively closed system can vary. Initial conversion of a primary magmatic CO_2 fluid to CH_4 with subsequent partitioning will fractionate the CH_4 to lighter $\delta^{13}C$ signatures. Later CH_4 associated with the total conversion of the CO_2 fluid will produce $\delta^{13}C$ signatures closer to the initial magmatic value. Therefore, a wide range of $\delta^{13}C_{CH_4}$ values, as observed in Table 4 (−10‰ to −30‰), can be produced during CH_4 generation (Voytov 1992). The majority of CH_4 found in igneous rocks show signatures between −20‰ and −28‰ (Table 4), indicating fractionation from the initial magmatic fluid at temperatures between 300 °C and 400 °C (Bottinga 1969). Plotting $\delta^{13}C$ values for CH_4 to C_3H_8 from hydrocarbon-bearing fluids at Khibina (Fig. 4), it is apparent that the $\delta^{13}C$ values become depleted with increasing carbon number (C*n*). Comparing these values with published $\delta^{13}C$ values for hydrocarbon gases in oil fields and geothermal (thermogenic) springs, it is clear that these fluids were not derived from biogenic processes (i.e. they have been generated by polymerization from simple hydrocarbon species with preferential enrichment of ^{12}C to ^{13}C with increasing C*n* rather than thermal cracking of complex organic species giving an inverse $\delta^{13}C$ distribution). This is also strong evidence for an abiogenic origin. The hydrogen isotope signatures observed in these hydrocarbon-bearing rocks are also in accordance with the theory that hydrocarbons were produced by Fischer–Tropsch synthesis. The signatures of these hydrocarbons commonly show depleted δD_{CH_4} values of around −135‰ and a δD_{H_2} values of −600‰ (Devirts *et al.* 1993; Sherwood-Lollar *et al.* 1993).

Table 5. *Carbon number ratios*

Rock type	C_2/C_1	C_3/C_2	C_4/C_3	C_5/C_4
Khibina				
Nepheline Syenite	0.01	0.22	0.21	—
Rischorrite	0.08	0.24	0.41	—
Ijolite	0.04	0.26	0.27	0.55
Kukisvumchorr Mine	0.07	0.25	0.18	0.28
Rasvumchorr Mine	0.05	0.13	0.24	0.18
Rasvumchorr Mine	0.07	0.08	0.1	—
Khibinite	0.03	0.05	0.01	—
Rischorrite	0.04	0.05	0.05	—
Lovozero				
Juvite	0.07	0.08	0.59	—
Syenite	0.05	0.21	0.06	—
Foyaite	0.03	0.47	0.19	—
Ilímaussaq				
Syenite	0.03	0.09	0.18	—
Naujaite	0.12	0.18	0.28	—
Foyaite	0.1	0.18	0.28	—
Lujavrite	0.13	0.07	0.05	—
Sodalite & nepheline	0.12	0.15	0.31	0.13
Arfvedsonite	0.09	0.13	0.38	0.03
Sodalite	0.12	0.21	0.32	—
Sodalite	0.1	0.15	0.25	0.25
Arfvedsonite	0.1	0.13	0.24	0.11
Nepheline	0.11	0.13	0.24	0.12
Eudialyte	0.3	0.15	0.25	0.12
Strange Lake				
fresh pegmatite	0.08	0.18	0.34	0.34
altered pegmatite	0.1	0.14	0.41	0.34
Canadian Shield				
Sudbury N	0.006	0.29	—	
Sudbury CCS	0.06	0.07	0.42	
Elliot Lake	0.002	0.18	1.5	
Timmins	0.06	0.15	0.32	
Red Lake – Dickenson	0.03	0.56	0.08	
Red Lake – Campbell	0.03	0.45	—	
Fennoscandian Shield				
Juuka	0.01	0.01	—	

The carbon ratios are calculated from the bulk gas analyses shown in Table 2 (vol%) from Khibina and Lovozero (Russia), Ilímaussaq (Greenland), Strange Lake (Quebec) and the Canadian and Fennoscandian shield rocks.

Textural observations provide further evidence that most of the hydrocarbons in alkaline and basic rocks have been generated by Fischer–Tropsch type reactions. Catalysts for the Fischer–Tropsch synthesis in industry include magnetite, Fe-silicates and zeolites. The often-observed close relationship between the hydrocarbon-bearing fluids and the magnetite, Fe-silicates and hydrated Na/K silicates (i.e. zeolites) in the rock assemblages suggests this is also occurring in nature (Salvi & Williams-Jones 1997; Potter *et al.* 1999).

Implications for the generation of abiogenic hydrocarbons in igneous rocks

The occurrence of hydrocarbons in igneous rocks may not be as rare as previously thought. Over the last 20 years there has been a proliferation of reports on the presence of hydrocarbons in igneous rocks found in ultrabasic, basic, peralkaline granites and silica-undersaturated igneous rocks from plutonic to hypabyssal and volcanic levels. Most of these occurrences have a biogenic or thermogenic origin, but many

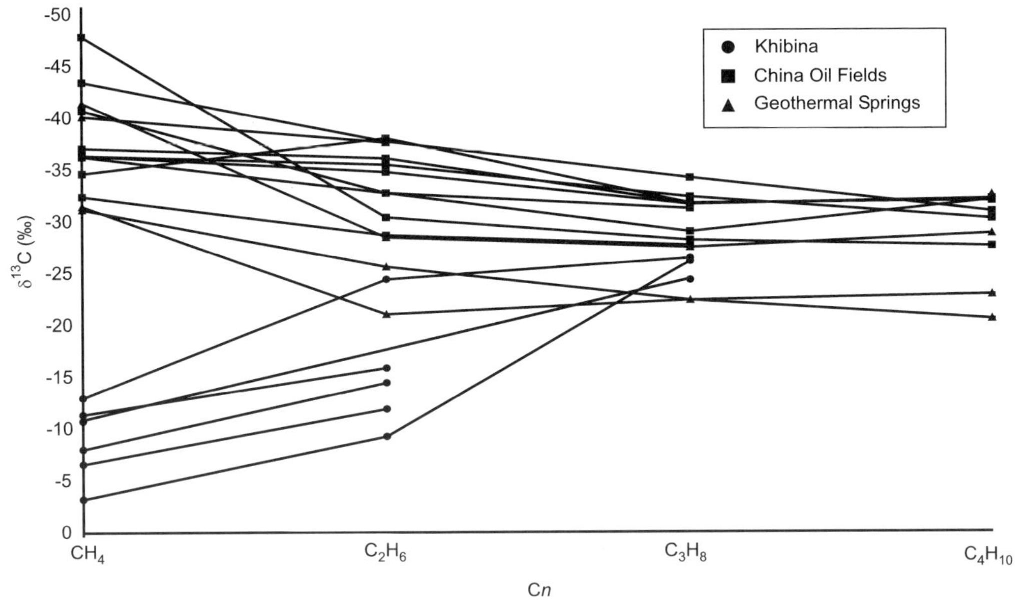

Fig. 4. A plot showing the distribution of $\delta^{13}C$ values with increasing carbon number (*Cn*) for the abiogenic hydrocarbon gases at Khibina (Voytov 1992) and for comparison, biogenic gases from the China oil fields (Dai *et al.* 1996; Chen *et al.* 2000) and various geothermal springs, USA (Des Marais *et al.* 1981).

have an abiogenic origin. A large proportion of these reports relates the hydrocarbons to post-magmatic alteration of the assemblage, in particular, serpentinization of ultrabasic and basic rocks.

Although a model involving near-solidus re-equilibration at low oxygen fugacities of an initially exsolved CO_2–H_2O fluid can explain the generation of hydrocarbon-dominated fluid compositions, the melt compositional requirements are such that they are only met with in very rare situations, such as in some hyperagpaitic nepheline syenite intrusions. Alternatively, the model for Fischer–Tropsch synthesis of hydrocarbons in igneous rocks, as proposed by Salvi & Williams-Jones (1997), Potter *et al.* (1998) and Potter (2000), is considered to be more generally applicable to predict the presence of hydrocarbons in a variety of igneous rocks. The operation of this process may become important in searching for future natural gas resources in these rocks. Fischer–Tropsch synthesis not only produces abundant CH_4, but, depending on the temperature at which the reactions take place, various other complex higher hydrocarbons. One obvious example of a potential reservoir for hydrocarbons formed by this process is the CH_4-clathrates trapped at the sea floor. Serpentinization of ocean floor basalts by circulating seawater and hydrothermal fluids has the potential to produce large volumes of CH_4 and higher hydrocarbons via Fischer–Tropsch reactions from magmatic CO_2 fluids. The continuing advances in ocean floor exploration may make these hydrocarbons an economically viable resource.

The Fischer–Tropsch model for hydrocarbon generation in igneous rocks can theoretically be applied to many magmatic bodies of all compositional types. Hydrothermally altered Fe-rich assemblages are common features in many intrusive bodies. Therefore, this model might suggest that hydrocarbons should be found in most igneous bodies. This is clearly not the case. It has been observed in industry that the Fe catalysts for Fischer–Tropsch reactions can be poisoned by the presence of sulphur (Anderson *et al.* 1965; Szatmari 1989). A comparative study of fluids present in the serpentinized carbonatitic intrusion, Sokli in Finland (Potter 2000) revealed the presence of little or no hydrocarbons in the samples. The majority of Sokli samples contain abundant sulphides which may have inhibited any Fischer–Tropsch type reaction from occurring. The extensive serpentinization (and phlogopitization) of the assemblage also suggests that large volumes of H_2O infiltrated the complex, potentially washing out any hydrocarbons generated. Therefore, in order for Fischer–Tropsch synthesis to occur, sulphur should not be present and, for the hydrocarbons to be retained within the assemblage, hydrothermal alteration should not be pervasive with large volumes of H_2O migrating through the complex.

Conclusions

Awareness among scientists of the presence of hydocarbons in igneous rocks has been steadily on the increase over the past 20 years. Although in most cases, the hydrocarbons have been derived from biogenic sources, there has been a recent proliferation of reports on the presence of hydrocarbons formed by abiogenic proceses. These are commonly found in basic and SiO_2-undersaturated alkaline igneous rocks and in hydrothermal plumes associated with basic rocks. From where these hydrocarbons were derived and how they were generated still remains debatable. In the 1970s it was proposed that these hydrocarbons were derived directly from the mantle and that large abiogenic hydrocarbon reservoirs might be stored in the deep crust, having flowed up from the mantle. However, more recent studies have determined that this is unlikely and that these hydrocarbons were generated during late or post-magmatic processes. Two models have been proposed: a late-magmatic equilibrium model and a post-magmatic disequilibrium model. It has been shown that a CH_4-rich fluid can evolve from a magmatic CO_2-rich fluid in a closed system at late-magmatic conditions (below 500 °C). This model can only be applied to igneous rocks with extreme compositions with low solidus temperatures and wide crystallization ranges (i.e. SiO_2-undersaturated alkaline rocks). A post-magmatic model has also been suggested, where hydrocarbons are generated during post-magmatic hydrothermal alteration of the mineral assemblages. This involved Fischer–Tropsch type reactions between an early CO_2-rich fluid and H_2 generated from the hydrothermal reactions. The Fischer–Tropsch reactions are catalysed in the presence of Fe-rich minerals abundant in the basic and alkaline igneous complexes (e.g. magnetite, aegirine). Textural, fluid inclusion, isotopic and bulk gas data from a variety of complexes support this model. Therefore, although the late-magmatic model may still be valid for generation of hydrocarbons in the SiO_2-undersaturated alkaline complexes, the Fischer–Tropsch model is favoured and can be applied to a wide variety of igneous environments. This model suggests that abiogenic hydrocarbons may be more common than previously anticipated, with possible implications for natural gas resources in the future.

The authors would like to thank A. Rankin and P. Treloar at the School of Earth Sciences and Geography, Kingston University, Kingston-upon-Thames, UK, whose help and support in research into this topic has made this paper possible. We would also like to thank C. Cornford and L. Larsen for helpful reviews for improving this manuscript.

References

ABRAJANO, T. A., STURCHIO, N. C., BOHLKE, J. K., LYON, G. L., POREDA, R. J. & STEVENS, C. M. 1988. Methane hydrogen gas seeps, Zambales ophiolite, Philippines—deep or shallow origin. *Chemical Geology*, **71**(1-3), 211–222.

ABRAJANO, T. A., STURCHIO, N. C., KENNEDY, B. M., LYON, G. L., MUEHLENBACHS, K. & BOHLKE, J. K. 1990. Geochemistry of reduced gas related to serpentinization of the Zambales ophiolite, Philippines. *Applied Geochemistry*, **5**, 625–630.

ANDERSEN, T. 1986. Magmatic fluids in the Fen Carbonatite Complex, S.E. Norway: Evidence of mid-crustal fractionation from solid and fluid inclusions in apatite. *Contributions to Mineralogy and Petrology*, **93**, 491–503.

ANDERSEN, T. & BURKE, E. A. J. 1996. Methane inclusions in shocked quartz from the Gardnos breccia, South Norway. *European Journal of Mineralogy*, **8**, 927–936.

ANDERSON, R. B. 1984. *The Fischer–Tropsch synthesis.* Academic Press, New York.

ANDERSON, R. B., KARN, F. S., KELLY, R. E. & SCHULTZ, J. F. 1965. *Sulfur poisoning of fixed beds of iron catalysts in the Fischer–Tropsch synthesis.* U.S. Dept. of the Interior, Bureau of Mines, Bulletin, **628**.

BELOKON, T. V., FRIK, M. G., KOBLOVA, A. Z. & KUCHINA, YE. L. 1995. Biomarker distributions in oil and rock organic matter in Russian petroliferous regions. *Geochemistry International*, **32**(9), 126–141.

BOTTINGA, Y. 1969. Calculated fractionation factors for carbon and hydrogen isotope exchange in the system calcite–carbon dioxide–graphite–methane–hydrogen–water vapour. *Geochimica et Cosmochimica Acta*, **33**, 49–64.

BOTZ, R., STÜBEN, D., WINCKLER, G., BAYER, R., SCHMITT, M. & FABER, E. 1996. Hydrothermal gases offshore Milos Island, Greece. *Chemical Geology*, **130**, 161–173.

CESARE, B. 1995. Graphite precipitation in C–O–H fluid inclusions: Closed system compositional and density changes and thermobarometric implications. *Contributions to Mineralogy and Petrology*, **122**, 25–33.

CHARLOU, J. L., BOUGALT, H., APPRIOU, P., NELSEN, T. & RONA, P. 1991. Different TDM/CH_4 hydrothermal plume signatures: TAG site at 26°N and serpentinized ultrabasic diapir at 15°05′N on the Mid-Atlantic Ridge. *Geochimica et Cosmochimica Acta*, **55**, 3209–3222.

CHARLOU, J. L., FOUQUET, Y., BOUGAULT, H., DONVAL, J. P., ETOUBLEAU, J., JEAN-BAPTISTE, P.,

DAPOIGNY, A., APPRIOU, P. & RONA, P. A. 1998. Intense CH_4 plumes generated by serpentinization of ultramafic rocks at the intersection of the 15°20′N fracture zone and the Mid-Atlantic Ridge. *Geochimica et Cosmochimica Acta*, **62**(13), 2323–2333.

CHEN, J., XU, Y. & HUANG, D. 2000. Geochemical characteristics and origin of natural gas in Tarim Basin, China. AAPG Bulletin, **84**(5), 591–606.

DAI, J., SONG, Y., DAI, C. & WANG, D. 1996. Geochemistry and accumulation of carbon dioxide gases in China. *AAPG Bulletin*, **80**(10), 1615–1626.

DARLING, W. G. 1998. Hydrothermal hydrocarbon gases: 2, Application in the East African Rift system. *Applied Geochemistry*, **13**(7), 825–840.

DARLING, W. G., GREISSHABER, E., ANDREWS, J. N., ARMANNSON, H. & O'NIONS, R. K. 1995. The origin of hydrothermal and other gases in the Kenyan Rift Valley. *Geochimica et Cosmochimica Acta*, **59**(12), 2501–2512.

DES MARAIS, D. J., DONCHIN, J. H., NEHRING, N. L. & TRUESDELL, A. H. 1981. Molecular carbon isotopic evidence for the origin of geothermal hydrocarbons. *Nature*, **292**, 826–828.

DEVIRTS, A. L., GAGAUZ, F. G., GRINENKO, V. A., LAGUTINA, YE.P., PEREVERZOV, V. V. & SHUKOLYUKOV, YU. A. 1993. Origin of hydrogen in Kempirsay-intrusion ultramafites. *Geochemistry International*, **30**(2), 139–144.

GERLACH, T. M. 1980. Chemical characteristics of the volcanic gases from Nyiragongo Lava Lake and the generation of CH_4-rich fluid inclusions. *Journal of Volcanic and Geothermermal Research*, **8**, 177–189.

GIARDINI, A. A., MELTON, C. E. & MITCHELL, R. S. 1982. The nature of the upper 400 km of the earth and its potential as the source for non-biogenic petroleum. *Journal of Petroleum Geology*, **5**(2), 173–190.

GIZE, A. & MACDONALD, R. 1993. Generation of compositionally atypical hydrocarbons in CO_2-rich geologic environments. *Geology*, **21**, 129–132.

GOLD, T. 1979. Terrestrial sources of carbon and earthquake outgassing. *Journal of Petroleum Geology*, **1**(3), 3–19.

GUNTER, B. D. & MUSGRAVE, B. C. 1971. New evidence on the origin of methane in hydrothermal gases. *Geochimica et Cosmochimica Acta*, **35**, 113–118.

HOLLOWAY, J. R. 1984. Graphite–CH_4–H_2O–CO_2 equilibria at low grade metamorphic conditions. *Geology*, **12**, 455–458.

HORITA, J. & BERNDT, M. E. 1999. Abiogenic methane formation and isotopic fractionation under hydrothermal conditions. *Science*, **285**, 1055–1057.

HUIZENGA, J. M. 1995. *Fluid evolution in shear zones from the late Archean Harare–Shamva–Bindura greenstone belt (NE Zimbabwe)*. PhD thesis, Proefschrift Vrije Universiteit, Amsterdam.

HUIZENGA, J. M. 2001. Thermodynamic modelling of C–O–H fluids. *Lithos*, **55**, 101–114.

IKORSKI, S. V. 1991. Hydrocarbon gases in alkali intrusions. *Geochemistry International*, **28**(6), 17–23.

IKORSKI, S. V., NIVIN, V. A. & PRIPACHKIN, V. A. 1993. New information about natural gases of alkaline massifs. *Isotopenpraxis*, **28**, 289–291.

JEFFREY, A. W. A. & KAPLAN, I. R. 1988. Hydrocarbons and inorganic gases in the Gravberg-1 well, Siljan ring, Sweden. *Chemical Geology*, **71**, 237–255.

KARLSEN, D. A., NEDKVITNE, T., LARTER, S. R. & BJØRLYKEE, K. 1993. Hydrocarbon composition of authigenic inclusions: Application to elucidation of petroleum reservoir filling history. *Geochimica et Cosmochimica Acta*, **57**, 3641–3659.

KARZHAVIN, V. K. & VENDILLO, V. P. 1970. Thermodynamic equilibrium and conditions for existence of hydrocarbon gases in a magmatic process. *Geochemistry International*, **7**, 797–803.

KELLEY, D. S. 1996. Methane rich fluids in the oceanic crust. *Journal of Geophysical Research*, **101**(B2), 2943–2962.

KELLEY, D. S. & FRÜH-GREEN, G. L. 2001. Volatile lines of descent in submarine pluonic environments: Insights from stable isotope and fluid inclusion analyses. *Geochimica et Cosmochimica Acta*, **65**(19), 3325–3346.

KERR, R. A. 1990. When a radical experiment goes bust. *Science*, **247**, 1177–1179.

KHARMALOV, YE. S. *et al.* 1981. Origin of carbonatites of the Kovdor deposit. *International Geology Review*, **23**(8), 865–880.

KIYOSU, Y., ASADA, N. & YOSHIDA, Y. 1992. Origin of light hydrocarbon gases from the Matsukawa geothermal area in Japan. *Chemical Geology*, **94**, 321–329.

KOGARKO, L. N., KOSZTOLANYI, CH. & RYABCHIKOV, I. D. 1987. Geochemistry of the reduced fluid in alkali magmas. *Geochemistry International*, **24**(7), 20–27.

KONNERUP-MADSEN, J. *et al.* 1981.

KONNERUP-MADSEN, J. 1988. Abiogenic hydrocarbon gases, *In*: SANTOSH, M. (ed.) *Fluid Inclusions*. Special Memoirs of Geological Society of India, 13–24.

KONNERUP-MADSEN, J. 2001. Non-biogenic hydrocarbon gases associated with solidification of the Ilímaussaq alkaline complex, South Greenland. *Geology of Greenland Survey Bulletin*, **190**, 163–170.

KONNERUP-MADSEN, J. & ROSE-HANSEN, J. 1982. Volatiles associated with alkaline igneous rift activity: Fluid inclusions in the Ilímaussaq intrusion and the Gardar granitic complexes (South Greenland). *Chemical Geology*, **37**, 79–93.

KONNERUP-MADSEN, J., DUBESSY, J. & ROSE-HANSEN, J. 1985. Combined Raman microprobe spectometry and microthermometry of fluid inclusions in minerals from igneous rocks of the Gardar Province (South Greenland). *Lithos*, **18**, 271–280.

KONNERUP-MADSEN, J., LARSEN, E. & ROSE-HANSEN, J. 1979. Hydrocarbon-rich fluid inclusions in minerals from the alkaline Ilímaussaq intrusion, South Greenland. *Bulletin de Minéralogie*, **102**, 642–653.

KONNERUP-MADSEN, J., ROSE-HANSEN, J. & LARSEN, E. 1981. Hydrocarbon gases associated with alkaline igneous activity: evidence from composition of fluid inclusions. *Rapport Grønlands Geologiske Undersøgelse*, **103**, 99–108.

KROT, A. N., POSUKHOVA, T. V., GUSEVA, YE. V., GALIMOV, E. M., BOTKUNOV, A. I., RAMENSKAYA, M. YE. & OGLOBINA, A. I. 1994. Origin of garnets containing hydrocarbon inclusions in the Mir kimberlite pipe. *Geochemistry International*, **31**(1), 122–128.

LANCET, M. S. & ANDERS, E. 1970. Carbon isotope fractionation in the Fischer–Tropsch synthesis and in meteorites. *Science*, **170**, 981–983.

LARSEN, R. B., BROOKS, C. K. & BIRD, D. K. 1992. Methane bearing, aqueous, saline solutions in the Skaergaard intrusion, East Greenland. *Contributions to Mineralogy and Petrology*, **112**, 428–437.

MATHEZ, E. A. 1987. Carbonaceous matter in mantle xenoliths: Composition and relevance to the isotopes. *Geochimica et Cosmochimica Acta*, **51**, 2339–2347.

MOROGAN, V. & LINDBLOM, S. 1995. Volatiles associated with the alkaline-carbonatite magmatism at Alnö, Sweden: A study of fluid and solid inclusions in minerals from the Långarsholmen Ring Complex. *Contributions to Mineralogy and Petrology*, **122**, 262–274.

NEAL, C. & STANGER, G. 1983. Hydrogen generation from mantle source rocks in Oman. *Earth and Planetary Science Letters*, **60**, 315–321.

NIVIN, V. A., DEVIRTS, A. L. & LAGUTINA, YE. P. 1995. The origin of the gas phase in the Lovozero massif based on hydrogen-isotope data. *Geochemistry International*, **32**(8), 65–71.

PETERSILIE, I. A. 1962. Origin of hydrocarbon gases and dispersed bitumens of the Khibina alkalic massif. *Geochemistry International*, **1**, 14–29.

PETERSILIE, I. A. & SØRENSEN, H. 1970. Hydrocarbon gases and bituminous substances in rocks from the Ilímaussaq alkaline intrusion, South Greenland. *Lithos*, **3**, 59–76.

PETERSILIE, I. A., IKORSKI, S. V., SMIRNOVA, L. I., ROMANIKHIN, A. M. & PROSKURYAKOVA, E. B. 1961. Application of gas logging to the investigation of natural gases and bitumens in the Khibina intrusive massif. *Geochemistry International*, **10**, 945–962.

POMMIER, B., REYMOND, J. P. & TEICHNER, S. J. 1985. Fischer–Tropsch synthesis on oxidised supported iron catalyst. *Zeitschrift physische chemie Nueue Folge*, **144**, 203–222.

PORFIR'EV, V. B. 1974. Inorganic origin of petroleum. *AAPG Bulletin*, **58**(1), 3–33.

POTTER, J. 2000. *The characterisation and origin of hydrocarbons in the alkaline igneous complexes in the Kola Alkaline Province*. PhD thesis, Kingston University, London, UK.

POTTER, J., RANKIN, A. H., TRELOAR, P. J., NIVIN, V. A., TING, W. & NI, P. 1998. A preliminary study of methane inclusions in alkaline igneous rocks of the Kola igneous province, Russia: implications for the origin of methane in igneous rocks. *European Journal of Mineralogy*, **10**, 1167–1180.

POTTER, J., RANKIN, A. H. & TRELOAR, P. J. 1999. The relationship between CH_4 and CO_2 inclusions and Fe–O–S mineralization in intrusions of the Kola Alkaline Province. *In*: STANLEY, C. J. *et al.* (eds) *Mineral deposits: Processes to processing*, Vol. 1. Balkema, Rotterdam, 87–90.

ROEDDER, E. 1984. *Fluid Inclusions*. Mineralogical Society of America, Reviews in Mineralogy, **12**.

SAKATA, S., SANO, Y., MAEKAWA, T. & IGARI, S. I. 1997. Hydrogen and carbon isotopic composition of methane as evidence for biogenic origin of natural gases from the Green Tuff Basin, Japan. *Organic Geochemistry*, **26**(5/6), 399–407.

SALVI, S. & WILLIAM-JONES, A. E. 1992. Reduced orthomagmatic C–O–H–N–NaCl fluids in the Strange Lake rare-metal granitic complex, Quebec/Labrador, Canada. *European Journal of Mineralogy*, **4**, 1155–1174.

SALVI, S. & WILLIAMS-JONES, A. E. 1997. Fischer–Tropsch synthesis of hydrocarbons during subsolidus alteration of the Strange Lake peralkaline granite, Quebec/Labrador, Canada. *Geochimica et Cosmochimica Acta*, **61**(1), 83–99.

SAMSON, I. M. & WILLIAMS-JONES, A. E. 1991. C–O–H–N–salt fluids associated with contact metamorphism, McGerrigle mountains, Quebec: A Raman spectroscopic study. *Geochimica et Cosmochimica Acta*, **55**, 169–177.

SAMSON, I. M., LUI, W. & WILLIAMS-JONES, A. E. 1995. The nature of orthomagmatic hydrothermal fluids in the Oka carbonatite, Quebec, Canada: Evidence from fluid inclusions. *Geochimica et Cosmochimica Acta*, **59**(10), 1963–1977.

SCHIDLOWSKI, M. 1982. Content and isotopic composition of reduced carbon in sediments. *In*: HOLLAND, H. D. & SCHLIDOWSKI, M. (eds) *Mineral deposits and the evolution of the biosphere*. Springer-Verlag, New York, 103–122.

SCHOELL, M. 1983. Genetic characterisation of natural gases. *AAPG Bulletin*, **67**(12), 2225–2238.

SCHOELL, M. 1988. Multiple origins of methane in the earth. *Chemical Geology*, **71**(1-3), 1–10.

SHERWOOD-LOLLAR, B., FRAPE, S. K., WEISE, S. M., FRITZ, P., MACKO, S. A. & WELHAN, W. A. 1993. Abiogenic methanogenesis in crystalline rocks. *Geochimica et Cosmochimica Acta*, **57**, 5087–5097.

SHERWOOD-LOLLAR, B., WESTGATE, T. D., WARD, J. A., SLATER, G. F. & LACRAMPE-COULOUME, G. 2002. Abiogenic formation of alkanes in the earth's crust as a minor source for global hydrocarbon reservoirs. *Nature*, **416**, 522–524.

SIMONEIT, B. R. T. 1988. *Organic matter in hydrothermal systems—maturation, migration and biogeochemistry*. Pergamon, Oxford.

SUGISAKI, R. & MIMURA, K. 1994. Mantle hydrocarbons: Abiotic or biotic. *Geochimica et Cosmochimica Acta*, **58**(11), 2527–2542.

SZATMARI, P. 1989. Petroleum formation by Fischer–Tropsch synthesis in platetectonics. *AAPG Bulletin*, **73**(8), 989–998.

TING, W., BURKE, E. A., RANKIN, A. H. & WOOLLEY, A. R. 1994. Characterisation and petrogenetic significance of CO_2, H_2O and CH_4 fluid inclusions in apatite from the Sukulu Carbonatite, Uganda. *European Journal of Mineralogy*, **6**, 787–803.

VARD, E. & WILLIAMS-JONES, A. E. 1993. A fluid inclusion study of vug minerals in dawsonite-altered phonolite sills, Montreal, Quebec: Implications for HFSE mobility. *Contributions to Mineralogy and Petrology*, **113**, 410–423.

VLIERBOOM, F. W., COLLINI, B. & ZUMBERGE, J. E. 1986. The occurrence of petroleum in sedimentary rocks of the meteor impact crater at Lake Siljan, Sweden. *Organic Geochemistry*, **10**, 153–161.

VOYTOV, G. I. 1992. Chemical and carbon-isotope fluctuations in free gases (gas jets) in the Khibiny. *Geochemistry International*, 14–24.

WELHAN, J. A. & CRAIG, H. 1983. Methane, hydrogen and helium in hydrothermal fluids at 21°N, East Pacific Rise. *In*: RONA, P. A., BOSTRÖM, K., LAUBIER, L. & SMITH JNR, K. L. *Hydrothermal processes at seafloor spreading centres*. Plenum, New York, 391–409.

WELHAN, J. A. & LUPTON, J. E. 1987. Light hydrocarbon gases in Guaymas Basin hydrothermal fluids: Thermogenic versus abiogenic origin. *AAPG Bulletin*, **71**(2), 215–223.

ZAKRZHEVSKAYA, N. G. 1964. Origin of gases in the Khibiny apatite deposits. *Geochemistry International*, 128 131.

ZHANG, Y. G. & FRANTZ, J. D. 1992. Hydrothermal reactions involving equilibrium between minerals and mixed volatiles 2. Investigations of fluid properties in the CO_2–CH_4–H_2O system using synthetic fluid inclusions. *Chemical Geology*, **100**, 51–72.

Coupled mineral–fluid evolution of a basin and high: kaolinization in the SW England granites in relation to the development of the Plymouth Basin

AGAMEMNON PSYRILLOS[1,2], STUART D. BURLEY[3,4], DAVID A. C. MANNING[1,5] & ANTHONY E. FALLICK[6]

[1] *Department of Earth Sciences, Oxford Road, University of Manchester, Manchester M13 9PL, UK*
[2] *Present address: B'Parodos Andrea Miaouli 9, Katerini 60100, Greece*
[3] *Development Geoscience, BG Group, 100 Thames Valley Park Drive, Reading RG6 1PT, United Kingdom*
[4] *Basin Dynamics Research Group, University of Keele, Keele ST5 5BG, United Kingdom*
[5] *Present address: School of Civil Engineering and Geosciences, University of Newcastle upon Tyne, Newcastle upon Tyne NE1 7RU, United Kingdom (email: david.manning@ncl.ac.uk)*
[6] *SUERC, East Kilbride, Glasgow, G75 0QU, Scotland*

Abstract: A new genetic model is proposed for the formation of the St Austell kaolin deposits, incorporating geological, isotopic, paragenetic and microthermometric evidence from kaolin–quartz veins combined with a reconstruction of the thermal evolution of the Cornubian pluton during the Mesozoic. Fluid inclusions in quartz, paragenetically associated with kaolin, document that the kaolinization took place at temperatures between 50 °C and 100 °C, indicating that the kaolinization is a low-temperature hydrothermal event coincident with the oil generation window.

Kaolinization occurred prior to the unroofing of the pluton, during the Late Jurassic to Early Cretaceous. The kaolinization is thus contemporary with the major Early Cretaceous uplift that affected the Cornubian massif as a consequence of rifting in the offshore Western Approaches. Geological, isotopic and geochemical considerations argue strongly against the involvement of unmodified meteoric waters in the kaolinization process. The most plausible fluid types for the kaolinization are either basinal brines expelled from Permo-Triassic sediments of the adjacent offshore Plymouth Basin, or highly evolved meteoric waters that circulated through the sediments enclosing the pluton. The kaolinization process converted large volumes of fractured granite to a porous quartz–kaolin rock matrix.

Sedimentary basin evolution has been linked to many types of mineralization (e.g. Cathles & Smith 1983; Eugster 1984; Garven 1985; Burley *et al.* 1989; Metcalfe *et al.* 1994). In this paper, we show that the Mesozoic offshore sedimentary basins of the Western Approaches Basin are the most probable source of low-temperature fluids that caused widespread kaolinization in the Cornubian granite batholith, using the St Austell pluton as an example. The fluids were highly saline NaCl–$CaCl_2$ brines, and precipitated quartz at temperatures of between 50 °C and 100 °C. These temperatures of formation, in the context of the thermal evolution of the adjacent offshore basins, suggest that the widespread kaolinization took place in the late Jurassic to early Cretaceous under a significant sediment cover. Migration of aqueous fluids from the now offshore Mesozoic sedimentary basins was probably associated with hydrocarbon migration, although most hydrocarbons have subsequently been lost through uplift, trap breaching and erosion.

In this paper, kaolinization in the St Austell pluton is treated as a process of fluid : rock interaction between a feldspathic rock (granite) and aqueous solutions, a geochemical system comparable to the diagenesis of clastic sediments in a sedimentary basin environment. This approach is used to re-examine the genesis of the St Austell kaolin deposits, focusing on kaolin–quartz-bearing mineral veins, and incorporating new paragenetic, fluid inclusion and stable isotope evidence. The end result of the fluid : rock interaction was to produce a quartz–kaolinite assemblage similar in mineralogy to many deep burial quartz arenite reservoirs in the North Sea (Harris 1989).

From: PETFORD, N. & MCCAFFREY, K. J. W. (eds) 2003. *Hydrocarbons in Crystalline Rocks*. Geological Society, London, Special Publications, **214**, 175–195. 0305-8719/03/$15

Geological context

The granite batholith of Cornwall is a 200 km-long and 40 km-wide plutonic body of late Variscan age, spatially associated with a wide range of mineralization phenomena, including extensive kaolin deposits formed by *in-situ* alteration of the granitic parent rocks. Some of the most extensive kaolin deposits are located in the St Austell pluton, towards the middle of the batholith, some 15 km from the flanks of the offshore Mesozoic Plymouth Basin (Fig. 1, inset map).

The post-consolidation evolution of the St Austell pluton is characterized by distinct episodes of vein mineralization (Jackson *et al.* 1989; Bristow & Exley 1994). These include so-called 'synbatholith mineral deposits' comprising quartz–tourmaline veins associated with the greisenization of the host granites, Sn–W mineralization and intrusion of rhyolite dykes. Fluid inclusion studies of these quartz veins and related greisen alteration halos indicate formation temperatures in excess of 250 °C (e.g. Jackson *et al.* 1977; Shepherd *et al.* 1985) relating their genesis to late magmatic fluids. Subsequently, a variety of 'post-batholith' hydrothermal vein mineralizations and alterations affected the granites, many of which are oriented north–south and termed 'cross-courses' as they intersect the earlier, predominantly east–west vein mineralization (Scrivener *et al.* 1994). The cross-courses include quartz–haematite veins associated with illitization of the granites formed at *c.*140 °C (Psyrillos *et al.* 2001), sulphide and fluorite–quartz-bearing veins formed between 100–170 °C (Shepherd & Scrivener 1987; Gleeson *et al.* 2000) and auriferous calcite–quartz veins formed at temperatures of around 110 °C (Scrivener *et al.* 1982).

Widespread kaolinization is the last mineralization event to have affected the granites, as determined from geological relationships (Bristow & Exley 1994; Psyrillos 1996). Investigations on the origin of the Cornish kaolin deposits have focused on two contrasting hypotheses. The first argues that the deposits are hydrothermal, associated with late magmatic fluids as an extension of the quartz–tourmaline veins and greisenization of the host granites (Collins 1878, 1909; Exley 1959; Bristow 1968;

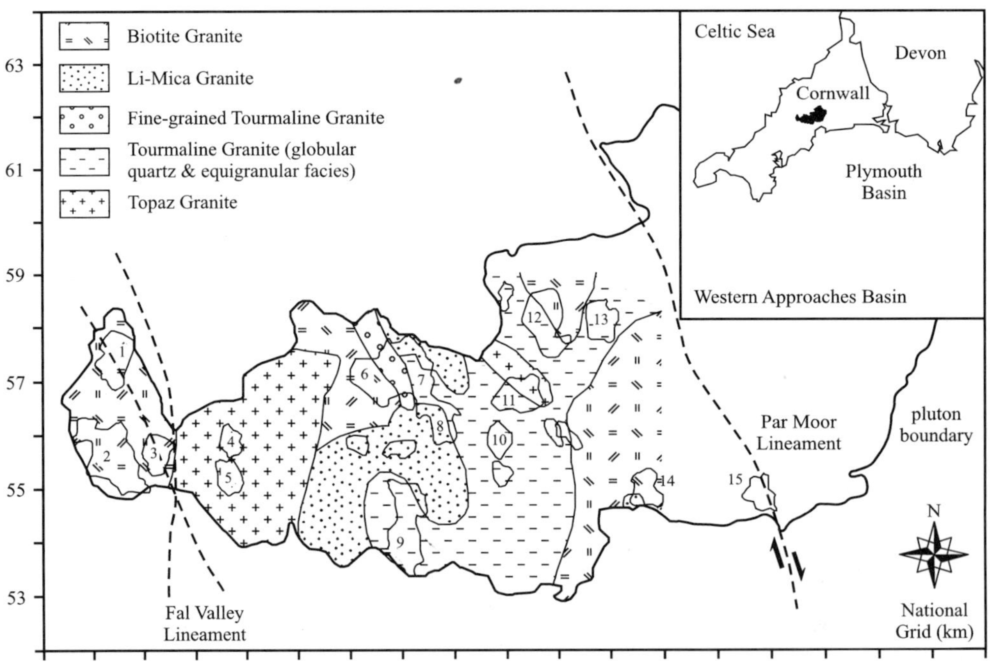

Fig. 1. Geological map illustrating the distribution of granite types in the western lobe of the St Austell pluton (Manning *et al.* 1996). The locations of two major tectonic lineaments are illustrated (Dangerfield *et al.* 1980; Bristow 1993). The boundaries of china-clay pits are shown for the year 1993. 1 Wheal Remfry, 2 Melbur, 3 Virginia, 4 Teviscoe, 5 Trethosa, 6 Dorothy, 7 Littlejohns, 8 Great Longstone, 9 Blackpool, 10 Wheal Martyn, 11 Gunheath, 12 Goonbarrow, 13 Rocks, 14 Baal, 15 Bodelva.

Bray & Spooner 1983). This hypothesis is supported by the occurrence of the kaolin deposits as inverted funnels, which extend to at least 250 m from the present day upper surface of the granites (Jackson *et al.* 1989). Kaolinization of the granite is spatially associated with zones of most intense quartz–tourmaline vein systems formed at temperatures in excess of 250 °C, and Exley (1976) documented that kaolinite 'crystallinity' increases in the proximity of some veins. However, this association may simply reflect the circulation of kaolinizing fluids along pre-existing discontinuities within the pluton, such as the stockworks of quartz–tourmaline veins (Psyrillos *et al.* 1998).

The alternative interpretation for the origin of the kaolinization advocates that the kaolin deposits formed in response to weathering of the granites, after the pluton was unroofed and exposed to tropical weathering conditions in the Mesozoic and early Tertiary (Hickling 1908; Coon 1911; Konta 1969; Bristow 1993). This hypothesis is strongly supported by the stable isotope composition of kaolin (Sheppard 1977), widely accepted to reflect isotopic equilibrium with a low-temperature meteoric fluid. Alderton & Rankin (1983) also support this concept on the basis of low-temperature, low-salinity fluid inclusions identified in the parent granites. To rationalize these contrasting hypotheses, Bristow & Exley (1994) invoke a multistage model involving an earlier high-temperature event associated with veining overprinted by low-temperature weathering. This theme is summarized by Alderton (1993), who concluded that the kaolinization took place from warm (100 °C) dilute waters of meteoric derivation towards the latter stages of magmatic–hydrothermal activity, with a probable overprint during deep inferred Tertiary weathering.

Methods

Kaolin vein samples and kaolinized granite samples were collected from quarries in the western lobe of the St Austell pluton to determine the mineralogy, paragenesis and stable isotope composition of kaolin and quartz. Kaolin separates of the >10 μm, 5–10 μm, 2–5 μm, 0.5–2 μm and <0.5 μm size fractions were produced. Mineralogical compositions were determined by XRD (CuK_α X-ray source). The terms 'kaolin' and 'kaolinite' follow AIPEA recommendations (Guggenheim *et al.* 1997): 'kaolin' is used to describe the entire 'kaolin' subgroup, whilst the term 'kaolinite' is only used when referring to the specific polytype.

Polished wafers were prepared according to the method of Barker & Reynolds (1984) for study of fluid inclusions in quartz. Quartz crystals were photographed using scanning electron microscopy (SEM) employing BSE and CL detectors to determine the relationship between fluid inclusion assemblages (FIAs) and the internal growth fabrics of quartz. Microthermometry was conducted, following the guidelines of Guscott & Burley (1993), using a FLUID Inc. modified USGS gas flow heating–freezing stage. The effects of vapour-bubble and ice metastability were extensively used to define precisely homogenization temperatures (T_h) and the melting temperatures of ice (T_{mi}) during microthermometry (cycling technique; Goldstein & Reynolds 1994) enabling temperatures to be measured to an accuracy of ±0.1 °C.

Conventional (chemical extraction) measurements of the stable isotope composition of kaolin were conducted at the Scottish Universities Environmental Research Centre (SUERC). Oxygen was extracted using the method of Clayton & Mayeda (1963) modified for ClF_3 and $\delta^{18}O$ was determined with a 'VG-SIRA 10' mass-spectrometer. Hydrogen isotope analyses were conducted according to the method of Bigeleisen *et al.* (1952) using a 'VG Micromass 602B' mass-spectrometer. The stable isotope composition of quartz was measured *in situ* with high spatial resolution, using a Fison's Instruments VG Isolab® 54 ion microprobe and following the method of Lyon *et al.* (1994). The instrument measures the ^{16}O and ^{18}O isotopes simultaneously, while isotope fractionation inside the instrument is controlled to a significant degree so that reproducibility on the standards is 1.5‰ (Saxton *et al.* 1995). This value is an order of magnitude less accurate than conventional bulk methods, but is compensated by *in-situ* analysis with high spatial accuracy. Ion-probe analyses were conducted along traverses on the same crystals to those used for fluid inclusion analysis. The analysed sample volume is that of a sputter crater with sample quantities of 3 picomoles. OH produced at the sputtered spot is detected by the instrument and the analysis is rejected, thus enabling the identification of analyses 'contaminated' by fluid inclusions or OH-bearing minerals (e.g. kaolin). Trials of $\delta^{18}O$ measurements using an ultra-pure synthetic quartz of known composition yielded a standard deviation of 1.4‰ for analysis measured over four days and less than 1‰ on any one day (Lyon *et al.* 1994). The reproducibility of results in this study from a quartz standard was ±0.6‰. All isotope data are reported relative to V-SMOW.

Results

Geology of the kaolin veins

Throughout the western lobe of the St Austell pluton, kaolinized granites are cut by networks of kaolin-bearing veins (Bristow 1977; Bristow & Exley 1994). The veins occur in the form of thin, less than 1 cm-wide, parallel-sided structures, considered to be contemporary with the kaolinization of the parent granites (Bristow & Exley 1994). These veins are particularly abundant in the Goonbarrow and Rocks quarries, as well as in the Littlejohns and Dorothy quarry complex. Cross-cutting relationships indicate that they post-date all other hydrothermal vein structures in the St Austell pluton, including the quartz–tourmaline and quartz–haematite veins (Bristow 1993; Bristow & Exley 1994; Psyrillos 1996). Kaolin veins form conjugate sets of steeply dipping planar structures, sometimes featuring striations indicative of shearing. The orientations of kaolin veins recorded in the Littlejohns/Dorothy pit complex show a systematic NNW–SSE strike with steep dip angles greater than 70° (Fig. 2) and are thus related to the so-called 'cross-course vein population' (Jackson *et al.* 1989; Alderton 1993; Scrivener *et al.* 1994). In some cases, kaolin veins occupy the cores of pre-existing quartz–tourmaline and quartz–haematite veins, cutting across the vein via joints to enter the host granite or to run along the margin of the pre-existing vein.

Kaolin vein petrography and mineralogy

Kaolin veins are simple structures, composed of monomineralic kaolin of the kaolinite polytype with textural characteristics comparable to those of finely crystalline kaolin found in the kaolinized granites (Fig. 3a). Occasionally, kaolin veins contain quartz in the form of narrow (up to 3 mm wide) white or pale grey zones restricted to the centre of the vein fills (Fig. 3b), with well-developed crystal faces and *c*-axes perpendicular to the vein wall. The crystals are less than 300 µm in length and contain abundant fluid inclusions aligned along growth bands (Fig. 4a). Although these quartz zones occur in the centre of kaolin veins, SEM observations show that kaolin crystals are not enclosed in quartz.

SEM/CL imaging of quartz from the cores of the kaolin veins reveals several generations of concentric growth zones *c*.10 µm in width. A 'generation' is defined as a correlatable sequence

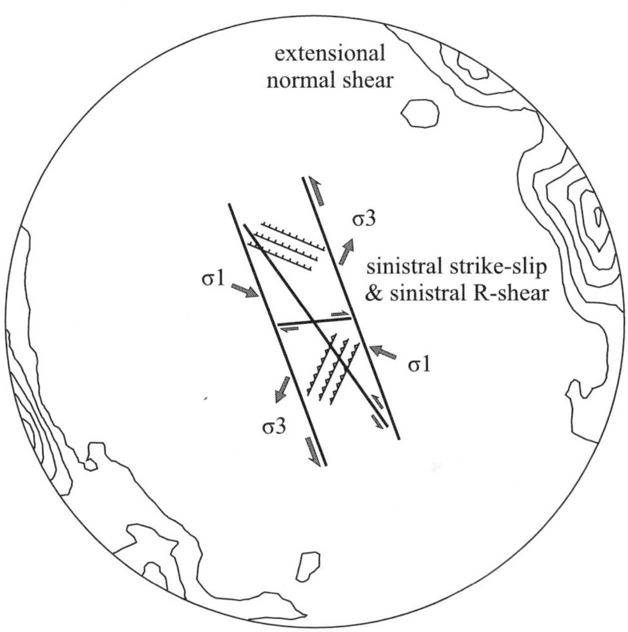

Fig. 2. Contoured stereonet illustrating the distribution of kaolin quartz vein poles in Littlejohns and Dorothy pits. The vein orientations coincide well with the theoretical fault orientations observed in a sinistral strike–slip zone, as described by Ramsay & Huber (1987) (minimum concentration 2.5%, contour interval 2.5%, maximum concentration 16.2%, 63 poles plotted).

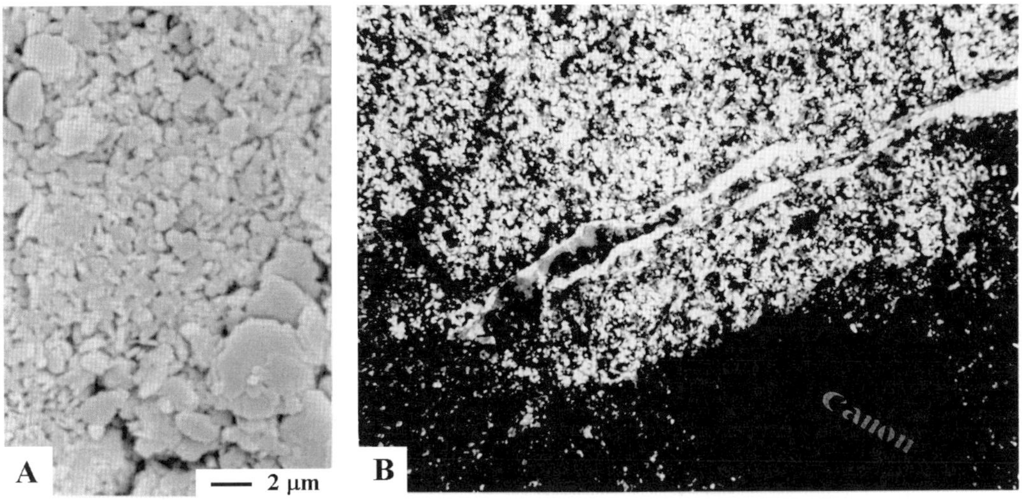

Fig. 3. (**a**) SEM/SEI micrograph illustrating kaolin crystals inside a kaolin vein. (**b**) Field photograph of a kaolin vein cross-cutting kaolinized tourmaline granites in Dorothy pit. The vein features white kaolin rims; pale grey quartz forms a narrow zone, *c*.3–5 mm wide, in the centre of the structure.

of distinguishable concentric growth zones that are bounded by a time event (such as an erosion or non-precipitation event). A 'zone' is defined as a contrast in CL emission that is assumed to be related to a property (CL emission centre) of the host quartz crystal. Figure 4b serves as a typical example of CL fabrics observed. The direction of crystal growth, as defined by the orientation of the concentric growth zones, is from the rim of the vein band towards the centre of the vein. Tightly spaced, non-luminescent growth bands with widths of 1–2 μm oriented perpendicular to the growth zones occur in areas of the crystals with abundant fluid inclusions. The quartz crystals show no evidence of brecciation and are cross-cut by non-luminescent, annealed microfractures aligned subparallel to the growth direction of the crystals. Sector zonation is also developed in places. From CL observations, three generations of concentric growth zones are identified in all the quartz crystals and constitute consistent growth events:

(1) The earliest generation is identified adjacent to the inner core of the quartz crystal and consists of concentric, brightly luminescent zones alternating with thin zones of low luminescence emission. Sector zonation is also developed in the latest concentric zones.

(2) The intermediate generation contains zones of low intensity luminescence emission with widths of 10–20 μm. The latest zones have uniform luminescence properties such that the alternating zones are difficult to distinguish because of the low CL contrast, and exhibit sector zonation. Areas with abundant fluid inclusions feature tightly spaced, poorly luminescing bands, oriented perpendicular to the growth zones. These bands are absent in the latest zones that contain no fluid inclusions.

(3) The latest generation of CL zonation overgrows all previous generations of concentric zones, although these are not crosscut by annealed microfractures observed in the CL image. The concentric CL zones have uniform, high-intensity luminescence emission and poor contrast between adjacent zones. Well-developed sector zonation is also present.

Fluid inclusion petrography and microthermometry

Three primary FIAs were defined by matching CL growth zone fabrics to the presence of fluid inclusions. FIA 1 is defined as the assemblage of primary fluid inclusions associated with the earliest concentric growth zones of the first generation and is located near the centre of the quartz crystals (Fig. 4b). FIA 2 consists of fluid inclusions associated with the late concentric growth zones of the first CL growth zone generation. FIA 3 is defined as the population of fluid inclusions hosted inside the early CL zones of the second generation. These inclusions are associated with the low intensity luminescence bands aligned perpendicular to the growth zones. No FIA was defined for the youngest concentric growth zones and annealed microfractures. The

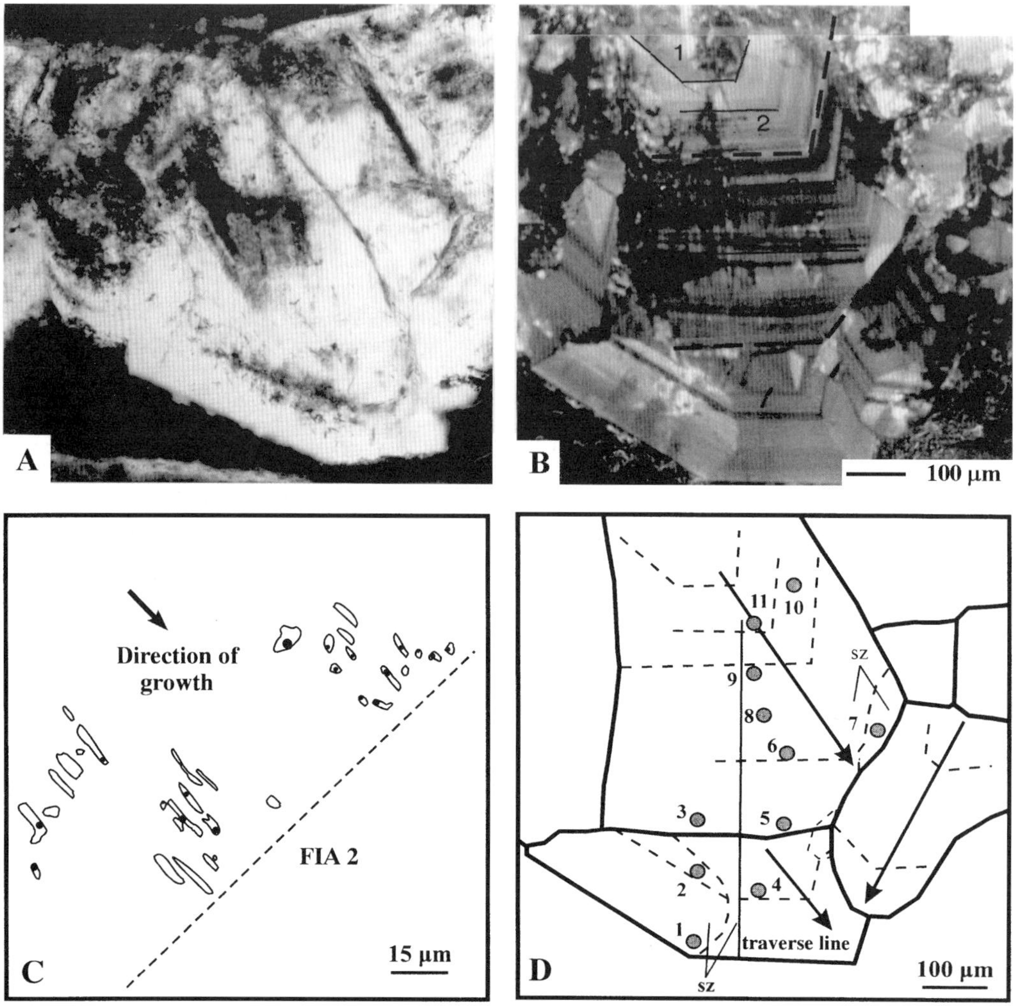

Fig. 4. (**a**) Optical microscope micrograph illustrating the quartz crystals selected for the fluid inclusion study of the kaolin veins. Areas with high abundances of fluid inclusions appear opaque. (**b**) SEM/CL micrograph of the quartz crystals illustrated in Figure 4a. Crystals feature concentric growth zones indicating a direction of growth towards the centre of the vein. Areas with abundant fluid inclusions exhibit poorly luminescing bands aligned perpendicular to the growth zones. Dashed lines indicate the boundaries of the three growth zone groups distinguished. The boundaries of primary FIAs are marked with solid lines. (**c**) Line drawing illustrating the morphology and liquid–vapour ratios of fluid inclusions of FIA 2 from the quartz crystal shown in Figures 4a and 4b. (**d**) Line drawing illustrating the topography of ion probe analysis points on the quartz crystal shown in Figures 4a and 4b. CL fabrics such as sector zones and concentric growth zones are also presented.

three primary FIAs defined comprise simple aqueous fluid inclusions, with no petrographical evidence for the presence of either immiscible gases or solid particles in the inclusions. The inclusions do not show fluorescence, indicating the absence of aromatic hydrocarbons.

FIA 1 comprises inclusions with size range 4–10 μm varying from nearly orthogonal in shape to elongate with irregular outlines. Irregular inclusions have narrow 'necks' between peripheral parts of the inclusions. Visually estimated L:V ratios range between 0.8 and 1.0. (Liquid/Vapour Ratio $= V_l/(V_l + V_v)$, where V_l = volume of liquid and V_v = volume of vapour bubble.) All-liquid inclusions have variable sizes. Large, all-liquid inclusions with sizes between 15 and 20 μm are also observed. Two-phase inclusions also exhibit variable sizes with small inclusions (2–4 μm) containing vapour bubbles. Inclusions of ***FIA 2*** are similar in size to those of FIA 1. Most of the inclusions are elongated/tabular and are aligned perpendicular to the quartz growth zones (Fig. 4c). Oval-shaped, slightly elongated inclusions also occur. Many inclusions display irregular shapes with narrow 'necks'. L:V ratios are variable and range between 0.8 and 1.0, although the relation between the inclusion size and the degree of fill is not systematic. All-liquid inclusions as large as 18 µm are present. Fluid inclusions of ***FIA 3*** are concentrated in areas of the crystals displaying low intensity luminescent lines oriented parallel to the crystal growth direction, while the inclusions are aligned parallel to the low intensity luminescent lines. Fluid inclusion sizes are similar to those reported for the previous two FIAs. Typically, inclusion size range is 2–6 µm, with some larger examples reaching a maximum of 15 µm. Inclusion shapes are variable; elongate elliptical or oval inclusions with rugged outlines are more usual, while irregular inclusions with angular outlines also occur. L:V ratios are variable between 0.8–1.0. As with the previous primary FIAs, all liquid inclusions exhibit variable sizes. Additionally, all-liquid inclusions are petrographically associated with two-phase inclusions exhibiting variable L:V ratios.

Microthermometric analyses were preceded by cooling to −5 °C to establish whether the all-liquid inclusions would nucleate vapour bubbles; none did, due to metastability. Table 1 summarizes the microthermometric measurements obtained. Homogenization temperatures (T_h) were measured for fluid inclusion assemblages 2 and 3. T_h was measured from two-phase inclusions and shows significant variation within the same FIAs. T_h values of both FIAs examined show substantial scatter. FIA 2 yields values from 117 °C to 189 °C (mean 152 °C, $\sigma = 22.5\%$). Values for FIA 3 are 125–205 °C (mean 146 °C, $\sigma = 22.3\%$).

The temperature of final ice melting (T_{mi}) was accurately measured because the phase change of ice melting was easily observed. All three FIAs yield comparable results; T_{mi} values range from −23 °C to −25 °C, giving consistent salinities of 24.0 wt% to 25.6 wt% NaCl equivalent. Freezing runs of inclusions from the three FIAs produced identical results. The inclusions required significant cooling at −110 °C to −100 °C before crystallizing clear ice. The first evidence of melting was visible at −50 °C to −55 °C (eutectic temperature T_e) when the inclusions exhibit a granular appearance because of the presence of hydrohalite. The inclusions become clear, containing ice crystals and water after a further increase in temperature (temperature of hydrohalite melting T_{mh}). At this temperature (−27 °C to −36 °C) the outlines of single ice crystals in the inclusions are visible.

Stable isotope composition of kaolin

Stable isotope analysis of kaolin was carried out using clay-separates produced from samples of kaolin veins, kaolin infilling the porosity of pre-existing veins, and kaolinized granites (Table 2). Both 5–10 µm and 0.5–2 µm size fractions were extracted, representing coarsely crystalline kaolin vermiform aggregates and finely crystalline kaolin, respectively (Psyrillos *et al.* 1999). The mineralogy of the clay separates is completely dominated by kaolin, with kaolinite being the only polytype identified. The coarser size fractions sometimes contain minor quartz impurities. The 0.5–2 µm fractions systematically yield very weak or no diffraction peaks of mica and quartz, thus showing the highest purity. Materials extracted from kaolin veins and pore-filling kaolin do not contain mica, although weak quartz diffraction peaks are recorded. XRD patterns of kaolin extracted from the kaolinized granites include minor hydrothermal muscovite, especially in the coarser size fractions, which is intergrown with kaolin in vermiform aggregates (Psyrillos *et al.* 1999).

Fractions of 5–10 µm size yield slightly variable $\delta^{18}O$ values (Table 2), averaging 18.1‰ ($\sigma = 1.5‰$, n = 9), whilst the 0.5–2 µm size fractions are more tightly clustered around an average of 19.1‰ ($\sigma = 0.9‰$, n = 9). Vein and pore-filling kaolin samples yield the most consistent results, averaging 19.8‰ ($\sigma = 0.4‰$, n = 8). The $\delta^{18}O$ compositions of the 5–10 µm size fraction

Table 1. *Summary of microthermometric data from fluid inclusions in quartz cores in kaolin veins*

FIA	T_h (°C)	mean T_h	σ (%)	n	T_e (°C)	T_{mh} (°C)	T_{mi} (°C)	n
1	—	—	—	—	−51	−31 to −35	−25	5
2	117 to 189	152	22.5%	11	−53	−27 to −34	−24	4
3	125 to 205	146	22.3%	15	−52	−30 to −36	−24	6

(–) Not measured.
T_{mh} temperatures were only measured for a few relatively large inclusions from every FIA (σ = standard deviation).

Table 2. *Kaolin stable isotope compositions measured using conventional techniques*

Sample		Size-fraction (μm)	$\delta^{18}O$ V-SMOW	δD V-SMOW
T4	kaolinized biotite granite from outside the tourmaline breccia (Wheal Remfry)	5–10	19,8	−53
		0.5–2	20,3	−58
105	kaolinized biotite granite clast from the breccia pipe (Wheal Remfry 192300, 057400)	5–10	17,9	
		0.5–2	19,3	
115	kaolinized biotite granite (Melbur 192400, 055900)	5–10	17,9	
		0.5–2	20,4	
138	kaolinized tourmaline granite (Goonbarrow 200900, 058000)	5–10	19,5	−66
		0.5–2	19,2	−69
205	kaolinized topaz granite (Treviscoe 194700, 056000)	5–10	17,7	
		0.5–2	19,1	
206	kaolinized topaz granite (Treviscoe 194700, 056000)	5–10	18,8	−64
		0.5–2	18,3	−58
215	kaolinized fine tourmaline granite (Blackpool 198000, 053700)	5–10	18,2	
		0.5–2	17,8	
222	kaolinized biotite granite (Goonbarrow 200600, 058400)	5–10	18,6	
		0.5–2	19,2	
229	kaolinized topaz granite (Rostowrack 194700, 055200)	5–10	14,5	
		0.5–2	18,3	

Sample		$\delta^{18}O$ V-SMOW	δD V-SMOW
116	pore-filling kaolin (Melbur 192400, 055900)	19,3	−65
131	kaolin from kaolin vein (Dorothy 197900, 056700)	19,7	−65
132	kaolin from kaolin vein (Dorothy 197900, 056700)	19,5	
140	kaolin from kaolin vein (Goonbarrow 200900, 058000)	20,2	
214	pore-filling kaolin (Blackpool 198000, 053700)	20,5	−58
236	pore-filling kaolin (Littlejohns 198400, 057000)	19,6	
304	kaolin from kaolin vein (Goonbarrow 201000, 058000)	20,1	−60
329	pore-filling kaolin (Littlejohns 198400, 056500)	19,3	−42

specimens are generally lowest, while 0.5–2 μm size fractions yield slightly higher values and the vein/pore-filling kaolins yield the highest $\delta^{18}O$ values. Differences in average $\delta^{18}O$ values between 0.5–2 μm and vein/pore filling kaolin fractions are statistically unimportant, as values fall within the ±0.5‰ error of the method. The average of the 5–10 μm size fractions deviates from the average of the 0.5–2 μm fractions by 1‰, while the standard deviation is 1.5‰. However, 5–10 μm fractions always contain significant mica and lesser quartz contamination. It is considered that the three types of samples consist a single population in terms of the $\delta^{18}O$ composition, with variability caused by impurities and not the paragenesis or textural types of the samples.

Unlike $\delta^{18}O$ values, δD values show some variability, from −58‰ to −69‰, and also include two analyses with values of −53‰ and −42‰. The overall mean δD is -60 ± 7.4‰ (1σ, $n = 11$), indicating a modest geological scatter over and above the analytical precision of ±5‰. Despite the limited scatter of δD compositions, the two elevated δD compositions cannot be attributed to impurities; the δD composition of sample 329 (Table 2, finely crystalline pure kaolin filling cavities of a quartz–haematite vein) is −42‰ with no impurities identified by XRD.

Oxygen isotope composition of quartz

Ion-probe measurement of $\delta^{18}O$ values was performed for the quartz crystal illustrated in Figure 4, using the analysis points in the crystal schematically illustrated in Fig. 4d. Measured quartz $\delta^{18}O$ compositions show significant variability with values between 18.4‰ and 26.8‰. The average of all the analyses is 23.6‰, similar to the average $\delta^{18}O$ value obtained from powdered quartz using conventional fluorination techniques (23.2‰; Psyrillos 1996). Variation of the quartz $\delta^{18}O$ composition across the crystal shows no systematic change in $\delta^{18}O$ composition with distance from vein core (Fig. 5). However, the correlation of sampling points with CL fabrics shows that two pairs of analysis points (numbers 2/4 and 6/7) are located

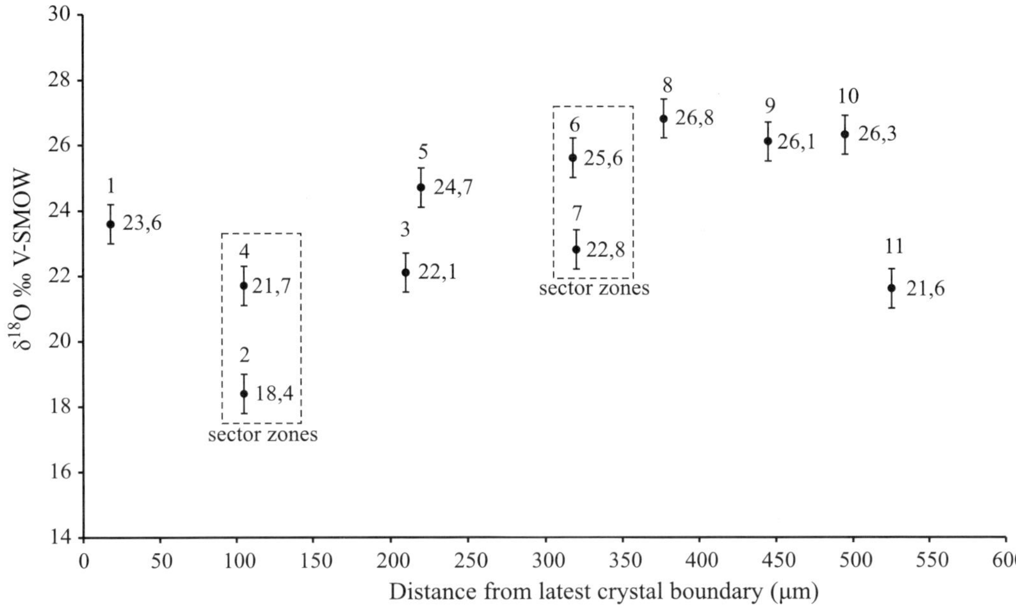

Fig. 5. Oxygen stable isotope compositions obtained *in situ* from the quartz crystal illustrated in Figure 4. Distances were measured along the traverse line illustrated in Figure 4d. Data points across adjacent sector zones show significant variation, although they were obtained along the same growth zones.

in equivalent growth zones but in adjacent sector zones. In the first pair the difference of $\delta^{18}O$ values is 3.3‰ and in the second 2.8‰, thus exceeding analytical scatter. This type of $\delta^{18}O$ distribution is reported in quartz cements from karst and fault breccias by Onasch & Vennemann (1995), who interpret the phenomenon as evidence of non-equilibrium partitioning of oxygen stable isotopes in the same concentric growth zone but across adjacent sector zones.

Interpretation

Temperature of quartz precipitation and fluid salinities

Quartz cores in the centre of kaolin veins contain fluid inclusions with constant salinities in all the FIAs examined. The freezing point depression T_{mi} yields salinities of 25 wt% equiv. NaCl (Potter *et al.* 1978), representing a brine seven times more saline than seawater. The eutectic temperature of the inclusions is around −52 °C, suggesting that the fluid composition is best described by the NaCl–$CaCl_2$–H_2O chemical system (Crawford 1981; Shepherd *et al.* 1985). The consistency of fluid inclusion salinities contrasts with the variability of the available T_h data and the observed L:V ratios. The spread of T_h values within individual inclusion populations is 50 °C to 60 °C, and may be explained either as entrapment from a heterogeneous fluid or thermal resetting. However, fluid inclusion assemblages also comprise single-phase, all-liquid inclusions paired with two-phase inclusions. The existence of all-liquid inclusions may be attributed either to necking-down after a phase change, e.g. the nucleation of vapour bubbles, or to metastability. If all-liquid inclusions were metastable, large inclusions with relatively consistent L:V ratios should be petrographically associated with smaller all-liquid inclusions. The fluid inclusion petrography characteristics, the variable T_h temperatures and the consistent salinities, as well as the presence of inclusion pairs consisting of all-liquid and two-phase inclusions, all comply with necking-down after the nucleation of vapour bubbles within the inclusions (Goldstein & Reynolds 1994). The limited number of T_h measurements and the lack of any T_h measurements for FIA 1 reflects a deliberate choice, as there is no merit in subjecting the sample to repeated heating cycles, causing decrepitation of fluid inclusions, simply to obtain inconsistent T_h data.

Fluid inclusion assemblages that have undergone necking-down after a phase change do not preserve entrapment temperatures, hence T_h values have no geological significance. However, the presence of vapour bubbles in some of the fluid inclusions provides a reliable lower limit for the temperature of entrapment and quartz precipitation. Inclusions trapped below a temperature of approximately 50 °C always exist as metastable all-liquid inclusions (Goldstein & Reynolds 1994). The presence of vapour bubbles provides unequivocal evidence that the primary fluid inclusions were trapped, and quartz precipitated at elevated temperatures of at least 50 °C.

Paragenetic relationship of kaolin and quartz

Kaolin veins are generally considered to be contemporary with the pervasive kaolinization of the St Austell granites (Bristow 1993; Bristow & Exley 1994). Kaolin crystals in the veins display the same textural characteristics as finely crystalline kaolin infilling feldspar dissolution porosity in the altered granites, suggesting they represent kaolin precipitation in open fracture porosity. An alternative interpretation, that moving fluids transported kaolin crystals from the kaolinized granites into open fractures, is unlikely, as the vein fills do not contain contaminant minerals from the wall rock granites. The association of quartz cores with kaolin veins has not been reported previously in the pluton and merits close attention. Quartz crystals inside kaolin veins are perpendicular to the vein walls and exhibit CL fabrics indicative of growth towards the vein centre. This growth arrangement indicates that quartz precipitation occurred within open fractures. Vein walls and individual quartz crystals display no evidence of shearing, such as brecciation or microfractures oriented subparallel to the vein walls. Quartz precipitation therefore initiated on the margins of the dilated fractures and resulted in the occlusion of fracture porosity. Regardless of the vein dilation mechanism, the precipitation of quartz must have taken place under near-hydrostatic pressures in a structural regime comparable to the seismic valving model of Sibson (1994).

The petrographical relations testify that a quartz precipitation event post-dated kaolin precipitation in the kaolin veins. However, it is not possible to assess the genetic links, if any, between kaolin and quartz, as there are no observations of kaolin enclosed in quartz. The oxygen stable isotope compositions of kaolin (Sheppard 1977; Psyrillos 1996) and quartz are also inappropriate to test the genetic relationship between the two minerals, as the isotopic composition of quartz is affected by non-equilibrium partitioning. It is thus assumed that kaolin and quartz precipitation in the kaolin veins are not contemporary, so the paragenetic sequence is summarised: quartz–haematite veins → kaolinization → quartz in kaolin veins.

Paragenetic and geological implications

The interpretation of the fluid inclusion assemblages has significant implications regarding the temperature at which the kaolinization took place. Necking-down after a phase change (nucleation of vapour bubbles) affects fluid inclusion populations uniquely during uplift and decrease of the ambient temperature. Small temperature decreases are sufficient to cause the nucleation of vapour bubbles in fluid inclusions while the necking process is still in progress. This is consistent with the geological evolution of the St Austell pluton, which is considered to have undergone continuous uplift since its emplacement in the Late Carboniferous–Early Permian (Alderton 1993).

An independent evaluation of the temperature at which the kaolinization took place can be made even though there is no definitive genetic link between quartz and kaolin in the kaolin veins. Paragenetic and geological relations indicate the quartz–haematite hydrothermal veins and the quartz cores of kaolin veins bracket the kaolinisation event. The quartz cores thus define a lower temperature limit of 50 °C for the pervasive kaolinization in the St Austell pluton. The depth at which the kaolinization took place cannot be accurately defined because of the lack of meaningful quartz T_h data. However, the minimum temperature of 50 °C combined with the salinity of the inclusions (25 wt% NaCl equiv.) yields a minimum depth of quartz precipitation of *c.*1 km, assuming a hydrostatic load, an average rock density of 2.65 gr/cm^3 and a present-day thermal gradient of 40 °C/km (as for the Carnmenellis pluton, Lee 1986). An upper temperature limit of around 100 °C for the kaolinization in the St Austell pluton is deduced by Psyrillos *et al.* (1998) on the basis of geochemical considerations regarding silica mineral authigenesis in the kaolinized granites.

Isotopic composition of fluids involved in the kaolinization

In his seminal study of the genesis of the Cornubian kaolin deposits, Sheppard (1977) documents that kaolin stable isotope compositions plot within a very narrow $\delta^{18}O$–δD field, which is considered to represent a single population

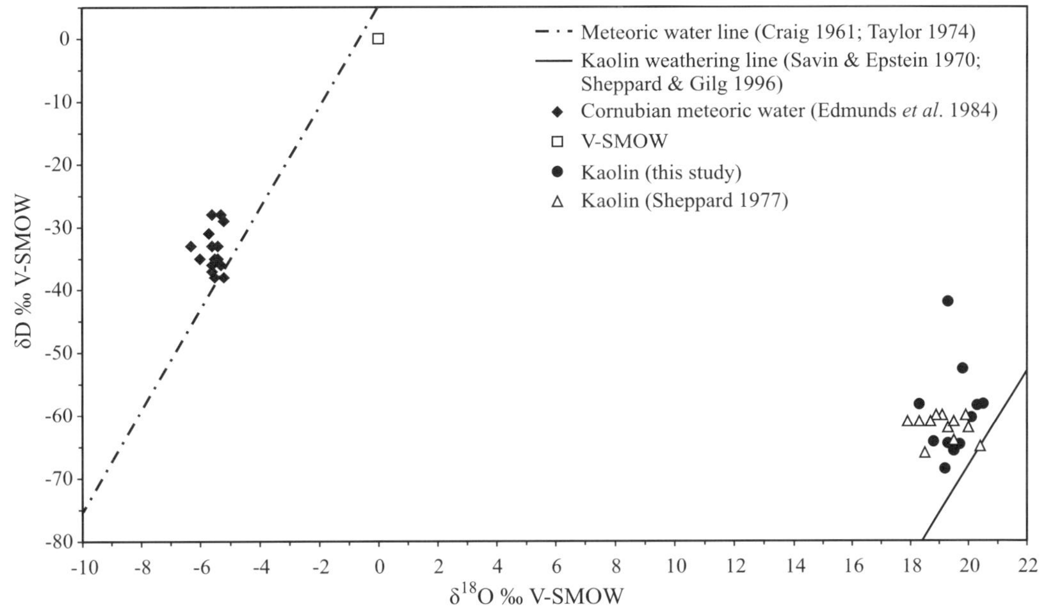

Fig. 6. Plot of kaolin $\delta^{18}O$–δD compositions from the kaolin deposits of SW England.

that has undergone no post-formation isotopic exchange of oxygen and hydrogen. He concludes that the kaolinization occurred as a result of meteoric water ingress to the pluton on the basis of: (1) the close proximity of isotope composition to the kaolin weathering line, (2) the limited scatter of isotope compositions, (3) the observation that kaolin $\delta^{18}O$ and δD compositions comply with equilibrium isotope fractionation from present-day meteoric waters at surface temperatures.

Figure 6 is a plot comparing the $\delta^{18}O$–δD data pairs obtained during this study with those of Sheppard (1977). Both data sets are similar in their proximity to the kaolin weathering line and in their absence of any correlation between $\delta^{18}O$ and δD, although isotope compositions from this study show a marked deviation towards δD compositions higher than those reported by Sheppard (1977). In this study, independent fluid inclusion and paragenetic evidence show that kaolinization must have occurred at a minimum temperature of 50 °C. The distribution of $\delta^{18}O$–δD isotope compositions may be the result of hydrogen stable isotope re-equilibration (Sheppard & Gilg 1996), in which the oxygen isotope composition is unaffected whilst hydrogen isotopes attempt to reach equilibrium with fluids other than those responsible for kaolin precipitation. Experimental evidence and theoretical considerations on the kinetics of stable isotope exchange in clay minerals indicate that hydrogen exchanges faster than oxygen (e.g. O'Neil & Kharaka 1976; Suzuoki & Epstein 1976; Fortier & Gilleti 1991). It is plausible that the abnormally elevated δD compositions identified either preserve an early isotopic signature, or record modification of the original δD composition, ignoring any effect of salinity on stable isotope partitioning.

Regardless of hydrogen stable isotope re-equilibration, the genetic interpretation of stable isotopes in kaolin must be reconsidered in the light of the independent determination of the kaolinization temperature in excess of 50 °C but below 100 °C. Figure 7a illustrates the oxygen fractionation curve in the kaolin–water system, using the equation of Sheppard & Gilg (1996) with a $\delta^{18}O$ value of 19.5‰. The $\delta^{18}O$ compositions of water in equilibrium with kaolin, at temperatures between 50 °C and 100 °C, range from 0‰ to 6.5‰. For hydrogen (Fig. 7b), assuming that hydrogen re-equilibration has not taken place, the curve for kaolin δD −60‰ represents the average kaolin δD value. In this context the δD compositions of water in equilibrium with kaolin range from −31‰ to −36‰. If the elevated δD compositions reported herein preserve an original hydrogen isotopic signature that has been subsequently reset, the curve for kaolin δD −42‰ yields water δD compositions in equilibrium with kaolin from −13‰ to −19‰.

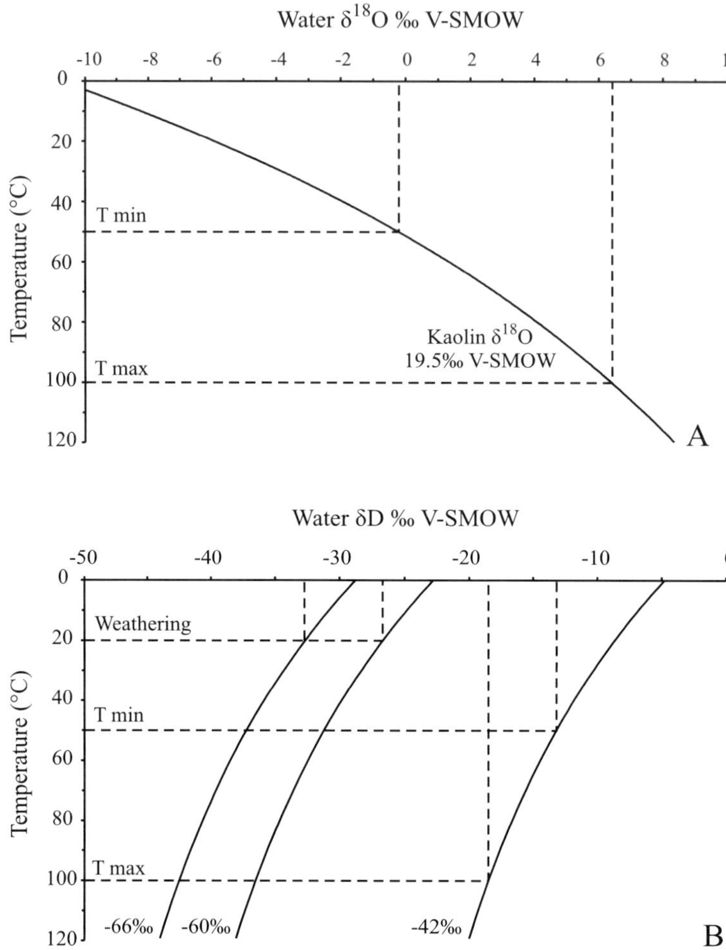

Fig. 7. (**a**) Oxygen stable isotope fractionation curve for kaolin ($\delta^{18}O$ 19.5‰). (**b**) Hydrogen stable isotope fractionation curves for kaolin. The curve for δD −42‰ is considered to represent equilibrium of kaolin with mineralizing fluids at temperatures between 50 °C and 100 °C. The curves for δD −66‰ and −60‰ represent the re-equilibrated kaolin δD composition in equilibrium with present-day meteoric waters.

Basin evolution and kaolin genesis: a new genetic model for kaolinization

Nature and sources of aqueous fluids

The previous discussion of the probable stable isotope composition of the fluids involved in the kaolinization of the St Austell pluton granites argues against the involvement of meteoric fluids in the kaolinization. Meteoric water $\delta^{18}O$ compositions generally range from −60‰ to 0‰, with the highest $\delta^{18}O$ meteoric water compositions encountered in areas with high-temperature climates, located at near-equatorial latitudes (Craig 1961; Faure 1986). Meteoric waters with $\delta^{18}O$ up to +5‰ are also encountered in closed basins, where extreme evaporation prevails at near-equatorial latitudes (Faure 1986).

Palaeogeographic reconstructions of NW Europe restore Cornubia to near-equatorial latitudes during the Late Carboniferous (Westphalian *c*.0°N and Stephanian *c*.10°N; Ziegler 1990), prior to the emplacement of the batholith. Ever since, Cornubia has followed a northerly migration path towards increased latitudes; by Rotliegend times it drifted to *c*.20°N and during the Late Triassic to 30–35°N (Ziegler 1990). Changes in latitude must be reflected in the isotopic composition of meteoric waters. The $\delta^{18}O$ of present-day meteoric waters in the Cornubian peninsula varies from −4‰ to −6‰ with an average between −5.0‰ and −5.5‰

(Edmunds *et al.* 1984). Therefore, meteoric fluids enriched in ^{18}O can only be associated with arid climatic conditions prevalent in Cornubia during the Permian–Triassic (Alderton 1993). Other likely sources of fluids with high $\delta^{18}O$ compositions are the sedimentary basins flanking the Cornubian peninsula. Bjørlykke *et al.* (1986), Burley & MacQuaker (1992), Egeberg & Aagaard (1989) and Warren & Smalley (1994) provide $\delta^{18}O$ compositions of present-day sedimentary brines from the North Sea ranging between −4‰ and +10‰.

In terms of the inferred water δD compositions, candidate fluids include present-day meteoric waters (−28‰ to −40‰, Edmunds *et al.* 1984), meteoric water in areas of extreme evaporation conditions (Faure 1986), as well as brines from sedimentary basins (0‰ to −45‰, Bjørlykke *et al.* 1986; Burley & MacQuaker 1992; Egeberg & Aagaard 1989; Warren & Smalley 1994).

Psyrillos *et al.* (1998) show that the granite kaolinization may result from a range of fluids undersaturated with respect to carbonates and sulphates, provided that the initial fluid $\log a(K^+/H^+)$ is lower than approximately 1.8 at 50 °C, 2.0 at 75 °C and 2.1 at 100 °C. This excludes waters from carbonate aquifers and meteoric waters from areas with extreme evaporative conditions, leaving highly modified meteoric waters and/or sedimentary brines from siliciclastic aquifers as candidate fluids.

Meteoric water descending the pluton would undergo chemical and isotopic modification due to the interaction with the host rocks. The character and extent of modifications depend on the composition of the host rocks, the prevailing temperatures and the time available for the interactions to take place. The country rocks of the St Austell pluton consist of metamorphosed schists and slates of Lower Devonian age, interbedded with limestones (Edmonds *et al.* 1975). The mineralogical composition of the country-rocks is thus similar to that encountered in a clastic sedimentary sequence in the diagenetic realm. Although the permeability characteristics of the host-rocks are controlled by fracture porosity, the fluid–rock interaction processes are comparable to those taking place during burial diagenesis.

Mesozoic sedimentary basins of the Western Approaches, flanking the Cornubian peninsula, are the alternative potential fluid source. After the emplacement and consolidation of the Cornubian batholith in the Late Carboniferous, the Cornish peninsula underwent erosion and uplift through to the Permian–Triassic, but may have had a Mesozoic cover prior to present-day exposure (Roberts 1989; Evans 1990; Dart *et al.* 1995; Burley & Cornford 1998), at least on the margins of the batholith. The location of the St Austell pluton is *c.*15 km north of the Plymouth Basin margin (Fig. 1). The basin contains a thick sequence of Permian–Triassic sediments of the Sherwood Sandstone Group and the Mercia Mudstone Group, consisting of sandstones, anhydritic claystones and siltstones. Up to 9 km thick (BIRPS & ECORS 1986), these sediments were deposited on the pre-Permian basement during the rapid fault-controlled subsidence of the basin and are overlain by Upper Cretaceous and Eocene sediments (Evans 1990). Further out in the English Channel, south of the Alderney Fault Zone, additional thick Jurassic–Cretaceous sediments (>2 km) are preserved (Ruffell 1995). Reconstruction of the burial history of these basins from offshore well data thus suggests that up to 10 km of Mesozoic sediments, mostly Permian–Triassic, were deposited in the Western Approaches, now only partially preserved because of middle Jurassic, Cretaceous and Tertiary uplift events (Evans 1990; Ruffell 1995).

Thus the Mesozoic basins immediately south and adjacent to the pluton provide a potential source of enormous volumes of formation waters. The chemical and stable isotope composition of formation fluids from the Permian–Triassic of the Plymouth Basin would have been comparable to present-day brines associated with the Permian–Triassic of the North Sea or the English Channel, with moderate to high salinities (20–25 wt% solute) and compositions dominated by alkalis and alkali earth metals. Psyrillos *et al.* (1998) show that formation waters from Permian–Triassic aquifers are geochemically capable of precipitating the St Austell kaolin assemblage, unlike waters from Jurassic aquifers. Moreover, salinities of fluid inclusions (25 wt% NaCl equiv.) in quartz in the kaolin veins are consistent with the salinities of formation fluids from Permian–Triassic aquifers.

Timing of fluid migration

Convective groundwater flow in and around the Cornubian batholith was driven by granite-produced radiogenic heat (Sams & Thomas-Betts, 1988a,b). These authors suggest that fluid convection cells are controlled by the extent of erosion of the plutonic body. If the pluton is not exposed on the surface, then surface fluids flow downwards around the flanks of the pluton and convect upwards inside the pluton

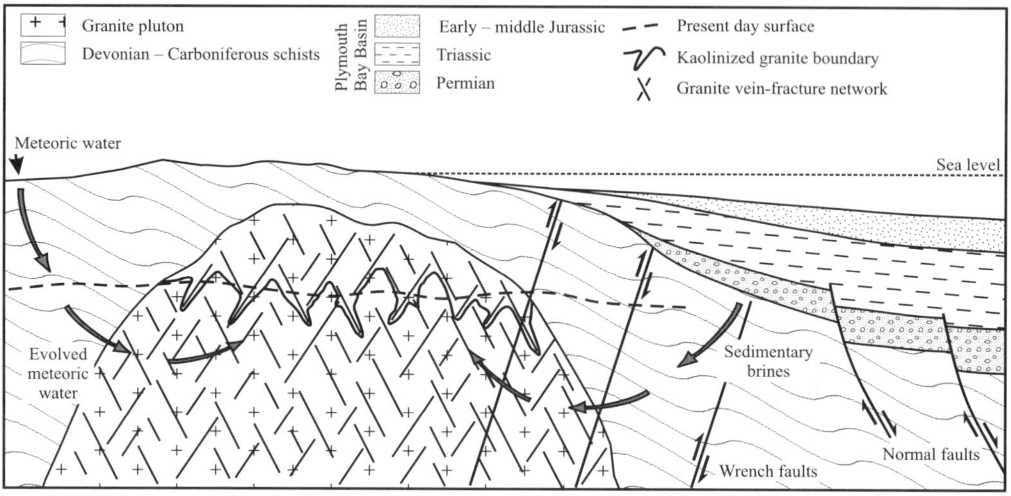

Fig. 8. Schematic illustration showing alternative sources of mineralizing fluids involved in the kaolinization of the St Austell pluton granites. Sedimentary brines from the thick Permian–Triassic sequence of the Plymouth Bay Basin and/or evolved meteoric fluids with a flow history inside the pluton's country rocks, were channelled inside the pluton via wrench faults that cross-cut the Cornubian peninsula. Once inside the pluton, fluid convected upwards using the dense network of fractures and hydrothermal veins, thus causing widespread kaolinization in the form of inverted funnels.

(Fig. 8). If a pluton is exposed, surface fluids follow an opposing route, down through the pluton and convecting upwards towards its flanks. An irreversible geological process, the erosion of the cover to the pluton, thus controls fluid convection and flow.

Sedimentary brines, by contrast, may have been focussed in the St Austell pluton via large, dextral wrench faults of Variscan age (Fig. 8) that cross-cut the Cornubian peninsula (Hobson & Sanderson 1983) in accordance with the model of Jackson *et al.* (1989). These structures have a prolonged history of reactivation during the Mesozoic and Cenozoic (Hobson & Sanderson 1983; Gayer & Cornford 1992; Willis-Richards 1993). In the St Austell pluton, two major fault zones, the Fal Valley and the Par Moor lineaments (Fig. 1; Dangerfield *et al.* 1980; Bristow 1993) strike NW–SE and generally form the eastern and western boundaries of the area hosting the most extensive kaolinization. Psyrillos *et al.* (1998) suggest that kaolin veins in the St Austell pluton formed due to tectonic deformation, on the basis of geochemical considerations. This conclusion is also supported by the observed structural styles indicative of tectonic deformation (conjugate sets of extensional fractures and shear striations). The consistency of vein orientations in large areas such as the Littlejohns/Dorothy pit complex (Fig. 2) also supports a tectonic origin; if volumetric contraction of the granite took place during kaolinization, the veins would be expected to be randomly orientated, with no consistency between kaolinized areas separated by unaltered granite. Vein orientations (Fig. 2) conform to the stress-fields expected from sinistral strike-slip faults (Ramsay & Huber 1987).

It is conceivable that both fluid transport mechanisms were active during the kaolinization, thus leading to mixing of fluids in or around the pluton. The involvement of sedimentary brines and/or evolved meteoric waters implies that the kaolinization of the pluton occurred before the complete erosion of the top cover of the pluton.

Thermal evolution and age of the kaolinization

The regional geology of the Cornubian massif and the adjacent sedimentary basins constrain the age of the kaolinization. Jackson *et al.* (1989) suggest that the unroofing of the Cornubian batholith occurred during the Early Mesozoic. Other estimates of the unroofing of the Dartmoor pluton use the nature and distribution of the granite detritus in sediments of the western English Channel. This detritus suggests that the Dartmoor pluton was not unroofed before the early Jurassic (Cosgrove & Salter 1966), based on the distribution of kaolin in the post-Armorican sediments of southwestern England, or the Late Cretaceous (Groves 1931), based on the origin of rock fragments in sediments. A Late Cretaceous age for the unroofing of the Dartmoor pluton is thus considered to be more representative. If the Dartmoor pluton is used as an analogue for the St Austell pluton, then the kaolinization occurred before the Late Cretaceous.

To estimate a maximum age for the kaolinization, additional isotope dating information can be used. Bray & Spooner (1983) report K/Ar isotope dating of muscovite associated with quartz–tourmaline veins which yields a numerical average of 271.4 ± 4.5 Ma. Similarly, groundmass muscovite ages from rhyolite dykes yield a numerical average of 271.7 ± 1.4 Ma. Hydrothermal and magmatic muscovites from the present-day surface of the St Austell pluton yield similar Ar/Ar ages (Chesley *et al.* 1993), suggesting that hydrothermal and magmatic muscovites cooled below the closure temperature of muscovite to Ar diffusion at approximately the same time and not later than 270 Ma. Harrison & McDougall (1980) and Snee *et al.* (1988) suggest that muscovite closure to Ar diffusion occurs at 320 ± 50 °C.

The subsequent thermal history of the St Austell pluton and the Cornubian batholith is, by comparison, poorly constrained. Chen *et al.* (1996) provide the only evidence regarding the thermal evolution of the Carnmenellis pluton into the Mesozoic–Cenozoic, through the use of apatite fission track analysis (AFTA). The thermal modelling results of Chen *et al.* (1996) suggest that the present-day surface of the Carnmenellis pluton cooled to temperatures less than the AFTA partial resetting temperature (125 ± 25 °C) between 155 and 137 Ma. During this period the Carnmenellis pluton underwent rapid cooling at a rate of 2.8 °C/Ma, while subsequent cooling to the present day was at a rate of 0.5 °C/Ma.

The available thermo-chronological data are valid for the present-day surface of the Carnmenellis and St Austell plutons. The increased cooling rate of the pluton at the time and temperature of AFTA resetting is incorporated as rapidly decreasing temperature at 146 Ma (2.8 °C/Ma), while the subsequent thermal evolution of the pluton is slower (0.5 °C/Ma), decreasing asymptotically towards the present-day surface temperature of 15 °C (Fig. 9a). The temperature profile is correlated with the inferred temperature range of 50–100 °C for the kaolinization, assuming the geothermal gradients and the ambient temperatures in the pluton were controlled by hydrothermal fluid circulation in the granites. The inferred age of the kaolinization is between 110 Ma and 130 Ma, before the unroofing of the pluton. The kaolinization occurred during the onset of the rapid cooling event as recorded by AFTA (Chen *et al.* 1996).

The process of stable isotope re-equilibration in clay minerals slows with decreasing temperature, although the behaviour of stable isotopes at temperatures generally lower than 100 °C cannot be determined with confidence (Sheppard & Gilg 1996). Factors affecting stable isotope re-equilibration for a given mineral include temperature variation, time and particle size (O'Neil & Kharaka 1976; Cole & Ohmoto 1986; Sheppard & Gilg 1996). The process of hydrogen stable isotope re-equilibration in kaolin is considered to have occurred due to interaction with waters having stable isotope compositions similar to those of present-day meteoric water during the *c.*100 Ma after the unroofing of the pluton. Present-day groundwaters, found in some of the mines in the region, are meteoric in origin and as hot as 50 °C (Durrance *et al.* 1982; Edmunds *et al.* 1984, 1988).

Regional implications

This thermo-chronological reconstruction has significant implications for the interpretation of the regional geology during the kaolinization of the St Austell granites. According to Evans (1990) and Ruffell (1995), the sedimentary basins of the Western Approaches underwent rifting associated with the opening of the Atlantic during the Jurassic and Early Cretaceous. During this period, basins south of the Cornubian massif (Plymouth and St. Mary's basins) underwent fault-controlled subsidence whilst the northern basins (e.g. Bristol Channel and Celtic basins) underwent regional uplift, reflected in the thermal evolution of the Cornubian batholith (Chen *et al.* 1996).

The thermal evolution of the pluton is correlated with the geological evolution of the adjacent sedimentary basins. Figure 9b illustrates a subsidence curve for the base of the Triassic, constructed from part of seismic line SWAT 9 in the Plymouth Basin (Evans 1990). The basin formed due to fault-driven subsidence during the Triassic. Its subsequent evolution is characterized by thermal relaxation subsidence until the Late Jurassic, when rifting occurred (Evans 1990). The Plymouth Basin was located at the northern flank of the Southwest Channel basin and underwent substantial thermal uplift (Cochran 1983; Steckler 1985) resulting in erosion, expressed by an unconformity and the absence of Late Jurassic–Early Cretaceous

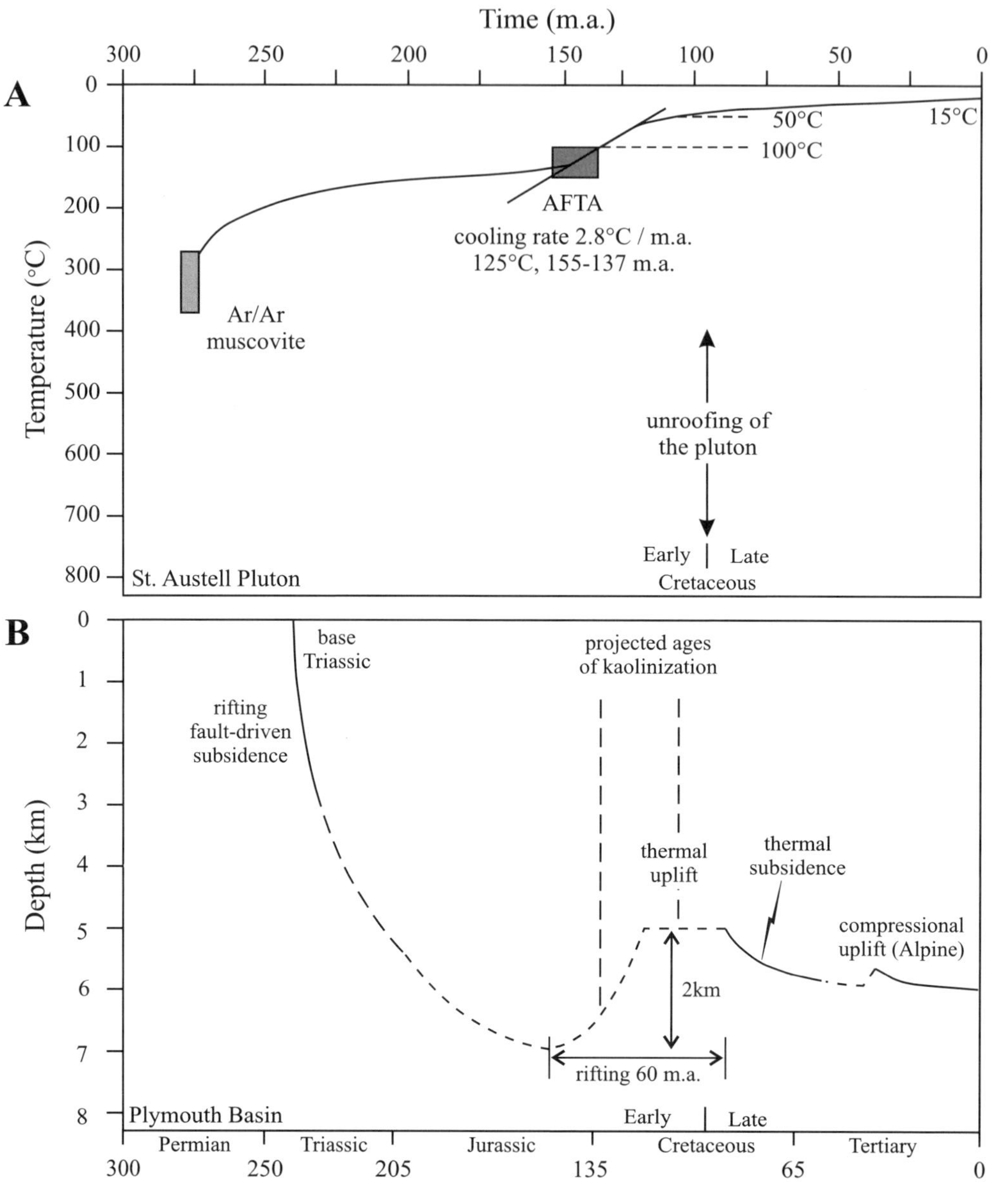

Fig. 9. (**a**) Temperature versus time curve for the present-day surface of the St Austell pluton (modified from Chen *et al.* 1996). The data used include a U/Pb monazite cooling age (725 ± 25 °C, Chesley *et al.* 1993) and the Ar/Ar cooling age of hydrothermal muscovite (320 ± 50 °C, Chesley *et al.* 1993). The AFTA age is valid for the present-day surface of the Carnmenellis pluton and represents cooling to less than *c.*125 °C (Chen *et al.* 1996). The rectangles represent the age and temperature uncertainties. AFTA thermal modelling yields an elevated cooling rate of 2.8 °C/Ma at 155–137 Ma and a subsequent cooling rate to the present day of 0.5 °C/Ma (Chen *et al.* 1996). (**b**) Generalized subsidence curve for the base of the Triassic in the Plymouth Basin constructed from seismic line SWAT 9 (Evans 1990). The basin underwent fault-controlled subsidence during the Early Triassic and subsequent thermal subsidence until the Middle Jurassic. During the Middle Jurassic to Early Cretaceous the basin underwent regional uplift related to rifting, associated with the opening of the Atlantic, and resulted in the erosion of up to 2 km of sediments (Evans 1990). The basin underwent continued thermal subsidence to the present day, except for the Tertiary inversion and uplift, which is associated with Alpine compression. The extrapolated ages of the kaolinization are projected on the subsidence curve and correspond well with the Mid-Jurassic to Early Cretaceous uplift.

strata in the northern parts of the basin. The total duration of rifting was approximately 50–60 Ma (Hillis 1988; Evans 1990). During this time regional uplift affected the northern basins of the Western Approaches as well as the Cornubian massif. According to Hillis (1988) the regional uplift had an axis of symmetry along the Cornubian massif due to the low density rocks of the batholith. Following the end of rifting, the Plymouth Basin underwent thermal subsidence that resulted in the deposition of Late Cretaceous and Tertiary strata towards the southern extremities of the basin. During the Tertiary the basins underwent inversion and minor uplift associated with Alpine compression (Hillis 1988; Evans 1990).

The extrapolated age of the kaolinization is projected in Figure 9b and is in good agreement with the Late Jurassic regional uplift event recorded in the Plymouth Basin. The chronological association of the kaolinization with basin uplift provides a geological framework for the kaolinization of the pluton and accounts for the observed tectonic features of the kaolin veins. Wrench faults in the Cornubian massif may have been reactivated and focussed basinal brines in the massif and the St Austell pluton from the Permian–Triassic sediments of the Plymouth Basin (Fig. 8). The rifting event in the southern parts of the Plymouth Basin during the Middle Jurassic to Early Cretaceous introduced a major thermal anomaly as predicted by models of rifting (Cochran 1983; Steckler 1985) and resulted in increased heat flow in the basin. This thermal anomaly promoted fluid movement within sediments of the Plymouth Basin and fluids may have migrated towards the northern extremities of the basin and into the Cornubian massif.

Basinal brine incursions in the Cornubian Massif

Basinal brines are linked with several different styles of extensive mineralization phenomena, such as the Northern Pennine orefield in the UK and the Mississipi Valley Type deposits of the central USA (e.g. Sawkins 1966; Alderton 1978). In the Cornubian peninsula, basinal brines are considered to be responsible for extensive base metal polymetallic mineralization associated with the 'cross-courses' (Shepherd & Scrivener 1987) and probably of Triassic age (Scrivener *et al.* 1994). Gleeson *et al.* (2000, 2001) provide further evidence of basinal brine involvement associated with cross-courses in the area of the Tregonning pluton, also suggesting that these brines originated from evaporitic systems in the Permian–Triassic sedimentary sequence of the Western Approaches basins.

Pervasive basinal brine migrations are here argued to have occurred into the Cornubian Massif for a much longer geological period than previously realized. Equally importantly, basinal brines are shown to be associated with widespread kaolinization of the granites, a style of mineralization fundamentally different to the polymetallic ores previously linked with basinal brines.

An elegant aspect of the model presented herein is that the inferred basinal brines had compositions similar to the ones associated with cross-courses and probably originated from the same aquifer systems, namely the thick Triassic sequence of the Plymouth Bay Basin. There is, however, a significant difference in the timing of mineralization. Basinal brine incursion associated with the kaolinization of the St Austell pluton occurred during the Mid-Jurassic to Early Cretaceous, many millions of years later than the generally accepted Triassic age of the cross-courses. At that time the Triassic sedimentary sequence in the Plymouth Basin had reached maximum burial depth with a potential supply of formation fluids available for mobilization during tectonic and thermal events. The rifting event of the late-Jurassic to Early Cretaceous provided both the tectonic instability and the regional thermal anomaly to expel deeply buried formation waters into the Cornubian Massif. The nature and origin of Triassic cross-course fluids in the area is by comparison poorly explained, as the neighbouring sedimentary basins had either a very thin sedimentary fill or had not even started their phase of fault-driven subsidence.

In addition to explaining the origin of kaolinization in a 'basement' area, this study demonstrates the value of integrating investigations of highs with basin studies. It shows that part of the geological history of the basin is recorded in the phenomena that are observed in an adjacent granite. This approach could be equally fruitful in other petroleum provinces.

Conclusions

The application of fluid inclusion methods and reassessment of the stable isotope composition of kaolin, demonstrates that the kaolinization of the St Austell granites is a low-temperature hydrothermal mineralization event, which occurred at temperatures between 50 °C and 100 °C.

A new genetic model for the formation of the St Austell kaolin deposits is proposed on the basis of geological, paragenetic and microthermometric considerations combined with a thermal reconstruction of the evolution of the pluton into the Mesozoic–Cenozoic. The model suggests that the kaolinization occurred prior to the unroofing of the pluton and probably during the Middle Jurassic to Early Cretaceous. The kaolinization is thus considered to be contemporary to the major Mid-Jurassic to Early Cretaceous uplift that affected the Cornubian massif as a consequence of rifting in the Western Approaches. The types of fluids involved in the kaolinization were either basinal brines from the Permian–Triassic of the nearby Plymouth Basin to the south and/or evolved meteoric water with a history of prolonged subsurface flow in the country rocks of the pluton. Thus the history of the pluton records events that occurred in the adjacent offshore basin, rather than the penetration of meteoric fluids during deep weathering (cf. Sheppard 1977).

This study forms part of a PhD thesis by A. Psyrillos, who acknowledges the financial support given by the University of Manchester (Boyd Dawkins Scholarship). The authors are grateful to ECCI Ltd (now Imerys Ltd – Pigments & Additives Group), and especially to J.H. Howe, for the invaluable assistance provided during fieldwork. The authors are grateful to S.M.F Sheppard and an anonymous referee for a constructive review of an earlier version of the manuscript. The two anonymous referees who reviewed the current version of the manuscript are also thanked for their critical remarks.

References

ALDERTON, D. H. M. 1978. Fluid inclusion data for lead-zinc ores from S. W. England. *Transactions of the Institute of Mining and Metallurgy*, **87**, B132–135.

ALDERTON, D. H. M. 1993. Mineralisation associated with the Cornubian Granite Batholith. *In*: PATTRICK, R. A. D. & POLYA, D. A. (eds) *Mineralisation in the British Isles*. Chapman & Hall, London, 270–354.

ALDERTON, D. H. M. & RANKIN, A. H. 1983. The character and evolution of hydrothermal fluids associated with the kaolinized St. Austell granite, SW England. *Journal of the Geological Society, London*, **140**, 297–309.

BARKER, C. E. & REYNOLDS, T. J. 1984. Preparing doubly polished sections of temperature sensitive sedimentary rocks. *Journal of Sedimentary Petrology*, **54**, 635–636.

BIGELEISEN, J., PERLMAN, M. L. & PROSSER, H. C. 1952. Conversion of hydrogenic materials to hydrogen for isotopic analysis. *Analytical Chemistry*, **24**, 1356–1357.

BIRPS & ECORS 1986. Deep seismic reflection profiling between England, France and Ireland. *Journal of the Geological Society, London*, **143**, 45–52.

BJØRLYKKE, K., AAGAARD, P., DYPVIK, H., HASTINGS, D. S. & HARPER, A. S. 1986. Diagenesis and reservoir properties of Jurassic sandstones from the Haltenbanken area, offshore mid-Norway. *In*: SPENCER, A. M., HOLTER, E., CAMPBELL C. J., HANSLIEN, S. J., NELSON, P. H. H. & ORMAASEN, E. G. (eds) *Habitat of hydrocarbons on the Norwegian continental shelf*. Norwegian Petroleum Society, Graham and Trotman Ltd., London, 275–286.

BRAY, C. J. & SPOONER, E. T. C. 1983. Sheeted vein Sn–W mineralisation and greisenisation associated with economic kaolinisation, Goonbarrow China Clay Pit, St. Austell, Cornwall, England: Geologic relationships and geochronology. *Economic Geology*, **78**, 1064–1089.

BRISTOW, C. M. 1968. The derivation of the Tertiary sediments in the Petrockstow Basin, North Devon. *Proceedings of the Ussher Society*, **2**, 29–35.

BRISTOW, C. M. 1977. A review of the evidence for the origin of the kaolin deposits in SW England. *In*: GALAN, E. (ed.) *Proceedings of the 8th International Kaolin Symposium and Meeting on Allunite, Madrid–Rome*.

BRISTOW, C. M. 1993. The genesis of the china clays of Southwest England—A multistage story. *In*: MURRAY, H., BUNDY, W. & HARVEY, C. (eds) *Kaolin genesis and utilisation*. Clay Minerals Society of America, Special Publication, **1**, 171–203.

BRISTOW, C. M. & EXLEY, C. S. 1994. Historical and geological aspects of the china-clay industry of Southwest England. *Transactions of the Royal Geological Society of Cornwall*, **21**, 247–314.

BURLEY, S. D. & CORNFORD, C. 1998. Carbonate cements constrain the burial history of the Portledge–Peppercombe Permian outlier, North Devon. *Geoscience in SW England*, **9**, 188–202.

BURLEY, S. D. & MCQUAKER, J. H. S. 1992. Authigenic clays, diagenetic sequences and conceptual diagenetic models in contrasting basin-margin and basin-center North Sea Jurassic sandstones and mudstones. *In*: *Origin, Diagenesis and Petrophysics of Clay Minerals in Sandstones*. SEPM Special Publication, **47**, 81–110.

BURLEY, S. D., MULLIS, J. & MATTER, A. 1989. Timing diagenesis in the Tartan Reservoir (UK North Sea): constraints from combined cathodoluminescence microscopy and fluid inclusion studies. *Marine & Petroleum Geology*, **6**, 98–120.

CATHLES, L. M. & SMITH, A. T. 1983. Thermal constraints on the formation of Mississippi Valley-type Lead zinc deposits and their implications for episodic basin dewatering. *Economic Geology*, **78**, 983–1002.

CHEN, Y., ZENTILLI, M. A., CLARK, A. H., FARRAR, E., GRIST, A. M. & WILLIS-RICHARDS, J. 1996. Geochronological evidence for post-Variscan cooling and uplift of the Carnmenellis granite, SW England. *Journal of the Geological Society, London*, **153**, 191–195.

CHESLEY, J. T., HALLIDAY, A. N., SNEE, L. W., MEZGER, K., SHEPHERD, T. J. & SCRIVENER, R. C. 1993. Thermochronology of the Cornubian batholith in southwest England: implications for pluton emplacement and protracted hydrothermal mineralisation. *Geochimica et Cosmochimica Acta*, **57**, 1817–1835.

CLAYTON, R. N. & MAYEDA, T. K. 1963. The use of bromine pentafluoride in the extraction of oxygen from oxides and silicates for isotopic analyses. *Geochimica et Cosmochimica Acta*, **27**, 43–52.

COCHRAN, J. R. 1983. Effects of finite extension times on the development of sedimentary basins. *Earth & Planetary Science Letters*, **66**, 289–302.

COLE, D. R. & OHMOTO, H. 1986. Kinetics of isotopic exchange at elevated temperatures and pressures. *In*: VALLEY, J. W., TAYLOR, H. P. JR & O'NEIL, J. R. (eds) *Stable Isotopes in High Temperature Geological Processes*. Mineralogical Society of America, Reviews in Mineralogy, **16**, 41–90.

COLLINS, J. H. 1878. *The Hensbarrow Granite District: a geological description and a trade history*. Lake & Lake, Truro, 56.

COLLINS, J. H. 1909. Geological features visible at Carpalla china clay pit. *Quarterly Journal of the Geological Society of London*, **65**, 155–161.

COON, J. M. 1911. China clay. *Transactions British Ceramics Society*, **12**, 81–92.

COSGROVE, M. E. & SALTER, D. L. 1966. The stratigraphic distribution of kaolinite in the post-Armorican formations of south-west England. *Proceedings of the Ussher Society*, **1**, 249–252.

CRAIG, H. 1961. Isotopic variations in meteoric waters. *Science*, **133**, 1702–1703.

CRAWFORD, M. L. 1981. Phase equilibria in aqueous fluid inclusions. *In*: Hollister, L. S. & Crawford, M. L. (eds) *Fluid Inclusions: Applications to Petrology*. Mineralogical Association of Canada, Short Course Handbook 6, 75–100.

DANGERFIELD, J., HAWKES, J. R. & HUNT, E. C. 1980. The distribution of lithium in the St. Austell granite. *Proceedings of the Ussher Society*, **5**, 76–80.

DART, C. J., MCCLAY, K. & HOLLINGS, P. N. 1995. 3D analysis of inverted extensional fault systems, southern Bristol Channel Basin, UK. *In*: BUCHANAN, J. G. & BUCHANAN, P. G. (eds) *Basin Inversion*. Geological Society, London, Special Publications, **88**, 393–413.

DURRANCE, E. H., BROMLEY, A. V., BRISTOW, C. M., HEATH, M. J. & FENMAN, J. M. 1982. Hydrothermal circulation and post-magmatic changes in the granites of SW England. *Proceedings of the Ussher Society*, **5**, 263–317.

EDMONDS, E. A., MCKOWN, M. C. & WILLIAMS, M. 1975. *South-west England*. British Regional Geology, HMSO, London.

EDMUNDS, W. M., ANDREWS, J. N., BROMLEY, A. V., KAY, R. L. F., MILDOWSKI, A., SAVAGE, D. & THOMAS, L. J. 1988. Granite-water interactions in relation to hot dry rock geothermal development. *In*: *Investigation of the Geothermal Potential of the UK*. British Geological Survey Report, 3–27.

EDMUNDS, W. M., ANDREWS, J. N., BURGESS, W. G., KAY, R. L. F. & LEE, D. J. 1984. The evolution of saline and thermal groundwaters in the Carnmenellis granite. *Mineralogical Magazine*, **48**, 407–424.

EGEBERG, P. K. & AAGAARD, P. 1989. Origin and evolution of formation waters from oil fields on the Norwegian shelf. *Applied Geochemistry*, **4**, 131–142.

EHRENBERG, S. N., AAGAARD, P., WILSON, M. J., FRASER, A. R. & DUTHIE, D. M. L. 1993. Depth-dependent transformation of kaolinite to dickite in sandstones of the Norwegian continental shelf. *Clay Minerals*, **28**, 325–352.

EUGSTER, H. P. 1984. Oil shales, evaporites and ore deposits. *Geochimica et Cosmochimica Acta*, **49**, 619–635.

EVANS, C. D. R. 1990. *The geology of the western English Channel and its western approaches*. British Geological Survey, UK Offshore Regional Report, HMSO, London.

EXLEY, C. S. 1959. Magmatic differentiation and alteration in the St. Austell granite. *Quarterly Journal of the Geological Society of London*, **114**, 197–230.

EXLEY, C. S. 1976. Observations on the formation of kaolinite in the St Austell granite. *Clay Minerals*, **11**, 51–63.

FAURE, G. 1986. *Principles of Isotope Geology*, 2nd edn. John Wiley & Sons, Chichester.

FORTIER, S. M. & GILLETI, B. J. 1991. Volume self-diffusion of oxygen in biotite, muscovite and phlogopite micas. *Geochimica et Cosmochimica Acta*, **55**, 1319–1330.

GARVEN, G. 1985. The role of fluid flow in the genesis of the Pine Point deposit, western Canadian sedimentary basin. *Economic Geology*, **80**, 307–324.

GAYER, R. A. & CORNFORD, C. 1992. The Portledge-Peppercombe Permian outlier. Read at the Annual Conference of the Ussher Society, January 1992, 15–18.

GLEESON, S. A., WILKINSON, J. J., SHAW, H. F. & HERRINGTON, R. J. 2000. Post-magmatic hydrothermal circulation and the origin of base metal mineralization, Cornwall, UK. *Journal of the Geological Society, London*, **157**, 589–600.

GLEESON, S. A., WILKINSON, J. J., STUART, F. M. & BANKS D. A. 2001. The origin and evolution of base metal mineralising brines and hydrothermal fluids, South Cornwall, UK. *Geochimica et Cosmochimica Acta*, **65**, 2067–2079.

GOLDSTEIN, R. H. & REYNOLDS, T. J. 1994. *Systematics of fluid inclusions in diagenetic minerals*. SEPM Short Course, 31.

GROVES, A. W. 1931. The unroofing of the Dartmoor granite and the distribution of its detritus in the sediments of southern England. *Quarterly Journal of the Geological Society, London*, **87**, 62–96.

GUGGENHEIM, S., ALIETI, A. ET AL. 1997. Report of the Association Internationale pour l'Etude des Argiles (AIPEA) nomenclature committee for 1996. *Clay Minerals*, **32**, 493–495.

GUSCOTT, S. C. & BURLEY, S. D. 1993. A systematic approach to reconstructing palaeofluid evolution from fluid inclusions in authigenic quartz overgrowths. *In*: PARNELL, J., RUFFELL, A. H. & MOLES, N. R. (eds) *Geofluids '93 Extended Abstracts.* Contributions to an International Conference on Fluid Evolution, Migration and Interaction in Rocks, 323–328.

HARRIS, N. B. 1989. Diagenetic quartz arenite and destruction of secondary porosity: an example from the Middle Jurassic Brent Sandstone of northern Europe. *Geology*, **17**, 361–364.

HARRISON, T. M. & MCDOUGALL, I. 1980. Investigations of an intrusive contact north-west of Nelson, New Zealand—I: Thermal, Chronological and Isotope constraints. *Geochimica et Cosmochimica Acta*, **44**, 1985–2003.

HICKLING, G. H. 1908. China clay: its nature and origin. *Transactions of the Institution for Mining Engineering*, **36**, 10–33.

HILLIS, R. R. 1988. *The geology and tectonic evolution of the Western Approaches Trough.* PhD thesis, University of Edinburgh.

HOBSON, D. M. & SANDERSON, D. J. 1983. Variscan deformation in Southwest England. *In*: HANCOCK, P. L. (ed.) *The Variscan Fold Belt in the British Isles.* Adam Hilger, Bristol, 108–129.

JACKSON, N. J., MOORE, J. MCM. & RANKIN, A. H. 1977. Fluid inclusions and mineralization at Cligga Head, Cornwall, England. *Journal of the Geological Society, London*, **134**, 343–349.

JACKSON, N. J., WILLIS-RICHARDS, J., MANNING, D. A. C. & SAMS, M. S. 1989. Evolution of the Cornubian ore field, South West England: Part II. Mineral deposits and ore-forming processes. *Economic Geology*, **84**, 1101–1133.

KONTA, J. 1969. Comparison of the proofs of hydrothermal and supergene kaolinisation in two areas of Europe. *Proceedings of the International Clay Conference, Tokyo*, **1**, 281–291.

LEE, M. K. 1986. Hot dry rock. *In*: DOWNING, R. A. & GRAY, D. A. (eds) *Geothermal Energy: the potential in the United Kingdom.* BGS-NERC, HMSO, London, 21–41.

LYON, I. C., SAXTON, J. M., MCKEEVER, P. J., CHATZITHEODORIS, E. & VAN LIERDE, P. 1995. Precision and reproducibility of in situ oxygen isotope ratio measurements on quartz obtained using an Isolab 54 ion microprobe. *International Journal of Mass Spectrometry Ion Process*, **151**, 1–16.

LYON, I. C., SAXTON, J. M. & TURNER, G. 1994. Isotopic fractionation in secondary ionisation mass spectrometry. *Rapid Communications Mass Spectrometry*, **8**, 837–843.

MANNING, D. A. C., HILL, P. I. & HOWE, J. H. 1996. Primary lithological variation in the kaolinised St. Austell Granite, Cornwall, England. *Journal of the Geological Society, London*, **153**, 827–838.

METCALFE, R., ROCHELLE, C. A., SAVAGE, D. & HIGGO, J. W. 1994. Fluid rock interactions during continental red bed diagenesis: implications for mineralisation in sedimentary basins. *Geological Society, London, Special Publication*, **78**, 301–324.

ONASCH, C. M. & VENNEMANN, T. W. 1995. Disequilibrium partitioning of oxygen isotopes associated with sector zoning in quartz. *Geology*, **23**, 1103–1106.

O'NEIL, J. R. & KHARAKA, Y. K. 1976. Hydrogen and oxygen isotope exchange reactions between clay minerals and water. *Geochimica et Cosmochimica Acta*, **40**, 241–246.

POTTER, R. W. II, CLYNNE, M. A. & BROWN, D. L. 1978. Freezing point depression of aqueous sodium chloride solutions. *Economic Geology*, **73**, 284–285.

PSYRILLOS, A. 1996. *Low-temperature hydrothermal mineralisation in the St. Austell pluton, Cornwall, England.* PhD thesis, University of Manchester.

PSYRILLOS, A., HOWE, J. H., MANNING, D. A. C. & BURLEY, S. D. 1999. Geological controls on kaolin particle shape and consequences for mineral processing. *Clay Minerals*, **34**, 193–208.

PSYRILLOS, A., MANNING, D. A. C. & BURLEY, S. D. 1998. Geochemical constraints on kaolinisation in the St. Austell granite, Cornwall, England. *Journal of the Geological Society, London*, **155**, 829–840.

PSYRILLOS, A., MANNING, D. A. C. & BURLEY, S. D. 2001. The nature and significance of illite associated with quartz–hematite hydrothermal veins in the St. Austell pluton, Cornwall, England. *Clay Minerals*, **36**, 585–597.

RAMSAY, J. G. & HUBER, M. I. 1987. *The techniques of modern structural geology. Vol. 2: Folds and Fractures.* Academic Press, New York, 527–530.

ROBERTS, D. G. 1989 Basin inversion in and around the British Isles. *In*: COOPER, M. A. & WILLIAMS, G. D. (eds) *Inversion Tectonics.* Geological Society, London, Special Publications, **44**, 123–129.

RUFFELL, A. 1995. Evolution and hydrocarbon prospectivity of the Britany Basin (Western Approaches Trough) offshore north-west France. *Marine & Petroleum Geology*, **12**, 387–407.

SAMS, M. S. & THOMAS-BETTS, A. 1988a. Models of convective fluid flow and mineralisation in Southwest England. *Journal of the Geological Society, London*, **145**, 809–817.

SAMS, M. S. & THOMAS-BETTS, A. 1988b. Three-dimensional numerical modelling of the conductive heat flow in southwest England. *Geophysical Journal of the Royal Astronomical Society*, **92**, 323–334.

SAVIN, S. M. & EPSTEIN, S. 1970. The oxygen and hydrogen isotope geochemistry of clay minerals. *Geochimica et Cosmochimica Acta*, **34**, 25–42.

SAWKINS, F. J. 1966. Preliminary fluid inclusion studies in the mineralization associated with the Hercynian granites of southwest England. *Bulletin of the Institute Mining and Metallurgy*, **714**, B109–B112.

SAXTON, J. M., LYON, I. C. & TURNER, G. 1995. High precision, *in situ* oxygen isotope ratio measurements obtained from geological and extra-terrestrial material using an Isolab 54 ion microprobe. *Analyst*, **120**, 1321–1326.

SCRIVENER, R. C., COOPER, B. V., GEORGE, M. C. & SHEPHERD, T. J. 1982. Gold-bearing carbonate veins in the Middle Devonian limestone of Hopes Nose, Torquay. *Proceedings of the Ussher Society*, **6**, 558.

SCRIVENER, R. C., DARBYSHIRE, D. P. F. & SHEPHERD, T. J. 1994. Timing and significance of crosscourse mineralization in SW England. *Journal of the Geological Society, London*, **151**, 587–590.

SHEPHERD, T. J. & SCRIVENER, R. C. 1987. Role of basinal brines in the genesis of polymetallic vein deposits, Kit Hill–Gunnislake area, SW England. *Annual Conference of the Ussher Society*, 491–497.

SHEPHERD, T. J., RANKIN, A. H. & ALDERTON, D. H. M. 1985. *Fluid Inclusion Studies*. Blackie, Glasgow, 93–113.

SHEPPARD, S. M. F. 1977. The Cornubian batholith, SW England: D/H and $^{18}O/^{16}O$ studies of kaolinite and other alteration minerals. *Journal of the Geological Society, London*, **133**, 573–591.

SHEPPARD, S. M. F. & GILG, H. A. 1996. Stable isotope geochemistry of clay minerals. *Clay Minerals*, **31**, 1–24.

SIBSON, R. H. 1994. Crustal stress, faulting and fluid flow. *In*: PARNELL, J. (ed.) *Geofluids: Origin, Migration and Evolution of Fluids in Sedimentary Basins*. Geological Society, London, Special Publications, **78**, 69–84.

SNEE, L. W., SUTTER, J. F. & KELLY, W. C. 1988. Thermochronology of economic mineral deposits: Dating the stages of mineralisation at Panasqueira, Portugal by high precision $^{40}Ar/^{39}Ar$ age-spectrum techniques on muscovite. *Economic Geology*, **83**, 335–354.

STECKLER, M. S. 1985. Uplift and extension at the Gulf of Suez: indications of induced mantle convection. *Nature*, **317**, 135–139.

SUZUOKI, T. & EPSTEIN, S. 1976. Hydrogen isotope fractionation between OH-bearing minerals and water. *Geochimica et Cosmochimica Acta*, **40**, 1229–1240.

TAYLOR, H. P. 1974. The application of oxygen and hydrogen isotope studies to problems of hydrothermal alterations and ore deposition. *Economic Geology*, **69**, 843–883.

WARREN, E. A. & SMALLEY, P. C. 1994. *North Sea formation waters atlas*. Geological Society of London, Memoirs, **15**.

WILLIS-RICHARDS, J. 1993. Controls on lithospheric strength, fracture opening and fluid movement in S.W. England. *In*: PARNELL, J., RUFFELL, A. H. & MOLES, N. R. (eds) *Geofluids '93 Extended Abstracts*. Contributions to an International Conference on Fluid Evolution, Migration and Interaction in Rocks, 128–131.

ZIEGLER, P. A. 1990. *Geological atlas of Western and Central Europe*, 2nd edn. Shell International Petroleum, Maatschappij B.V.

Fracture-dominated flow in the Borrowdale Volcanic Group at Sellafield, NW England: the identification of potential flowing features, development of conceptual models and derivation of effective parameters

P. J. DEGNAN[1], A. K. LITTLEBOY[1], U. McL. MICHIE[1], C. P. JACKSON[2] & S. P. WATSON[3]

[1] *United Kingdom Nirex Limited, Curie Avenue, Harwell, Didcot, Oxfordshire OX11 0RH, UK (e-mail: Paul.Degnan@nirex.co.uk)*
[2] *Serco Assurance, Building 424.4, Harwell, Didcot, Oxfordshire OX11 0RA, UK*
[3] *Geo-Analysis, 23 Tuckers Road, Faringdon, Oxfordshire SN7 7YQ, UK*

Abstract: Between 1989 and 1997 United Kingdom Nirex Limited (Nirex) studied in detail the geology and hydrogeology of a rock volume near Sellafield in Cumbria, NW England. The aim of the study was to determine the suitability, or otherwise, of the site as the location for a deep repository for intermediate-level and certain low-level radioactive wastes. An important factor in determining site suitability was the nature of groundwater flow in the potential repository host rock, the Borrowdale Volcanic Group. In the host rock, interpretation of borehole core, wireline logs and hydrogeological pumping test data indicated that groundwater flow was predominantly through a limited subset of discontinuities, mainly fractures, parts of which form networks of connected channels. Within this overall understanding of the nature of groundwater flow, there is a wide range of possible geometrical descriptions for the flow channels. Determination of one or more appropriate conceptualizations of the flow system must be soundly based on site characterization data as a prerequisite for any numerical modelling study. In the first part of this paper, details of the site characterization studies that were used to identify the location, orientation and mineralogical characteristics of discontinuities, and in particular the set of discontinuities referred to as Potential Flowing Features (PFFs), are provided. These features have either demonstrable present-day open porosity, or display evidence of geologically recent groundwater flow as part of the evolution of the current groundwater system. It is inferred that the PFFs observed in boreholes correspond to flowing features and that the borehole data can be used to infer the distribution and characteristics of flowing features that are present in the unobserved rock mass. On this basis, knowledge of the distribution, orientation and permeability associated with the PFFs provided the framework for developing conceptual models for groundwater flow. Numerical models were constructed to represent the flow. It is not computationally practicable to undertake regional scale groundwater flow and transport calculations in which small-scale variability is explicitly represented. Therefore upscaled effective parameters need to be derived as a precursor to running large-scale numerical model simulations. A summary of the Nirex upscaling procedure applied to the flowing feature network in the Borrowdale Volcanic Group is provided.

Between 1989 and 1997, United Kingdom Nirex Limited (Nirex) investigated the geology and hydrogeology of the rock volume around a site near Sellafield in NW England, to determine whether the site might be suitable as a location for an underground repository for solid intermediate level, and certain low level, radioactive waste (Fig. 1a). (In March 1997 Nirex was refused planning permission to build a proposed underground rock laboratory at Sellafield. Since that time Nirex has concentrated on consolidating the information gained from the Sellafield investigations and carrying out generic studies.) The host rock for the previously proposed repository would have been the Borrowdale Volcanic Group. This paper outlines some of the investigation methods that were applied to identify the nature of the groundwater pathways through the Borrowdale Volcanic Group, the conceptualization of those pathways and the derivation of effective properties. These tasks were required for the incorporation of Sellafield specific information into numerical models for groundwater flow. Although the investigation techniques and the characteristics of the fractures that carry groundwater flow have been studied in the context of a hydrogeological programme for radioactive waste disposal, the methodologies used by Nirex to characterize fluid pathways are likely to be relevant to studies in other

From: PETFORD, N. & MCCAFFREY, K. J. W. (eds) 2003. *Hydrocarbons in Crystalline Rocks*. Geological Society, London, Special Publications, **214**, 197–219. 0305-8719/03/$15 © The Geological Society of London.

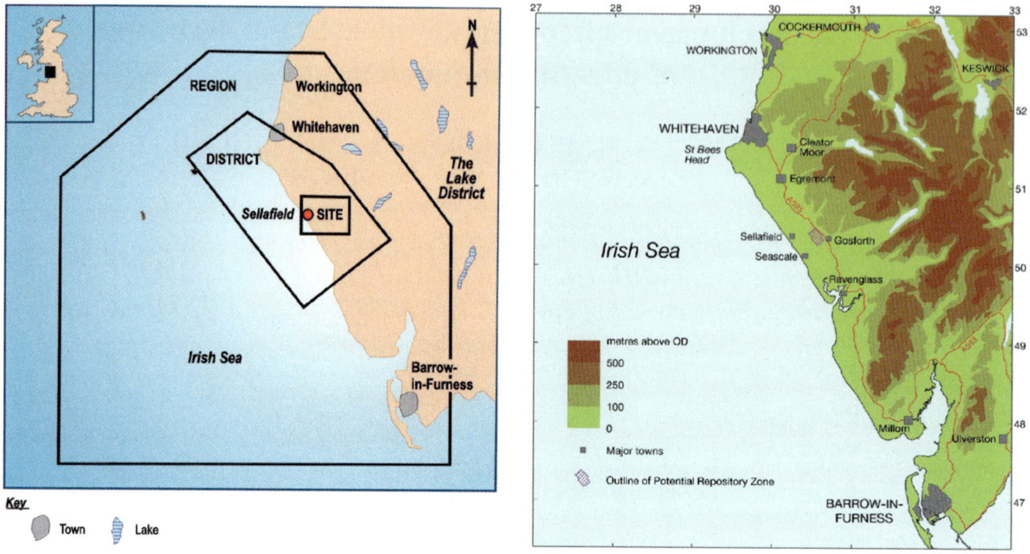

Fig. 1. (**a**) Sellafield location and Nirex earth science investigation areas. (**b**) Topography of the area around Sellafield.

disciplines, such as petroleum exploration, geothermal evaluation and hydrogeology, and for other types of rock. It is likely that the conceptual models also have widespread application in other crystalline terrains, although the parametrization of the hydrogeologically significant properties and the boundary conditions will clearly be site specific.

The geological research undertaken by Nirex covered an area (60 × 65 km) in NW England, with generalized investigations at large (regional) length scales and much more detailed studies in the relatively small volume immediately around the location of the potential underground repository, located on a 3–5 km wide coastal plain adjacent to the Irish Sea. From the relatively flat lying coastal plain, the land rises steeply to the east to a maximum height of just under 1000 m about 20 km inland, forming the Cumbrian hills and mountains of the Lake District (Fig. 1b).

Earth science information compiled by Nirex included an analysis of seismic survey data and results from airborne geophysical surveys, as well as results from new geological, geophysical, hydrological and hydrogeological surveys. As part of the detailed site characterization, Nirex drilled 29 deep boreholes, to a maximum of 2 km depth, and over 30 shallow boreholes to 100 m depth. Many of these boreholes were monitored over a number of years for groundwater pressure fluctuations in packered intervals (Nirex 1996a). This geological, geophysical and hydrogeological information provided support for an assessment of the long-term safety of the potential underground repository (Nirex 1997a).

A key aspect of the long-term safety of a potential repository is the way in which radionuclides transported in groundwater may return to the near-surface environment. Therefore an important component of the system studied at Sellafield is the nature of groundwater flow through the geosphere. Previous work by Nirex and others (e.g. Nirex 1997a; Nagra 1994; SKB 1999) has highlighted the importance of understanding in appropriate detail the nature of the groundwater pathways that may allow the transport of any long-lived radionuclides that may be released from a deep geological repository. Matrix and fracture flow both contribute to groundwater flow at Sellafield, but to different extents within different hydrogeological units. It is fracture flow that dominates in the Borrowdale Volcanic Group.

Geology and hydrogeology

The geological setting of the area around Sellafield is structurally a transitional zone between the western margin of the Lake District Massif of Lower Palaeozoic rocks and the adjacent, mainly offshore, East Irish Sea basin consisting largely of Mesozoic sedimentary rocks. A geological map of the site is shown in Figure 2. The basement rocks in the area include the

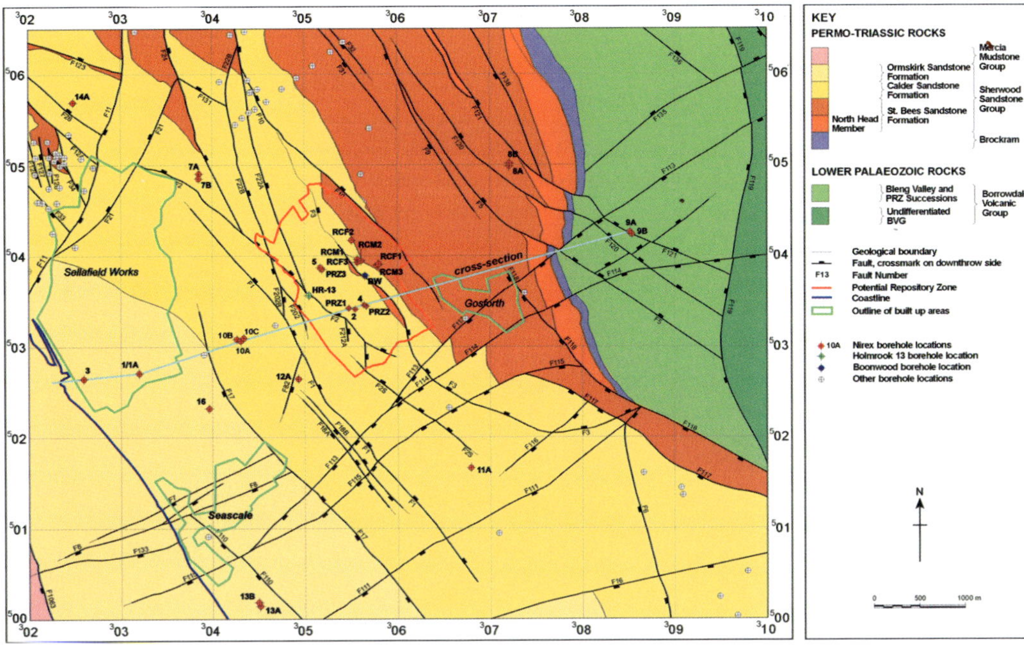

Fig. 2. Geological map of the Sellafield Site area (see Fig. 1a for location).

Ordovician age Borrowdale Volcanic Group, which was considered potentially suitable as the host rock for an underground repository. The Group comprises a several kilometres-thick sequence of basaltic, andesitic, dacitic and rhyolitic volcanic and volcaniclastic rocks, including lavas, ignimbrites, welded and non-welded tuffs, breccias and intrusive igneous rocks (Millward *et al.* 1994). The whole assemblage was intensely faulted and fractured as a

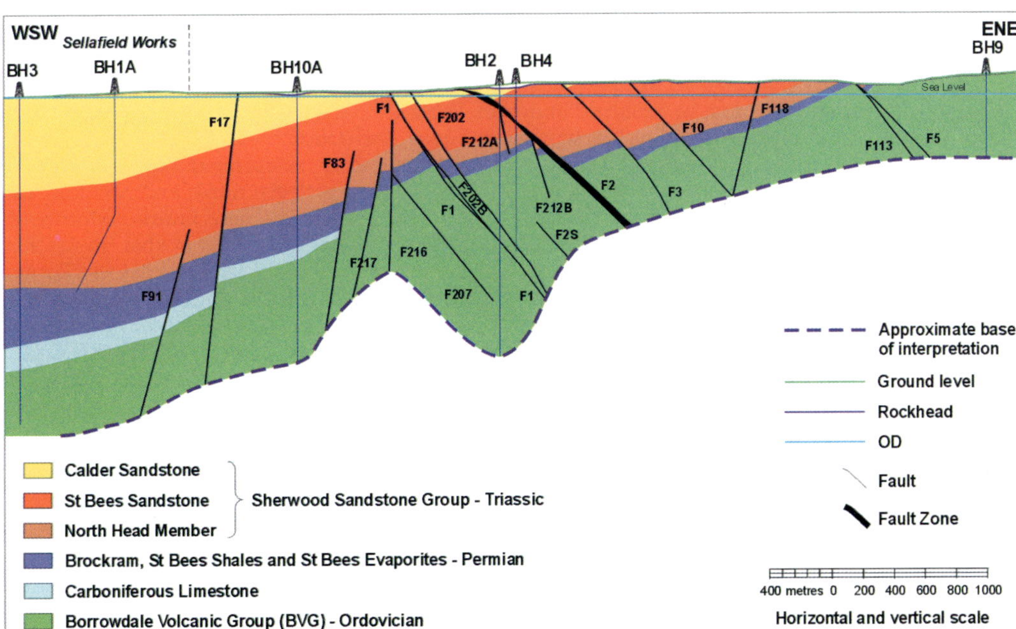

Fig. 3. SW–NE geological section through the Sellafield Site (position of cross-section indicated on Fig. 2).

Table 1. *Pre-Quaternary stratigraphy in the local Sellafield area*

TRIASSIC	Mercia Mudstone Group*		
	Sherwood Sandstone Group	Ormskirk Sandstone Formation	
		Calder Sandstone Formation	
		St Bees Sandstone Formation	
		North Head Member	
	Cumbrian Coast Group	St Bees Shale Formation	Moss Side Member
PERMIAN		St Bees Evaporite Formation	
	Appleby Group	Brockram	
UNCONFORMITY			
CARBONIFEROUS (Dinantian)	Chief Limestone Group	Urswick Limestone	
		Frizington Limestone	
		Martin Limestone	
UNCONFORMITY			
ORDOVICIAN	Borrowdale Volcanic Group	Fleming Hall Formation	Longlands Farm Member
			Sides Farm Member
			Town End Farm Member
		Brown Bank Formation	
			Seascale Hall Member
		Bleawath Formation	
		Broom Farm Formation	
		Moorside Farm Formation	

* Mercia Mudstone Group not present in Sellafield boreholes; offshore only.

result of contemporary volcano-tectonic activity. Subsequently, the rock mass was intruded with granitic rocks and further deformed and metamorphosed at low (sub-greenschist) grade during the tectonic events related to closure of the Iapetus Ocean.

Underneath the coastal plain at Sellafield and offshore, an unconformity surface on the basement rocks is overlain by younger sedimentary rocks. This surface dips at approximately 25° to the west and is down-faulted in that direction (Fig. 3), such that at the coast the top of the Borrowdale Volcanic Group is some 1600 m below the ground surface. In the area that was considered for the potential repository, the top surface of the Borrowdale Volcanic Group is at a depth of approximately 500 m below ground level. Late Palaeozoic and Mesozoic sedimentation is represented in the area by an on-lapping sequence of Dinantian-age Carboniferous Limestone (Barclay *et al.* 1994) and overstepping Permo-Triassic rocks of the Appleby, Cumbrian Coast and Sherwood Sandstone Groups (Barnes *et al.* 1994). Offshore the Mercia Mudstone Group is present. Finally, Quaternary deposits are present, up to 180 m thick and overlying the Mercia Mudstone Group offshore, and up to 100 m thick, overlying the Sherwood Sandstone Group onshore. These Quaternary sediments are sedimentologically heterogeneous, comprising glacial tills and glaciotectonized sands and gravels that are laterally variable in thickness and facies (McMillan *et al.* 2000). The lithostratigraphy in the local Sellafield area is shown in Table 1.

Today, the structural grain through the region is dominated by extensional fault zones, with faults trending between north and NW and also east to NE. Many of these faults are probably deep-seated crustal structures that have been active at several times since at least the Permo-Triassic. They have influenced early sedimentary basin development, for example, a NW–SE fault zone (the Fleming Hall Fault Zone) locally formed the eastern edge of Permian evaporites and other basinal facies. The structural framework and geological evolution of the area is more fully described by Michie (1996) and Akhurst *et al.* (1997).

The present-day hydrogeology of the area can be best understood by a knowledge of the hydrogeological properties of the various rocks, their structural features and through the disposition of distinct groundwater bodies (or 'regimes') characterized by their hydrochemistry, in particular their salinity. At shallow levels below the coastal plain, fresh meteoric groundwaters of Ca–HCO_3 type are present. Throughout much of the area investigated in detail a relatively sharp saline transition zone occurs, separating the Coastal Plain regime from the Hills and Basement regime. The latter is characterized by moderate salinity groundwaters with Br/Cl ratios that suggest that a significant component of the salinity is derived from a putative source to the east. In the west of the study area a third groundwater body exists, the Irish Sea basin regime. This has a highly saline groundwater component inferred to derive from the dissolution of Permo-Triassic halites by ancient meteoric recharge. The Mercia Mudstone Group, that is present offshore and which thickens to the west in the East Irish Sea basin, is thought to be the source of the observed brines. To the east of the site, the rather less concentrated saline groundwater flowing from the Hills and Basement regime is inferred to be flowing upwards as it encounters the transition zone with the denser brines of the Irish Sea basin regime. It is also believed to be mixing with these brines to create the salinity variations currently observed at depth.

Groundwater movement through the Sellafield area is primarily driven in a NE to SW direction by gravity acting on the elevated water table associated with the high topography of the Lake District uplands to the east of the site. At depth, flow is largely driven by gravity acting on differences in fluid density. These variations in density are the result of changes in the temperature and salinity of the groundwater, which both vary significantly across the area. Therefore, the coupling of groundwater flow to the transport of salinity and the transport of heat are important considerations. Furthermore, the groundwater system at Sellafield is dynamic. Over long time scales, the surface conditions are changing, for example, as a result of glaciation leading to overpressures and sea-level changes. The groundwater system changes in response, but the response time for parts of the system may be longer than the timescale of the changes, so that the system never reaches equilibrium with the surface conditions.

Investigative techniques

Introduction

A variety of discontinuities within the Borrowdale Volcanic Group have been identified in the Nirex boreholes and in outcrop including faults, open joints, altered intrusive margins, primary fabrics, lithological boundaries, internal disconformities and mineralized and vuggy veins of several ages. A discontinuity is not synonymous with a fracture. A fracture is used here in a generic sense, following Kulander *et al.* (1990) and Ameen (1995), as a dislocation feature that forms as a result of an applied stress exceeding rock strength. Faults, veins and joints are all different and particular types of fracture.

Due to the complex structural history in the basement, fracturing is pervasive and occurs on all length scales, whilst the igneous and metamorphic origin of the Borrowdale Volcanic Group means that the matrix of the intact rock between fractures is inherently of low permeability. Consequently, most of the groundwater flow takes place through a network of discontinuities predominantly comprised of fractures. However, evidence indicates that many of the fractures are tight in places and effectively sealed. Most fractures cross-cut and abut against others, forming networks on a variety of length scales. Consequently, individual open fractures or groups of open fractures may be relatively well connected hydraulically, or they can be effectively isolated from others. In the latter case, as with sealed fractures, they do not contribute to meaningful flow on a large scale.

Nirex has used various geological, geophysical and hydrogeological methods to differentiate between the total discontinuity population and the subset of fractures that are important for groundwater flow. In the following sub-sections the methods used to identify the total discontinuity population will be discussed first. This is followed by a summary of techniques used to characterize the fractures important for flow.

Borehole discontinuity analysis

The main source of data on discontinuity characteristics has been the deep boreholes. The borehole discontinuity data set is a compilation of high-resolution information derived from the interpretation of borehole imagery and detailed core logging. It provides a comprehensive record of discontinuity frequency and orientation.

Wireline geophysical logging was applied over the full length of all of the deep boreholes investigated by Nirex. The borehole imagery tools (FMS/FMI and BHTV/UBI) mapped electrical and acoustic properties and their variations in the borehole wall. Any planar features intersecting the borehole wall that have variable electrical or sonic properties, compared to the bulk of the rock mass, are recognized by their sinusoidal form in the images. Using automated algorithms and expert judgment, the orientation of a planar discontinuity (at the borehole scale) can be derived from the sine wave patterns. Although wireline techniques cannot always differentiate fractures from other types of discontinuity in the imagery logs, most of the features in the Borrowdale Volcanic Group identified from the imagery were eventually interpreted as fractures.

The location and nature of fractures and other discontinuities were also measured during core logging. This involved the systematic observation of the discontinuity type and its associated mineralization. Many of the deep boreholes were drilled using wireline triple core barrels to facilitate the extraction of intact 10 cm diameter cores and as a result most of the deep boreholes have extensive core coverage. On the whole, core quality and recovery (>99%) was excellent (Chaplow 1996). Upon examination, discontinuities in the cores were compared with features identified in the borehole imagery logs. In this way the depth of the cores was accurately assessed (the borehole trajectories having been fully surveyed) and the orientation of the discontinuities logged in the core was appraised. Comparison of core and borehole images also allowed any drilling-induced breaks in the core to be identified.

Approximately 32,500 fractures and other discontinuities were identified from the wireline imagery analysis and core logging of almost 30 km of borehole (for the Borrowdale Volcanic Group and other lithostratigraphical units). This information has provided an extensive dataset of fracture frequency and orientation. Visual examination and statistical interpretations of the spatial distribution of the total fracture data set indicates that the rock units can be subdivided into discrete intervals in the boreholes (structural domains) which have a common set of discontinuity characteristics, namely fracture frequency and orientation (Nirex 1996b). Not surprisingly, the locations of the structural domains were seen to be influenced by lithology, with the variation in discontinuity frequency and orientation often related to the degree of welding of the predominantly pyroclastic rocks. Intensely welded rocks typically have a relatively high discontinuity frequency. However, discontinuity frequency and orientation also have a general association with other features, such as proximity to major fault zones.

The integrated imagery and core discontinuity data provide fracture frequency estimates. Data from the Borrowdale Volcanic Group in three boreholes (RCF1, 2 and 3) at the potential repository site were used in a geostatistical analysis (Nirex 1997b). The results indicated that the density of fractures was fairly consistent for the three boreholes. Variograms indicate a structure range of about 50 m within which fracture properties are correlated, with each fracture set attaining a sill in a range from 20–70 m. At greater lags, generally around 100–120 m, the variance of fracture number and of particular fracture sets decreases (hole effect). This reflects a periodicity in the frequency estimates, which can be loosely correlated with lithological variations in the Borrowdale Volcanic Group. Low values of fracture density tend to correspond to the occurrence of pyroclastic breccias in a tuff-dominated sequence, whilst high values are more common in banded rhyolite sequences. The observed fracture density of 5–8 fractures/m is an underestimate of the true fracture density due to the high proportion of steeply dipping fractures. It should be noted that all of the borehole-derived data, both imagery and core, were potentially subject to sampling bias due to the near vertical orientation of the boreholes. A modified Terzaghi weighting method was used to compensate for this limitation (Nirex 1996b). After correction for orientation bias, the density is 10–20 m^{-1} in the boreholes, which compares well with the equivalent vertical densities measured during surface mapping.

Surface mapping

Nirex investigated the extent of statistical correlations between fracture characteristics of the Borrowdale Volcanic Group at outcrop within the western Lake District and fractures derived from the boreholes. This was carried out specifically to understand the two- and three-dimensional fracture relationships and also to establish whether fracture patterns and fracture densities observed locally at the surface can be used as analogues for fracture characteristics at depth. Surface outcrops of the Borrowdale Volcanic Group to the east of the boreholes were investigated in the Craghouse Park and Latterbarrow regions (Nirex 1997c) and in Blengdale. The work comprised aerial photo interpretation and detailed outcrop mapping.

The outcrop at all sites examined displays an interconnected network of fractures with sometimes complex intersection relationships. There was good correspondence between the orientations observed at the surface at Craghouse Park and Latterbarrow and at depth in the boreholes, but not with those mapped in the Blengdale area. The orientations of the fractures at Craghouse Park and Latterbarrow demonstrate that there are three main groups of fractures or sets which all share the same distinct, preferred orientations. This conclusion was substantiated by a study of the modal orientations demonstrating that the strikes of the fracture sets were NNW (340°), SE (105°) and NE (070°), and that they are all steeply dipping. The NNW-trending set can locally be divided further into two sub-sets, although these cannot be distinguished everywhere (Nirex 1997c).

Fracture lengths were measured using integrated aerial photographic and mapped outcrop data at several locations in the Craghouse Park and Latterbarrow areas. Fractal analysis was then carried out on the spatial and scaling distributions (Nirex 1997b). Although there are censoring problems at all the sites, and possibly truncation errors due to under-sampling, it was found that the distribution of lineament lengths for some fracture sets conformed closely to a power law distribution. Fracture lengths measured on scan lines ranged over about 2 orders of magnitude (0.2–20 m), with a single set at each site indicating a fractal dimension of about 0.7 and 0.82 (the sets had different orientations). A conclusion from the relatively simple analysis that was undertaken is that scale invariance of fracture size should not be assumed for all sets. The corresponding fractal dimension for a three-dimensional volume, using trace maps and aerial lineament mapping, gave fractal dimensions of approximately 2, in good agreement with the two-dimensional fractal dimension ($D_{3\text{-}D} = D_{2\text{-}D} + 1$) and indicating approximate self-similarity over the 0.1–1000 m range (Fig. 4).

A scan line survey was carried out at locations where there was appropriate three-dimensional access and where fracture traces could be identified on two adjacent surfaces with different orientations (Nirex 1997c). The method that was used involves establishing a scan line along the edge of two rock surfaces that intersect. Fractures which themselves intersect the edge scan line can be seen on both surfaces and their 'semi-trace lengths' are then measured on each. There is an assumption that if there was a significant difference in semi-trace lengths for a single fracture, then it would be reasonable to infer that the fractures were anisotropic in shape. For example, comparison of the ratio of vertical to horizontal semi-trace lengths measured at Latterbarrow suggest that fracture surfaces have variable degrees of anisotropy, generally being longer in the vertical than in the horizontal direction. It was recognized that this anisotropy could be important for modelling groundwater flow.

The site fracture density in the horizontal plane was measured (i.e. the average of the densities along two or more sub-horizontal scan lines). The results indicate a range of 7–25 m^{-1} and there is little systematic variation in this between the areas studied. The variability between the sites was found to be less than the variability at a site.

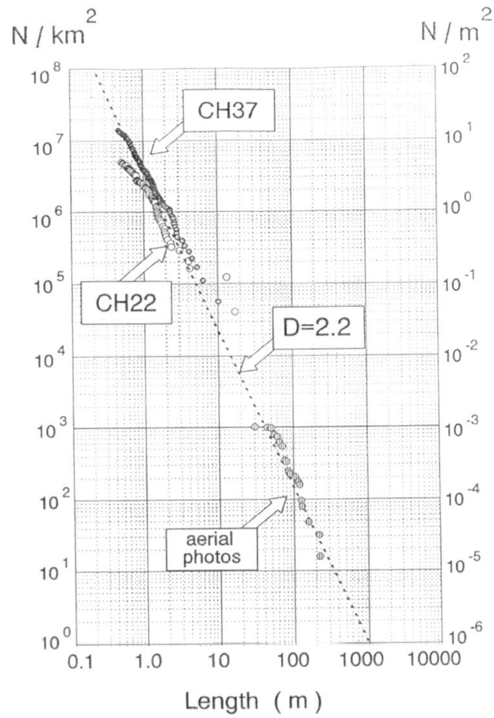

Fig. 4. Plot of length against number per km² for two representative surface mapped sites in the Craghouse/Latterbarrow area, and for the largest sample fractures in a photolineament interpretation at 1:2000. Data fall on a line establishing a power-law scaling relationship for trace-length in the range 0.5–250 m.

Flow zone identification

As a first attempt to differentiate the fractures that are important for groundwater flow from the total discontinuity population, a series of down borehole measurements was made to identify discrete horizons where inflows occurred. These measurements were made during production logging over large intervals of the boreholes, usually over a period of one or two days per borehole (Nirex 1997d). Intervals as deep as 1800 m were tested. Compressed air or nitrogen was introduced into the boreholes to induce a drawdown in the water level. The drawdown ranged from 0 m to 470 m, but was generally held at

200 m. During each test, a number of continuous logging runs were undertaken at different logging speeds in order to monitor evolving changes in parameter profiles. Three key fluid parameters were measured: fluid velocity, conductivity and temperature. Flow zones were inferred at locations where a sufficient inflow of fluid occurred to produce a measurable change to the logged parameter profile.

Generally only the 'down pass' production logging data were used in interpretations. After preliminary on-site inspection, information was reprocessed and reinterpreted off-site using rigorous averaging and recalculation procedures. Of the three measured parameters, the fluid conductivity profile was found to be more sensitive to changes than the temperature profiles, which were in turn more sensitive than flow velocity measurements made with the flow meter. In the Borrowdale Volcanic Group, many of the log responses were quite sharp, indicating inflow from discrete entry points (individual open fractures). However, in many cases it was not possible to identify a unique point because the changes in the profile were indistinct. This was seen to occur, for example, in intervals known to be crossed by fault damage zones and in such cases it is likely that a dense assemblage of fractures exists to form a flow zone.

There are some limitations in the interpretation of the flow logging data. In particular, the magnitude of variation in conductivity or temperature does not correspond to the volume or rate of fluid flow. Furthermore, the flow zone data can only be used to identify some of the locations of inflow, and the quality of inflow recognition will vary within a borehole and also between boreholes. This is because, for example, at the deepest measurement point in a borehole the water column is relatively static. Consequently, very limited inflows can make a measurable difference in fluid properties. However, near the top of a section, similar inflows may not be recognized if the cumulative flow beneath that point is significant. The variations in the drawdowns applied in the different boreholes and differences in borehole completion may also influence inflow recognition and make meaningful comparisons difficult. Despite the acknowledged limitations in the methodology used, there is value in recording the position of intervals that demonstrate the inflow of groundwater.

Wireline measurements

Flow-zone identification using production logging techniques is relatively time-consuming and expensive. For this reason, Nirex investigated the use of wireline-derived geophysical information to identify sections of potential inflow as an alternative or complementary technique. In the Borrowdale Volcanic Group there is the expectation that zones of fractured rock will in general display high porosity, low electrical resistivity and low seismic velocity, in comparison to more intact rock. Low bulk density has also been observed to coincide with faulting. In order to identify the location of more permeable zones within the Borrowdale Volcanic Group, wireline data derived from neutron porosity, bulk density, compressional seismic velocity and shallow resistivity measurements were analysed individually. They were also combined into a synthetic wireline property log (Nirex 1997e, f; Brereton *et al.* 1998) reflecting an overall rock characteristic that suggests hydrogeological significance.

A further wireline logging parameter was also investigated. Theoretical and empirical studies indicate that a relationship exists between fracture permeability and the attenuation of acoustic waveforms, particularly tube waves and Stoneley waves. Both qualitative and quantitative estimates of permeability from Stoneley wave analysis have been described in the literature as reliable indicators of open fracture occurrences (e.g. Burns 1991; Paillet 1991; Johnson *et al.* 1992; Wei *et al.* 1995). At Sellafield, the correspondence between Stoneley waves and the wireline property log was good in certain boreholes and was also found to correlate well with fracture features which had also been located with the FMI/FMS and BHTV/UBI logs. However, correspondence with flow zones was erratic and analysis of the wireline property log, the individual logs and the Stoneley wave events failed to unambiguously identify zones of high permeability or inflow. This is believed to be because the Stoneley wave events respond to open, but not necessarily connected fractures.

Potential flowing features

Subsequent to the standard core logging activities briefly described above, the mineralogical characteristics of fractures (both the wall rocks and fracture infills) and other mineralized features were systematically evaluated as part of a comprehensive programme to investigate mineral paragenesis, diagenetic events and the relationship of fracture mineralization to faulting history. One of the aims of the study was to investigate whether mineralogical evidence could be used to make inferences concerning groundwater evolution and past and present groundwater flow paths. Consequently, work undertaken by Nirex has sought to explore any linkages, genetic or statistical, between mineral paragenesis and currently flowing features observed in boreholes.

The mineralogical studies that have been reported (Milodowski *et al.* 2000) were carried out on samples derived from intervals in 21 of the 29 boreholes drilled (Fig. 5). The fracture logging strategy varied from borehole to borehole. Some boreholes were continuously logged and this was usually undertaken without any *a priori* knowledge of the location of flowing zones, in order to avoid any potential bias. However, where there were logistical or time constraints, fracture-targeting decisions had to be made and in such circumstances production logging information was sometimes used to concentrate the mineralogical studies in target sections identified from their vicinity to flowing areas.

Flow zones are not expected to account for all actual or potential discrete groundwater inflows, because of the limitations of the logging methods used, and because some points that should provide inflow to a borehole are not recognized because of clogging. Additionally, a fracture may be sealed adjacent to a borehole wall, but only centimetres away it could be a fully flowing feature. If the seal were to open, through mineral dissolution or tectonic activity for example, a new inflow would be seen. Because it is believed that certain pre-existing fractures have a greater likelihood to flow at some time in the future (on the timescale of relevance to Nirex, up to 1 Ma), the term Potentially Flowing Feature (PFF) was coined for those features that had certain clear diagnostic mineralogical and/or petrographical characteristics related to geologically recent (Quaternary) or current groundwater flow.

A PFF is a discontinuity in the rock mass that is potentially capable of conducting groundwater flow, whether it does so or not at the borehole at present. A PFF has been defined on the basis of six petrographical criteria (Milodowski *et al.* 1998) that indicate that it has demonstrable open porosity, at least at core scale (Fig. 6) and/or that it displays evidence of mineralization or rock water interaction that can be attributed directly to the development of the present-day groundwater system (Fig. 7).

The temporal relationships between the fracture minerals have also been deduced from examination of fracture cross-cutting relationships, petrographical fabric analysis, isotopic geochemistry ($^{86}Sr/^{87}Sr$, C, O and S) and fluid inclusion microthermometrical analysis. There was also limited radiometric (K–Ar, ^{39}Ar–^{40}Ar, ^{230}Th–U) dating and palaeomagnetic dating of appropriate fracture minerals (Nirex 1997g). This work led to the recognition of nine discrete Mineralisation Episodes, designated ME1 to ME9 (Table 2). Each episode has a particular mineral assemblage and the age relationships between them have been established. Of particular importance in the identification of PFFs are the most recent episodes, ME8 and ME9. The ME8 is a manganese or ferroan oxide mineral coating that especially occurs in the near-surface environment and so is not found to any great extent in the Borrowdale Volcanic Group at depth.

The ME9 is dominated by recent calcite that almost always occurs with fractures that are unsealed at the present day, and analysis of the distribution of the ME9 calcite has demonstrated that its presence closely corresponds with flow zones in the Borrowdale Volcanic Group. The presence of ME9 calcite is therefore a diagnostic feature of a PFF. Radiometric dating of a small number of samples of the late calcite mineralization indicates ages from 17,000 to greater than 300,000 years BP (before present). The coeval association of the late calcite cements in the near surface with the ME8 Fe and Mn oxyhydroxide mineralization episode, which has been independently dated as extending from the Tertiary (Miocene) and possibly up to the present day, indicates that the calcite mineralization phase may also extend from the late Tertiary up to the present day.

An interesting feature of the morphology of the ME9 calcite is that it varies in sympathy with the hydrochemistry of the groundwater system, in particular with the salinity. The evolution of crystal morphology can therefore be used to infer changes in the depth of the saline transition zone between the Coastal Plain Regime and the underlying more saline groundwater bodies.

Most of the PFFs in the Borrowdale Volcanic Group are fracture related. The PFFs tend to be highly clustured and their orientations are widely scattered (Nirex 1997h). In all the boreholes inspected for PFFs, moderate to steep NE-dipping features predominate, as can be seen from the gently SW-dipping girdle of PFF poles and the rose diagram in Figure 8. However, although there are orientation data for individual PFFs, the borehole data do not allow determination of the orientation of PFF clusters, which could be quite different from the orientations of the individual PFFs that comprise them.

There does seem to be a loose association of PFFs with faults. However, although PFFs are associated with many faults, not all faults are

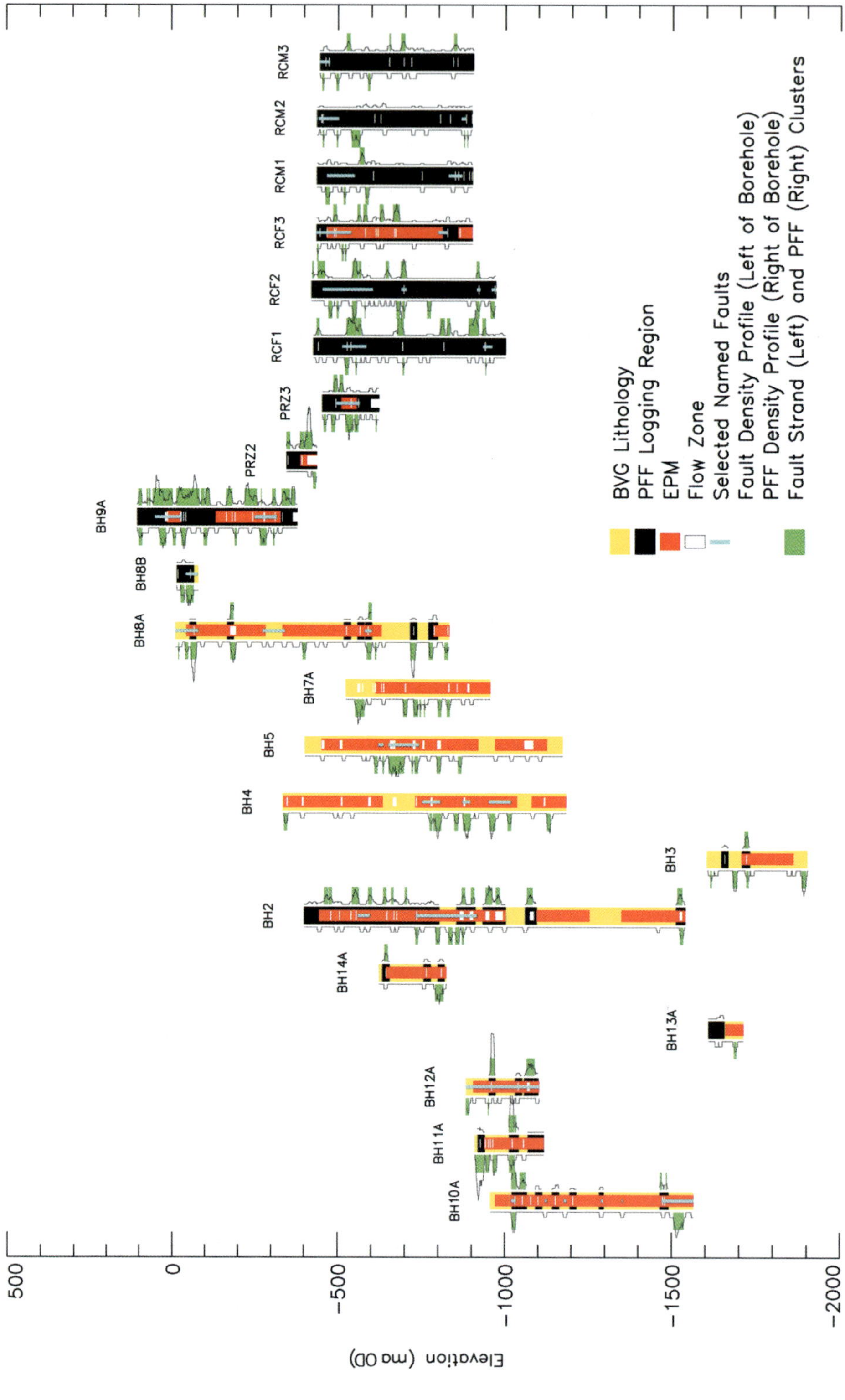

Fig. 5. The locations of PFF clusters, flow zones and major identified faults in the Borrowdale Volcanic Group, as logged in the Nirex deep boreholes (see Fig. 2 for borehole locations).

associated with PFFs. In generating PFFs, it may be that the presence of a fault may be less significant than the fact that certain of the early mineralization, especially ME6 calcite, has undergone dissolution (often with subsequent ME8 and 9 precipitation). It is probably this process, in tandem with a degree of fracture reactivation through changes in regional or local stress fields, that produced the actively flowing network, recognized in boreholes as PFFs,

The locations of clusters of PFFs correspond well with the locations of flow zones. Information about the frequency of PFF clusters and the distribution of their orientation, which was based on information about the density of individual PFFs and the distribution of their orientation, provided key features that were used in the development of the conceptual model for groundwater flow in the Borrowdale Volcanic Group.

Hydrogeological testing

Several different types of hydrogeological test were carried out to investigate permeability on a number of different scales. They were also used to provide insights into the hydrogeological significance of the PFFs. The most comprehensive set of tests was the Environmental Pressure Measurement tests (or EPMs). These were single packer drawdown tests on approximately 50 m-long contiguous intervals of borehole (Sutton 1996) carried out during pauses in drilling activity. A comparison was made between the distributions of measured hydraulic conductivity for intervals that contain and that do not contain PFF clusters. The analysis indicates that the distributions are different and intervals containing PFF clusters are generally more permeable than those without, although there is some overlap.

Fig. 6. Example of a PFF developed within a tensional jog fracture, displaying secondary porosity from anhydrite dissolution and partial infilling by euhedral ME9 calcite. The open fracture coincides precisely with a flow zone identified by production logging.

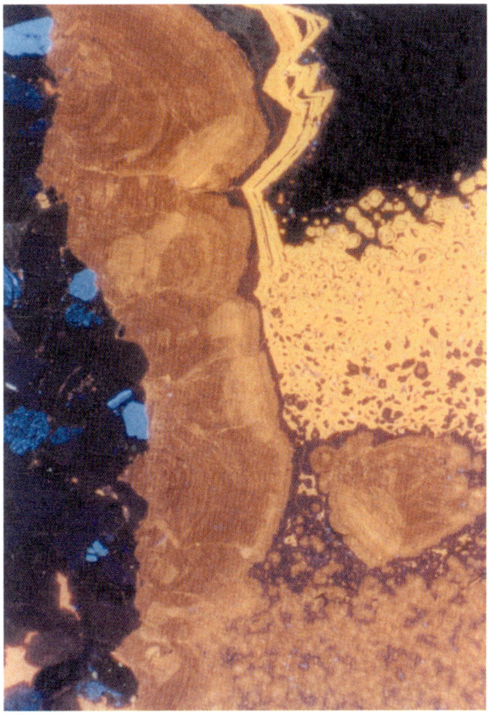

Fig. 7. Cathodoluminescence photomicrograph of a part of a PFF. It shows finely zoned bright luminescent late-stage ME9 calcite mineralization along the walls of an open fracture. The fracture wall is lined by an earlier generation of dull red luminescent ferroan calcite with equant morphology, showing fine but diffuse growth zoning. There is also a geopetal filling of fracture pore space by ME9 calcite precipitated from the water column. Field of view: 3 mm width.

Table 2. *Mineralization episodes in the Sellafield area*

Mineralizing episode	Principal associated minerals	Dominant type of mineralization
ME1	K-feldspar/adularia ±quartz, ±chlorite, ±albite, ±haematite	Silicate
ME2	**Quartz** ±epidote, ±calcite, ±chlorite, ±apatite, ±K-feldspar, ±albite, ±sericite, ±haematite	Silicate (and carbonate)
ME3	**Pyrite** ±traces of chalcopyrite, arsenopyrite, marcasite, galena, sphalerite, Bi-Se sulphosalts and quartz	Sulphide (and possibly silicate)
ME4	**Anhydrite** ±barite, ±fluorite, ±haematite, ±quartz, ±siderite (possibly), ±K-feldspar	Sulphate
ME5	Albite, K-feldspar, **kaolinite**, **illite**, ±haematite	Silicate
ME6 a	*Early ME6a*: ferroan/manganoan carbonate now replaced completely by specular haematite and **calcite** with abundant inclusions of Fe- and Mn-oxides. *Late ME6a*: **calcite** and haematite	Carbonate
ME6 b	Dolomite, **ferroan dolomite**, ankerite, ±siderite, ±quartz, ±anhydrite, ±ferroan calcite	±sulphate
ME6 c	**Calcite** (usually ferroan) ±barite, ±fluorite, ±haematite, ±pyrite, ±sphalerite, ±galena	±sulphate
ME7	**Illitic clay** and **haematite**	Silicate and oxide
ME8	**Mn-** and **Fe-oxides/oxyhydroxides**	Oxide
ME9	**Calcite** ±pyrite, ±anhydrite, ±gypsum	Carbonate ±sulphate ±sulphide

A series of 100 short interval tests, typically 1–2 m in length, were carried out contiguously over a 156 m-long section in the Borrowdale Volcanic Group of borehole RCF3 (Armitrage *et al.* 1996). Analysis of the results confirmed that for intervals not crossed by an identified PFF, the effective permeability determined from the tests is extremely low (Fig. 9) and is consistent with the permeability of the matrix (determined from measurements on cores), and that for intervals crossed by one or more identified PFFs the effective permeability is significantly larger on average than that of the matrix. Significantly, the permeability of a zone with a number of PFFs in it was more than the permeability that would be expected from the permeability of the same number of PFFs with their individual permeability measurements summed. Other pumping tests, over packered intervals approximately 20 m long, give results that confirm that clusters of PFFs are generally more permeable than intervals with only one PFF. Clustering of PFFs is therefore hydrogeologically significant.

A series of four major pumping tests was undertaken in isolated intervals of the main geological units within the RCF3 borehole. The Borrowdale Volcanic Group test involved inducing a long-term (88-day) drawdown, with pressure response monitoring in packered-off intervals in the pumped borehole and in surrounding boreholes (Nirex 1997i). Conclusions derived from the test include that the pressure response was anisotropic, i.e. the drawdown was not spherical around the 40 m pumped interval, but nor was it radial (Fig. 10). Another conclusion of the test was that on a length scale of less than hundreds of metres from the signal source, the groundwater flow through the Borrowdale Volcanic Group is controlled by the interconnection of many features of comparatively low transmissivity, as opposed to a smaller number of features with high transmissivities.

Conceptualization of the flowing feature network

An early step in the development of a numerical model for groundwater flow and transport is to specify the conceptual models of the flow system. A conceptual model is a brief, clear, simple and unambiguous description of the system. It defines the processes acting within the system, the parameters required to model those processes and the conditions on the boundaries of the system. The conceptual model for groundwater flow also identifies the hydrogeological units, that is the subdivisions of the

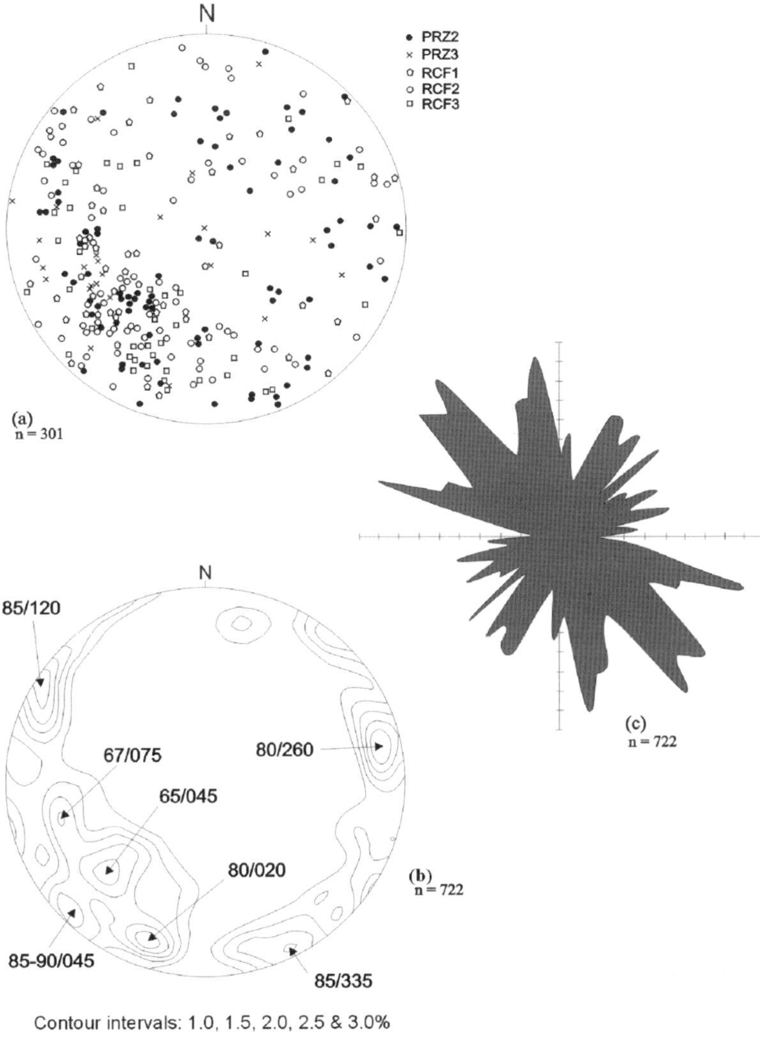

Fig. 8. Orientated PFF data for the Borrowdale Volcanic Group with a Terzaghi weighting for individual borehole orientations. (**a**) and (**b**) are lower hemisphere stereographic plots of poles to PFFs and contoured to show the principal orientation clusters (labelled). (**c**) Rose diagram of strike. n = 722

rock mass into parts within which the rocks have similar hydrogeological parameters, and also the way in which the groundwater flows within the different units. There may be several different conceptual models that are each consistent with the available data. Indeed, one of the tasks that should be undertaken is the generation of alternative models, in order to ensure that any conceptual model bias is minimized. The range of valid conceptual models generated at any one time should reflect the conceptual uncertainty regarding the description of the site. Reduction in the number of possible conceptual models was an iterative process of hypothesis testing using new information. Consequently, although the number of possible conceptual models for the Borrowdale Volcanic Group was initially large, over a number of years the number of conceptual models was reduced to the few described here.

Hydrogeological units

The geological structure and lithostratigraphy of the site provide the basic framework for identifying major hydrogeological units. For the Nirex 97 performance assessment (Nirex 1997a), the

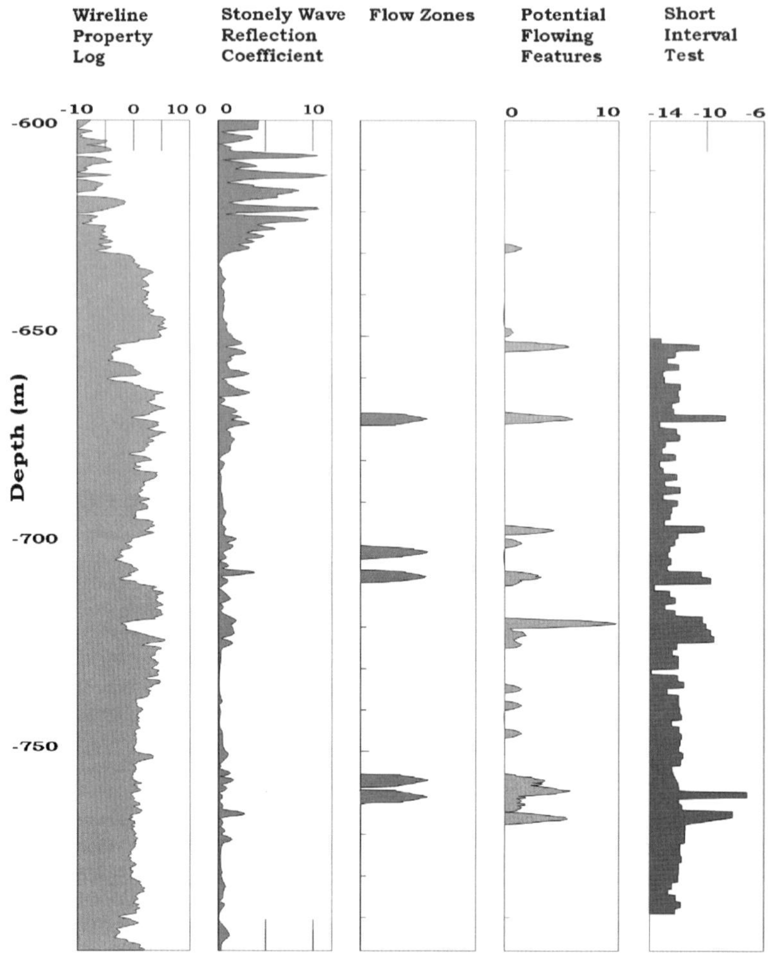

Fig. 9. Profiles of Wireline Property Log, Stoneley wave coefficient, location of flow zones, PFFs and Short Interval Test transmissivity (m^2/s) in a section of the Borrowdale Volcanic Group in borehole RCF3.

Borrowdale Volcanic Group was initially subdivided into two hydrogeological units. Where present within 50 m of the present-day ground surface, a near-surface Borrowdale Volcanic Group hydrogeological unit was defined (termed the Near-Surface Borrowdale Volcanic Group) and at depth a deep Borrowdale Volcanic Group hydrogeological unit was defined (termed the Undifferentiated Borrowdale Volcanic Group). In the near-surface unit, weathering effects, mineral dissolution and the possibility of fracture dilation associated with stress-relief were considered to have created different hydrogeological properties from the deep hydrogeological units.

Major faults were identified in seismic sections and in boreholes, and faulting on various length scales is believed to be pervasive in the Borrowdale Volcanic Group hydrogeological units. Faults and fault damage zones influence the groundwater flow system by modifying the fracture properties of the rock in the vicinity of the fault itself, and by juxtaposing units with differing properties. Because of the potential hydrogeological significance of faulting, it is possible to further subdivide the Borrowdale Volcanic Group into rock within damage zones associated with faults, and rock outside (Nirex 1997j) and conceptually this would be intuitively sensible. For the purposes of calibrating numerical models, the Undifferentiated and Near-Surface Borrowdale Volcanic Group hydrogeological units were divided into Faulted and Undifferentiated sub-units. This classification was further extended by creating additional hydrogeological sub-units which correspond to the major

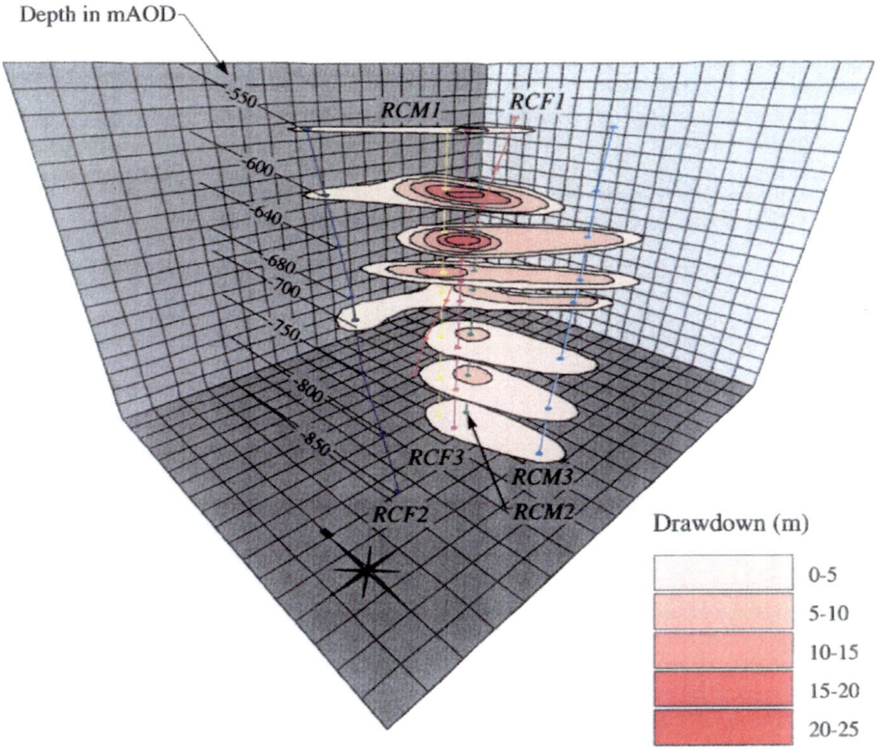

Fig. 10. Contoured drawdown at 2,110 hours into the long-term Borrowdale Volcanic Group hydrogeological test in borehole RCF3. The test section was a relatively permeable section at −640 to −680 m aOD (Sides Farm Member). Pressure responses were monitored in hydraulically isolated sections in 12 observation boreholes and in the test borehole. Vertical and horizontal scales are equal.

structural domains associated with lithostratigraphical units. More information on the hydrogeological subdivisions is provided in reference Nirex 1997a.

Conceptualization of groundwater flow

In the Borrowdale Volcanic Group, groundwater flow is predominantly through a sparse network of interconnected open discontinuities forming flowing features, which comprise a subset of the total fracture population. At the location in the borehole where they are observed, PFFs are considered to indicate the presence of a flowing feature within the local unobserved rock volume. Accordingly, the observations of PFFs frequency and orientation in boreholes have been used to develop a description of a network of flowing features within the Borrowdale Volcanic Group.

The PFFs in the Borrowdale Volcanic Group do not appear to have a random distribution in space, but instead show marked spatial clustering, suggesting that flowing features could also form clusters. As shown by pumping test analysis, clustering of PFFs appears to be significant hydrogeologically and the clusters of flowing features are thought to be indicative of a framework of groundwater flow on a larger scale. There are many possible geological controls on the occurrence of these flowing feature clusters, giving rise to a loose association of the clusters with fault damage zones. Gutmanis *et al.* (1998) discuss hydrogeological characteristics and fluid flow in faults, using the Sellafield investigations as the basis for their study. Figure 11 illustrates the PFF clustering directly. It shows how the density of PFFs varies down the logged intervals of Borehole 2. The density at each point was calculated from the number of PFFs in a 10 m interval centred on each point. Pronounced clusters are clearly present.

Figure 12 shows the clustering more indirectly. It presents a comparison of the distribution of the density of PFFs with that to be expected from a Poisson (random) distribution in space. The figure shows a histogram of the actual frequencies of 10 m intervals containing different

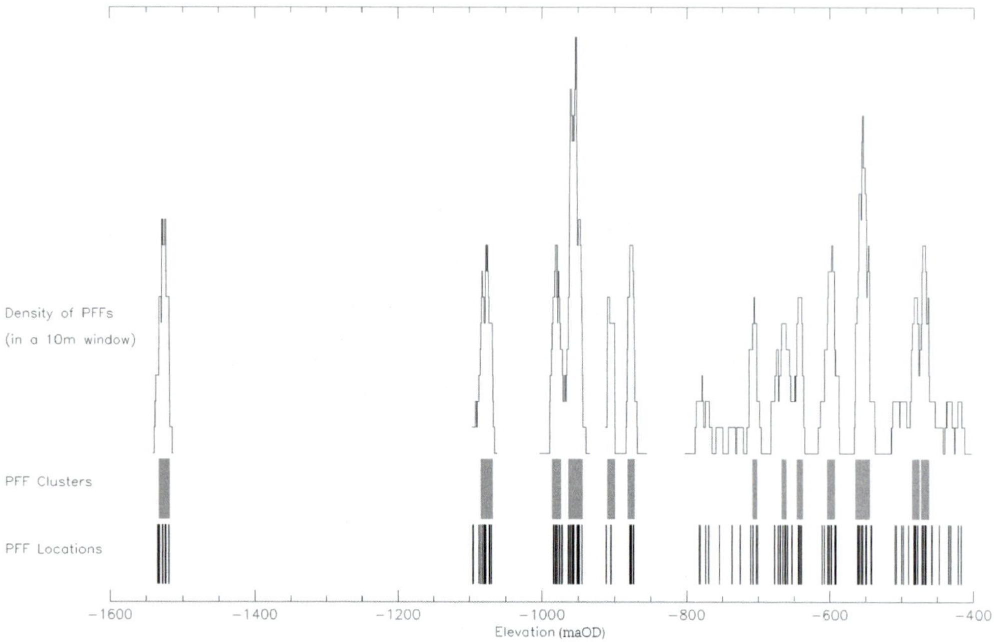

Fig. 11. The distribution of PFFs and PFF clusters down Borehole 2 as calculated using a 10 m moving window. *Note*: no logging between −1150 and −1500 m aOD.

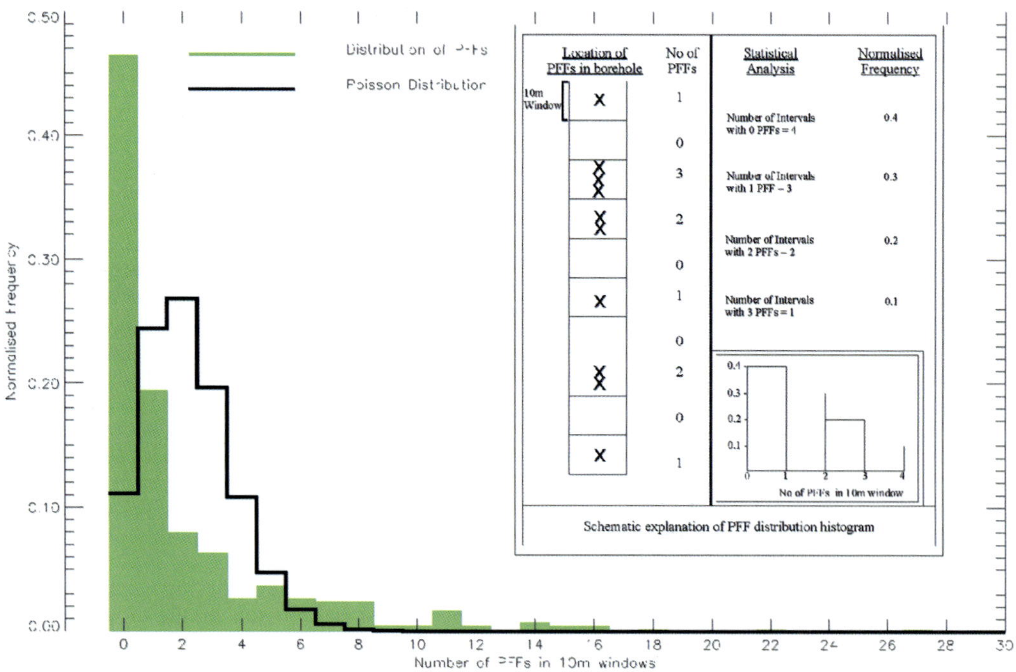

Fig. 12. Comparison of distribution of PFFs with that which would be obtained assuming a Poisson Distribution.

numbers of PFFs from the observed population of PFFs and the corresponding histogram for a Poisson distribution (with the same number of features). Only sections of borehole longer than 100 m that had been logged for PFFs were included in the analysis. Logged intervals of smaller length that had been targeted on specific features of interest were excluded from the analysis, in order to avoid biasing the analysis. It can be seen that there are many more sections of borehole not crossed by any PFFs than would be expected for a Poisson distribution. Also, there are fewer sections of borehole crossed by a small number of PFFs than would be expected, and there are more sections of borehole with many PFFs, than would be expected for a Poisson distribution.

The connectivity of the flowing feature clusters, which is an important aspect of the conceptual model, cannot be determined from the borehole samples, although some inferences have been made from the surface mapping. However, interpretation of the RCF3 pumping test indicates the likely extent of interconnection between flowing features. Even with the available data, alternative concepts for fracture interconnections are possible (Fig. 13). At one extreme, the flowing feature clusters (and indeed, the PFF clusters upon which they are based) might simply be isolated regions with a higher local density of flowing features and with a length scale of up to several hundred metres, connected to each other by a relatively sparse network of background flowing features. At the other end of the spectrum, the flowing feature clusters might form an extensive well-connected network of one-dimensional 'strings' and/or two-dimensional 'sheets'. The hydrogeological properties of the Borrowdale Volcanic Group might be expected to be quite different for the case in which the flowing feature clusters form a well-connected network, and the case in which they do not. Flowing feature clusters are connected hydrogeologically by the background flowing features that do not form part of flowing feature clusters, even if the flowing feature clusters do not form a well-connected network.

In order to address the uncertainty about the connectivity of flowing feature clusters within the Borrowdale Volcanic Group, two extreme conceptual models were taken forward in the upscaling calculations to derive effective regional-scale parameters (Nirex 1997a). The first model is one in which the flowing feature clusters form a well-connected network in their own right. The second is one in which the clusters are not well connected and, on a large length scale, flow is predominantly through the background flowing features. Clearly, any case intermediate between the two extreme conceptual models is possible.

Heterogeneity within the Borrowdale Volcanic Group

The hydrogeological properties of the Borrowdale Volcanic Group are heterogeneous at all length scales and may vary in different directions. At the scale of an individual flowing feature (fractions of a centimetre to many metres), there

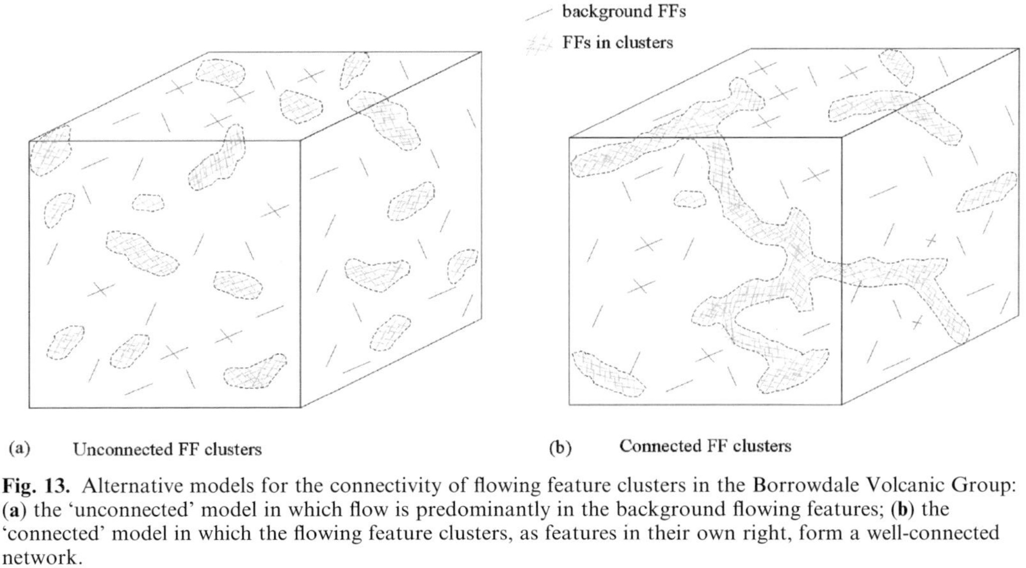

Fig. 13. Alternative models for the connectivity of flowing feature clusters in the Borrowdale Volcanic Group: (**a**) the 'unconnected' model in which flow is predominantly in the background flowing features; (**b**) the 'connected' model in which the flowing feature clusters, as features in their own right, form a well-connected network.

are variations in internal structure. In particular, the aperture and the nature of infilling minerals are likely to be heterogeneous. This could result in significant channelling of flow within the flowing features and to variations in transport properties (e.g. sorption) over the fracture surface. On longer length scales, there are variations in the hydrogeological properties from one flowing feature to another relating to, for example, lithology, proximity to structural features, effects of diagenesis, depth, etc. Because of such variations, the properties of flowing features within a region of given size can best be described by statistical distributions. The analysis undertaken by Nirex indicates that there is considerable variation in hydrogeological properties in the Borrowdale Volcanic Group between the deep boreholes. This leads to variation in the effective properties of large blocks of the Borrowdale Volcanic Group (order of several hundred metres). The assignment of anisotropic permeability values is considered appropriate for the Undifferentiated Borrowdale Volcanic Group, with horizontal permeability in a NE–SW direction likely to be less than permeability in a vertical direction, but greater than permeability in a horizontal NW–SE direction. Finally, there are heterogeneities and possibly, but less likely, anisotropy in the properties of the rock matrix. These property variations would not be significant for groundwater flow, which is predominantly through flowing features, but may have a significant effect upon radionuclide transport, because of their effect on rock matrix diffusion.

Derivation of effective properties

At Sellafield, as at any site, the hydrogeological properties exhibit variability on all length scales from kilometres down to centimetres or smaller. The detailed variation of the hydrogeological (and other) properties of rocks cannot be completely characterized with a practicable number of measurements. Rather, the variability and correlation structure of each property can be characterized to within an uncertainty that depends on, for example, the number and type of measurements and the magnitude of the variability and its correlation structure.

It is not computationally feasible to undertake regional-scale groundwater flow and transport calculations in which the small-scale variability, in particular, is explicitly represented. Such calculations would require many orders of magnitude greater computational resources than are currently available. Moreover, such calculations would provide a description of the groundwater system with far more detail than is necessary for most objectives. These issues are addressed through the use of effective parameters on the length scale of numerical models that are computationally practicable. The effective parameters are parameters that, in the context of the models, give a good representation of the behaviour of the system at an appropriate level of detail. For steady-state groundwater flow modelling, permeability is the key hydrogeological parameter of interest and the following information concerns the upscaling methodology applied for this parameter. For radionuclide transport calculations, fracture porosity is also required. Further information on the upscaling of this parameter is provided in Nirex 1997a.

The data that are required to determine effective parameters fall into two categories, geometrical data and hydrogeological data. In addition, knowledge of the physical processes that influence flow, and valid conceptual models, are required. A hydrogeological unit is generally a subdivision of the rock mass to which effective permeability could be assigned. Ordinarily, it is not possible to measure directly the effective parameters of large-scale hydrogeological units for use in regional-scale models. These parameters must therefore be calculated by a process of 'upscaling' on the basis of data derived on smaller length scales. Each effective parameter can be considered an average, in some sense, of the corresponding smaller length scale property. The appropriate average depends on the structure of the variability within the domain, and is generally not a simple arithmetic mean. In addition, because of the variability within a hydrogeological unit, it is not possible to characterize completely the properties of the hydrogeological unit with a finite number of measurements. There are therefore uncertainties about the distributions and correlations of the measurements. These uncertainties need to be systematically propagated through the upscaling calculations.

Groundwater flow in the hydrogeological units comprising the Borrowdale Volcanic Group is predominantly through the network of flowing features, represented at borehole intersections by PFFs. In view of the uncertainty about the connectivity of the flowing feature clusters in the Borrowdale Volcanic Group, the effective permeabilities used in discretized numerical models were derived by assessing the impact of the two end-member conceptual models in which (a) flowing feature clusters form a well-connected network in their own right, and (b) the flowing feature clusters are not well connected and, on the large length scale, flow is predominantly through the background flowing features.

The estimation of effective permeability for the Borrowdale Volcanic Group was derived in a number of steps. These steps are described in the following paragraphs.

Firstly the PFF clusters crossing each borehole were identified, using the mineralogical and petrological criteria noted previously. For analysis purposes, a cluster was defined as a region in a borehole (or more strictly, in the core), where the density of PFFs exceeded five times the average density for the chosen interval length. A moving window of 10 m length was used to assess the PFF density, centred on points with a few tens of centimetres separation. The mean spacing of PFF clusters, the mean spacing of background PFFs (i.e. that do not form clusters) and the mean spacing of PFFs within the PFF clusters were then calculated, together with estimates of the uncertainties in these parameters. Results indicate that the mean spacing of PFF clusters (as measured in a vertical section) is about 100 m, the mean spacing of background PFFs is about 10 m and the mean spacing of PFFs in a PFF cluster is about 1 m. These values were taken to correspond to the mean spacings for equivalent flowing features in the unobserved rock mass.

The next stage was the determination of the distribution of transmissivity in the flowing feature clusters and background flowing features, from the measurements of permeability (hydraulic conductivity) from EPM pumping tests on borehole sections containing PFFs. Although the EPM tests are very comprehensive, not all the boreholes were logged for PFFs. Therefore, in some cases where PFFs had not been logged, flow zones associated with higher than average permeability were tentatively taken to correspond to clusters of PFFs. This approach may have introduced a slight bias towards higher transmissivities into the distribution of flowing feature clusters and a slight bias towards lower transmissivity into the distribution for background flowing features. However, the beneficial effects of having a larger data set on which to base the statistical upscaling analysis were judged to more than outweigh the detrimental effects of the possible bias.

The mean and standard deviation of the logarithm (to base 10) of the permeability for both background and clustered PFFs were calculated. Given the spacing of flowing feature clusters (about 100 m), which is significantly longer than the typical length of EPM pumping test intervals (50 m), a reasonable approximation to the distribution of transmissivity T', of flowing feature clusters on the small to medium length scale is given by the distribution for

$$T' = l_{\mathrm{EPM}} k_{\mathrm{EPM}},$$

where l_{EPM} is the length of an EPM interval (taken to be 50 m) and k_{EPM} is the permeability measured in an EPM crossed by a PFF cluster.

Each EPM interval may be crossed by several background PFFs. On average there were five PFFs per typical 50 m EPM interval length. The distribution of transmissivity for the background flowing features was inferred by matching the distribution of effective permeability determined from numerically simulated EPMs, to the distribution of permeability determined from EPMs crossed by background flowing features only.

The flowing features, away from direct observation, were taken to have the same orientation distribution as the PFFs. This was also taken to describe the orientation of the flowing feature clusters. Consequently, there should be some anisotropy in the effective permeability.

For the conceptual model for flow in the deep Borrowdale Volcanic Group, in which flowing feature clusters form a network in their own right, the permeability was initially upscaled using the following analytical expression:

$$\mathrm{k}_{\mathrm{ei}} = \frac{C_{Gi} G_i \hat{T}}{b},$$

where b is the average spacing of the transmissive features; C_{Gi} is a factor that represents the combined effects of modelling uncertainty and the connectivity and variability of the features. The prior estimate for C_{Gi} is 1. In cases in which detailed numerical analysis were undertaken, the value of C_{Gi} was calibrated; G_i is a geometric factor, which accounts for the orientation of the features relative to the direction of interest; $\hat{T}$ is the geometric mean transmissivity of the features.

The upscaling analysis was supported by numerical calculations using the discrete fracture network programme NAPSAC (Hartley *et al.* 1996). In these calculations the flowing feature clusters were represented as being planar and realizations of networks of flowing feature clusters in a 2 km cube were generated. The distribution of the lengths of the flowing feature clusters was taken to be a truncated power-law fractal distribution based on the observed distribution of the lengths of features from the surface outcrop mapping.

Each fracture was tessellated into domains with sides of approximately 40 m, within which transmissivity was taken to be constant. In this approximation, smaller length scale variability was taken to be incorporated within the medium length scale variability, and this was taken to have a correlation length of approximately 40 m. The effective regional-scale permeabilities obtained from the numerical calculations were in good agreement with the values calculated using the simple analytical formula with a prior estimate for C_{Gi} of 1. The results from NAPSAC were then used to calibrate C_{Gi} in the analytical expression in order to derive the effective permeability.

For calculating the effective permeability for the alternative conceptual model in which flowing feature clusters do not form a connected network, NAPSAC realisations of networks of background flowing features in a 200 m cube were generated. This volume is sufficient because the flowing features are much smaller than the flowing feature clusters. The length scale of the small-scale variability was taken to be about 3 m and so each fracture was tessellated into domains with sides of approximately 3 m, within which transmissivity was taken to be constant. Again, the numerically derived effective permeability measures were in good agreement with the analytical results.

The uncertainties in the regional-scale parameters for both models were then determined using the First-Order Second-Moment Method (Dettinger & Wilson 1981) to propagate the uncertainties in the various parameters in the analytical expression through to the uncertainty in the permeability for each conceptual model. Finally, an overall estimate of the uncertainty about the effective regional-scale permeability of the Borrowdale Volcanic Group was obtained by combining the uncertainties for the two extreme conceptual models. The resulting probability distribution function has a mean that is the mean of the PDFs of the two extreme cases and an uncertainty which is significantly larger than the uncertainty for either of the two cases.

Finally, it was noted that the Borrowdale Volcanic Group is highly faulted and that the locations of fault damage zones are loosely associated with flow zones in boreholes, although distributions of damage zone properties are not necessarily different from the distributions in

Table 3. *Regional scale Borrowdale Volcanic Group (BVG) effective permeabilities in the Sellafield area and their uncertainties derived from upscaling*

Hydrogeological unit	Permeability		
	k (m^2)	m ($\log_{10}(k)$)	s ($\log_{10}(k)$)
Near Surface BVG	8.51×10^{-14}	−13.070	0.604
Faulted Near-Surface BVG	2.69×10^{-13}	−12.570	0.604
Undifferentiated BVG horizontal NE–SW (combined)	1.75×10^{-18}	−17.757	1.724
Undifferentiated BVG horizontal NW–SE (combined)	3.04×10^{-18}	−17.517	1.724
Undifferentiated BVG vertical (combined)	2.76×10^{-18}	−17.559	1.724
Faulted Undifferentiated BVG horizontal NE–SW (combined)	1.75×10^{-18}	−17.757	1.724
Faulted Undifferentiated BVG horizontal NW–SE (combined)	3.04×10^{-18}	−17.517	1.724
Faulted Undifferentiated BVG vertical (combined)	2.76×10^{-18}	−17.559	1.724
Undifferentiated BVG horizontal NE–SW (connected)	4.22×10^{-17}	−16.375	0.925
Undifferentiated BVG horizontal NW–SE (connected)	7.33×10^{-17}	−16.135	0.925
Undifferentiated BVG vertical (connected)	6.65×10^{-17}	−16.177	0.925
Faulted Undifferentiated BVG horizontal NE–SW (connected)	4.22×10^{-17}	−16.375	0.925
Faulted Undifferentiated BVG horizontal NW–SE (connected)	7.33×10^{-17}	−16.135	0.925
Faulted Undifferentiated BVG vertical (connected)	6.65×10^{-17}	−16.177	0.925
Undifferentiated BVG horizontal NE–SW (unconnected)	7.26×10^{-20}	−19.139	1.114
Undifferentiated BVG horizontal NW–SE (unconnected)	1.26×10^{-19}	−18.899	1.114
Undifferentiated BVG vertical (unconnected)	1.15×10^{-19}	−18.941	1.114
Faulted Undifferentiated BVG horizontal NE–SW (unconnected)	7.26×10^{-20}	−19.139	1.114
Faulted Undifferentiated BVG horizontal NW–SE (unconnected)	1.26×10^{-19}	−18.899	1.114
Faulted Undifferentiated BVG vertical (unconnected)	1.15×10^{-19}	−18.941	1.114

The table presents the mean, m, and the standard deviation, s, of the logarithm to base 10 of the permeability, k. For the BVG hydrogeological units, 'combined' indicates that the uncertainty in whether the flowing feature clusters, considered as features in their own right, form a well-connected network has been taken into account. Use of 'connected' indicates the conceptual model for flow in the BVG in which the flowing feature clusters are considered to form a well-connected network, and 'unconnected' indicates the conceptual model for flow in the BVG in which flow is predominantly through the background flowing features.

intact rock. Any large block of the deep Borrowdale Volcanic Group, with sides of the order of tens of metres and upwards, would be expected to contain faults. Therefore, the contribution of fault damage zones to the effective permeability has been taken into account through the link to upscaling the effects of flowing feature clusters and flowing features and it is not necessary to treat fault damage zones explicitly (although in sensitivity studies such explicit representations were simulated). Table 3 presents the final distribution for the effective permeability in the deep Borrowdale Volcanic Group after upscaling, for use in regional continuum models.

Conclusion

An important factor contributing to the development of robust and defensible simulations of groundwater flow in the Sellafield area was the successful integration of knowledge from a wide range of scientific disciplines and techniques, using the expertise of both modellers and site characterization earth scientists. A key message from the studies summarized here, and which may be of particular value to reservoir studies in the petroleum industry, is that a knowledge of the geometry of the 'complete' fracture network (as observed for example, from surface analogues or from the total discontinuity data set in a borehole) does not necessarily provide the appropriate spatial information required to model successfully fluid flow in fractured rock. This is because the parts of the fractures that actively contribute to flow are a small and particular sub-set of the total fracture population.

Another problem is, how to discriminate between the currently flowing fractures, or those that have a reasonable likelihood to flow, from those that are tightly sealed and unlikely to become active? Production flow logging, using temperature, fluid conductivity or fluid flow measurement, provides only a partial and spatially limited indication of the bulk fluid flow in the whole reservoir. At Sellafield, detailed mineralogical analysis allowed the identification of a particular late calcite found at depth infilling fractures of the Borrowdale Volcanic Group and in other formations. This calcite precipitated from groundwaters that are part of the current groundwater system, or one that has flowed in the geologically recent past (Pleistocene). The presence of the late calcite, with or without other diagnostic criteria, allows the identification of Potential Flowing Features and therefore provides a means to discriminate the sub-set of discontinuities that are flowing or which have the potential to flow, from ancient sealed fractures. More recent work (Rogers & Evans, 2002) has attempted to correlate the relationship between the orientation of fractures controlling groundwater flow in the Borrowdale Volcanic Group and the prevailing stress field, using the 'critically stressed fracture' approach of Barton *et al.* (1995). This approach has the potential to provide additional information that should further refine our ability to generate stochastic fracture networks that only include fracture geometries that are likely to be significant for flow.

As has been demonstrated in this paper, it is possible to identify the intersections with boreholes of the features carrying flow, such as flowing zones, or Potential Flowing Features. More problematic, however, is the characterization of the connectivity of the features carrying flow and their geometry, because these cannot be determined directly from measurements within a borehole. However, it is believed that insights are available from analysis of hydrogeological testing responses and from surface analogues.

It is important not to focus numerical modelling effort and interpretations on a single conceptual model when two or more conceptual models are consistent with the information available from a site. Recognition of the validity of more than one conceptual model permits future data-gathering activities to be more closely specified to reduce the number of allowable conceptual models. This should be an iterative process. Furthermore, the conceptual uncertainties that are inherent in the range of valid conceptual models at any one time need to be accommodated in any interpretation. Two end member conceptual models for flow in the Borrowdale Volcanic Group were developed for the Sellafield area. The first model is one in which the flowing feature clusters form a well-connected network. The second is one in which the clusters are not well-connected and, on a large length scale, flow is predominantly through the background flowing features.

It is not possible to parametrize numerical models of regional-scale groundwater flow and transport in which the detailed variability in flow properties is explicitly represented. Consequently, upscaling methodologies need to be applied in order to take empirical results measured at one scale and transform them for use at other scales. Detailed analysis of site characterization data provided the evidential basis for this upscaling work, which was a prerequisite for the detailed numerical flow and transport modelling Nirex carried out as part of the Sellafield investigations.

A great many individuals and companies contributed to the site characterization programme at Sellafield and this paper includes only a brief summary of some of the work carried out by the various contractors on behalf of Nirex. Some of the contributors directly involved in the work described here, with their affiliations at the time of the work, include T. Milodowski, R. Metcalfe and R. Barnes (British Geological Survey), J. Heathcote (Entec Ltd), T. Batchelor, J. Gutmanis, B. Lanyon and P. Ledingham (GeoScience Ltd), A. Bath, J. Black, H. Richards, K. Been and M. Brightman (Golder Associates), K. Cliffe, J. Porter and M. Shepley (AEA Technology), D. Sanderson, S. Dee and Z. Xing (Southampton University), and members of Sir Alexander Gibb & Co. Ltd and Schlumberger Ltd. We apologize for any omissions. The authors thank S. Rogers and an anonymous referee for their reviews of the paper and also thank their various colleagues for their comments.

References

AKHURST, M. C. and 24 others. 1997. *The geology of the west Cumbria district*. Memoir of the British Geological Survey, Sheets 28, 37 and 47 (England and Wales).

AMEEN, M. S. 1995. Fractography and fracture characterisation in the Permo-Triassic sandstones and Lower Palaeozoic Basement, West Cumbria, UK. *In*: AMEEN, M. S. (ed.) *Fractography: fracture topography as a tool in fracture mechanics and stress analysis*. Geological Society, Special Publications, **92**, 97–147.

ARMITRAGE, P., HOLTON, D., JEFFRIES, N. L., MYATT, B. J. & WILCOCK, P. M. 1996. *Groundwater flow through fractured rock at Sellafield*. European Commission Nuclear Science & Technology Report EUR 16870.

BARCLAY, W. J., RILEY, N. J. & STRONG, G. E. 1994. The Dinantian rocks of the Sellafield area. *Proceedings of the Yorkshire Geological Society*, **50**, 37–49.

BARNES, R. P., AMBROSE, K., HOLLIDAY, D. W. & JONES, N. S. 1994. Lithostratigraphical subdivision of the Triassic Sherwood Sandstone Group in west Cumbria. *Proceedings of the Yorkshire Geological Society*, **50**, 51–60.

BARTON, C. A., ZOBACH, M. D. & MOOS, D. 1995. Fluid flow along potentially active faults in crystalline rock. *Geology*, **23**, 683–686.

BRERETON, N. R., EVANS, C. J. & KINGDON, A. 1998. Hydrogeological and geotechnical rock property characterisation from geophysics. *Proceedings of the Yorkshire Geological Society*, **52**, 199–213.

BURNS, D. R. 1991. Predicting relative and absolute variations of *in situ* permeability from full waveform acoustic logs. *The Log Analyst*, **32**, 246–255.

CHAPLOW, R. 1996. The geology and hydrogeology of Sellafield: an overview. *Quarterly Journal of Engineering Geology*, **29**, Supplement 1, S1–S12.

DETTINGER, M. D. & WILSON, J. L. 1981. First-order analysis of uncertainty in numerical models of groundwater flow. 1. Mathematical development. *Water Resources Research*, **17**, 149–161.

GUTMANIS, J. C., LANYON, G. W., WYNN, T. J. & WATSON, C. R. 1998. Fluid flow in faults: a study of fault hydrogeology in Triassic sandstone and Ordovician volcaniclastic rocks at Sellafield, north-west England. *Proceedings of the Yorkshire Geological Society*, **52**, 159–175.

HARTLEY, L. J., HERBERT, A. W. & WILCOCK, P. M. 1996. *NAPSAC (Release 4) Summary Document*. AEA Technology Document AEA-D & R-0271.

JOHNSON, L. D., LUI, H. L., WINKLER, K. W. & LUTHI, S. M. 1992: Permeability estimation in naturally fractured fields by analysis of Stoneley waves. *The Log Analyst*, **33**, 493–494.

KULANDER, B. R., DEAN, S. L. & WARD, B. J. 1990. *Fractured core analysis: Interpretation, logging and use of natural and induced fractures in core*. Methods in Exploration Series **8**. American Association of Petroleum Geologists, Tulsa, Oklahoma, USA.

MCMILLAN, A. A., HEATHCOTE, J. A., KLINCK, B. A., SHEPLEY, M. G., JACKSON, C. P. & DEGNAN, P. J. 2000. Hydrogeological characterisation of the onshore Quaternary sediments at Sellafield using the concept of domains. *Quarterly Journal of Engineering Geology and Hydrogeology*, **33**, 301–323.

MICHIE, U. MCL. 1996. The geological framework of the Sellafield area and its relationship to hydrogeology. *Quarterly Journal of Engineering Geology*, **29**, S13–S27.

MILLWARD, D., BEDDOE-STEPHENS, B., WILLIAMSON, I. T., YOUNG, S. R. & PETTERSON, M. G. 1994. Lithostratigraphy of a concealed caldera-related ignimbrite sequence within the Borrowdale Volcanic Group of west Cumbria. *Proceedings of the Yorkshire Geological Society*, **50**, 25–36.

MILODOWSKI, A. E., GILLESPIE, M. R., NADEN, J., FORTEY, N. J., SHEPHERD, T. J., PEARCE, J. M. & METCALFE, R. 1998. The petrology and paragenesis of fracture mineralisation in the Sellafield area, west Cumbria. *Proceedings of the Yorkshire Geological Society*, **52**, 215–241.

MILODOWSKI, A. E., GILLESPIE, M. R. & KEMP, S. J. 2000. *A review of the fracture mineralogy and mineral distribution in the Sellafield boreholes*. British Geological Survey Report to Nirex WG/94/37.

NAGRA 1994. *Kristallin-I Safety Assessment Report*. Nagra Technical Report NTB 93-22.

NIREX 1996a. *Rock Characterisation Facility Longlands Farm, West Cumbria: Report on baseline groundwater pressures and hydrochemistry*. Nirex Science Report SA/96/006.

NIREX 1996b. *Structural atlas of the Sellafield boreholes: structural domains and statistics illustrated by graphical display of discontinuity data*. Nirex Science Report SA/96/005.

NIREX 1997a. *Nirex 97: An Assessment of the Post-Closure Performance of a Deep Waste Repository at Sellafield*. Nirex Science Report S/97/012.

NIREX 1997b. *Evaluation of the heterogeneity and scaling of fractures in the Borrowdale Volcanic Group in the Sellafield area*. Nirex Science Report SA/97/028.

NIREX 1997c. *The lithological and discontinuity characteristics of the Borrowdale Volcanic Group at outcrop in the Craghouse Park and Latterbarrow areas*. Nirex Science Report SA/97/029.

NIREX 1997d. *Locations of flow zones in Sellafield boreholes*. Nirex Science Report SA/97/073.

NIREX 1997e. *Spatial heterogeneity of the rock mass within the potential repository zones*. Nirex Science Report S/97/005.

NIREX 1997f. *Spatial heterogeneity of the rock mass properties*. Nirex Science Report SA/97/021.

NIREX 1997g. *Constraining the timing of movements on Sellafield faults*. Nirex Science Report SA/97/033.

NIREX 1997h. *Summary data compilation for the location, distribution and orientation of Potential Flowing Features in the Sellafield boreholes*. Nirex Science Report SA/97/031.

NIREX 1997i. *Detailed interpretation of the Borrowdale Volcanic Group phase of the Borehole RCF3 Pump Test*. Nirex Science Report SA/97/008.

NIREX 1997j. *Conceptualisation of the architecture and hydrogeology of faults at Sellafield*. Nirex Science Report SA/97/025.

PAILLET, F. L. 1991. Qualitative and quantitative interpretation of fracture permeability using acoustic waveform logs. *The Log Analyst*, **32**, 256–270.

ROGERS, S. F. & EVANS. C. J. 2002. Stress-dependent flow in fractured rocks at Sellafield, United Kingdom. *In*: LOVELL, M. & PARKINSON, N. (eds) *AAPG Methods in Exploration Series No. 13*, 241–250.

SKB 1999. *SR97—Main Report*. Swedish Nuclear Fuel and Waste Management Co. Stockholm.

SUTTON, J. S. 1996. Hydrogeological testing in the Sellafield area. *Quarterly Journal of Engineering Geology*, **29**, S29–S38.

WEI, Z. G., EGGER, P. & DESCOEUDRES, F. 1995. Permeability predictions for jointed rock mass. *International Journal of Rock Mineral Science & Geomechanics Abstracts*, **32**, 251–261.

Modelling fracture systems in extensional crystalline basement

C. A. E. SANDERS[1], L. FULLARTON[2] & S. CALVERT[3]

[1]*Petronas Carigali Sdn. Bhd., Petronas Twin Towers, 55000 Kuala Lumpur, Malaysia*
[2]*Shell UK Exploration and Production, 1 Altens Farm Road, Nigg, Aberdeen AB12 3FY, UK*
[3]*Midland Valley Exploration Ltd., 14 Park Circus, Glasgow G3 6AX, UK*

Abstract: Using observations from the extensional basin setting in Vietnam, conceptual models were developed to simulate and analyse fracture systems typical of crystalline basement in such structural settings. Information from field observations, seismic surveys and three-dimensional (3-D) structural modelling were integrated and used to build geologically realistic 3-D fracture networks. A major advantage of the method used in this study is that it allows a better understanding of the apparently chaotic fracture networks characteristic of such rocks, and of the processes responsible for fracturing. Several fracture generating processes are discussed and modelled, with emphasis on tectonic fractures and the relation to structural modelling. An example is presented highlighting the differences in fracturing in the hanging wall and footwall during lithospheric extension superimposed on a primary (igneous) cooling fracture network. Results suggest that during flexural uplift, the hanging wall is significantly more deformed than the footwall, implying the former is more prone to fracturing than the footwall for both kinematic and flexural isostatic processes.

Fractured basement has proved to be a successful petroleum play in the extensional offshore setting of SW Vietnam (Fig. 1 and Areshev *et al.* 1992). The unconventional nature of this play demands different exploration and production strategies compared with sedimentary plays. Typically a comprehensive exploration strategy is ill defined. Cartoon style concepts have been defined (e.g. Areshev *et al.* 1992; Tandom *et al.* 1997), but few attempts have been made to link the concepts in a quantitative way to the available data, and to test these concepts in a predictive manner.

The structurally high horst blocks form the usual target (Fig. 2). The seal on the fractured reservoir is formed by shale deposits overlying the basement. The basement is most probably charged with hydrocarbons from the side and above from the earliest clastic lacustrine deposits (Sladen 1997). The presence of an extensive connected fracture network in the basement then defines whether a fault block is a potential viable prospect.

In this paper we reassess some of the structural concepts that define the petroleum play. Modern methods of appraising fractured reservoirs rely on a number of data sets including those from cores, FMI, wireline, video and field observations. More often than not, the prediction of regional fracture networks requires a full understanding of these diverse data sets. Here we present a novel technique that incorporates all available data into a comprehensive model describing the range of fracture systems found in extensional settings in crystalline rocks. In the absence of primary porosity and permeability, the basement reservoir potential is entirely defined by fractures forming the secondary porosity and permeability. Fracture appraisal is therefore of prime importance for the successful development of these fields. Three basic questions arise from this, namely: (1) where are the fractures (fracture location and fracture density), (2) what is their geometry (orientation and interaction of the fractures, aperture, length etc.), and (3) how has their geological and fluid flow behaviour changed over time, if at all, in response to different stress regimes? Clearly the first two questions must be answered before the third can be properly addressed. Detailed knowledge of the spatial variation in density and orientation of the fracture systems is a first-order requirement in any appraisal of a hydrocarbon trap, and forms the main focus of this paper.

Technical background

This paper is based on several case studies (mostly relying on seismic surveys) from the Cuu Long basin offshore SW Vietnam and illustrates the techniques that have been applied. For reasons of confidentiality, only conceptual models that represent the tectonic setting are presented.

The main tools used were two-dimensional and three-dimensional structural modelling software. These were used to gain qualitative and quantitative analysis of deformation processes over time. In addition, discrete fracture networks (DFN) were modelled to simulate geologically realistic fracture systems. The modelling is constrained by data from seismic surveys, observations from fieldwork and wells. The workflow is designed

From: PETFORD, N. & MCCAFFREY, K. J. W. (eds) 2003. *Hydrocarbons in Crystalline Rocks*. Geological Society, London, Special Publications, **214**, 221–236. 0305-8719/03/$15

Fig. 1. Area of study location for fieldwork and case studies in SW Vietnam.

to predict fracture systems based on the geological understanding of the relationship between the fracture network and the geological history, and in particular the structural history of the area. This is in contrast to more conventional, purely stochastic fracture models, which make only limited use of the geological background. The workflow for prediction and understanding fracture systems involves three steps:

(1) The first is the identification of the geological process (or processes) that creates the fractures. It is commonly observed that more than one process can be involved in creating fracture sets over the geological history of the basement rock (e.g. Tandom *et al.* 1997).

(2) Once a geological process has been identified, the corresponding fracture system is simulated according to its generic and geometric characteristics and at the appropriate time in the deformation history.

(3) Then a final correlation can be made of the simulated fracture network to well data or other observations.

Geology of SW Vietnam

The crystalline basement in SW Vietnam is characterized by Mesozoic granitic to dioritic plutons and their volcanic equivalents (e.g. Areshev *et al.* 1992). Both onshore and offshore basement share the same geological history of subduction-related calc alkaline magmatism throughout the Mesozoic, but becoming more alkaline towards the end of the Cretaceous. This change in magma composition marks the transition from a convergent period to widespread extensional deformation that disrupted the granitoid-dominated crust and created the now-submerged Cuu Long basin. NW–SE extension occurred, probably during the Palaeogene, with local deposition of lacustrine and fluvial-brackish sediments. The fault activity decreased during the Oligocene, followed by a thermal sag basin with a transition to transgressional marine basin infill. Gradual basin infill with predominantly marine clastic sediments lasted until the present day

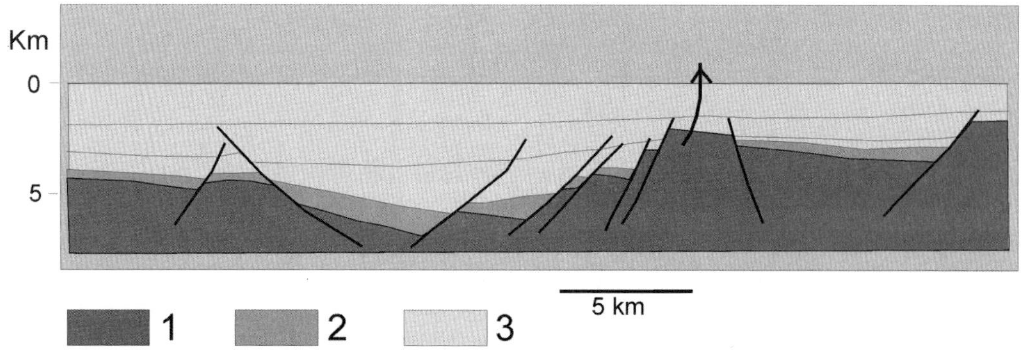

Fig. 2. Typical schematic section over the Cuu Long basin in SW Vietnam (modified after Sanders *et al.* 2001). The structurally highest horst blocks are usually targeted for fractured basement reservoirs. 1, crystalline basement; 2, Early Oligocene clastic lacustrine or fluviatile sediments (E horizon); 3, clastic lacustrine and marine basin infill (sand–shale).

with (re-)activation of faults probably due to sediment compaction (Sanders *et al.* 2001). There is no evidence for Neogene inversion similar to that found in the nearby Nam Con Son basin to the south (e.g. Matthews *et al.* 1997).

Field studies and fractures

Onshore exposures of granitic rocks in SW Vietnam and on the Con Dao islands provide an extremely useful analogue for the offshore, hydrocarbon-bearing basement. Field studies of onshore basement outcrops reveal three generations of fractures: (1) primary fractures related to the emplacement and cooling of the magma, (2) secondary fractures related to tectonic deformation, and (3) fractures related to exhumation. Each is discussed in turn below.

Primary fractures

Primary fractures or 'cooling fractures' are fractures formed during the emplacement, crystallization and cooling of the magma in the upper crust. At the present day these rocks can be well studied in undisturbed outcrops located outside the main deformation front, for example near Camly Park (Fig. 3). Primary fractures are

Fig. 3. Outcrop granites at Camly Park (SW Vietnam), showing cooling fractures in section view (**a**) and in top view (**b**). Note the regular spacing and absence of damage zones. Several fractures are filled with pegmatite minerals (late volatile granitic melts). See truck (in **a**) and hat (in **b**) for scale.

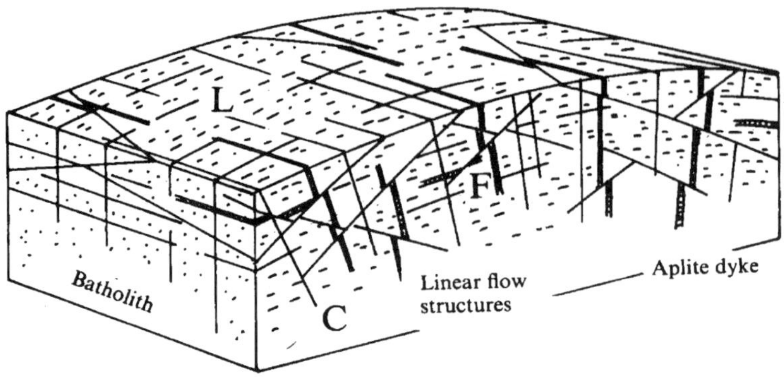

Fig. 4. Diagram showing primary fracture patterns in granitic rocks (after Cloos 1922).

characterized by evenly spaced joint sets often orthogonal to each other. (Figs 3 and 4). The fractures are predominantly dilational fractures (or type I fractures) often filled by pegmatites, volatile-rich melts formed by late-stage segregation of the crystallizing magma. As such, these fracture sets can be identified as having formed early in the history of the igneous basement, prior to the onset of regional extension. All intrusive rocks are expected to have primary fractures, but their spacing and orientation may vary greatly within the magma body. A strong correlation was observed between crystal size and fracture spacing in the field. For purposes of offshore exploration and production, the geometry of primary fracture sets is hard to predict, although some progress is being made (see Koenders & Petford 2003). However, the presence of these primary fracture sets may have a significant effect on the porosity and connectivity behaviour of hydrocarbon reservoirs.

Secondary fractures

Secondary fractures result from deformation of the rock mass due to tectonic forces. In the Cuu Long basin in Vietnam, this occurred predominantly during the extensional basin deformation. The fracture sets are commonly characterized by conjugate shear sets, with the smallest angle between the fractures in the direction of sigma 1 of the regional stress field.

Tectonic fractures are characterized predominantly by 'type II' or shear fractures, showing considerable shear displacement. With ongoing shear movement along the fracture planes, the fractures develop damage zones with gouge or fault breccia and at the same time increase in length (e.g. Price & Cosgrove 1990). When strain concentration continues, subseismic faults occur (Fig. 5). Rocks characterized predominantly by shear fractures generally have a high porosity and connectivity (compared to primary fractures), due to the greater length of the fractures, the width of the damage zones, and frequently occurring subseismic faults (as in Fig. 5b), which tend to act as corridors for fluid flow. The variation in density and orientation of the tectonic fractures is dominated by the history and intensity of the deformation processes.

Exhumation fractures

Further fracturing takes place during exhumation of the rocks to the surface. Prior to exhumation, rocks are under ambient *in-situ* stress regimes. Due to unloading associated with exhumation, the stress regime acting on the rock mass changes. The rocks carry residual stresses which at high levels in the crust result in fracturing of the rock parallel to the Earth's surface (Price & Cosgrove 1990). Associated with this are fractures at right angles. Exhumation fractures are thought to occur predominantly in the upper 50 m or so below the surface (Areshev *et al.* 1992). Exposure at the surface can result in the fractures forming channels for meteoric fluids that can greatly enhance the rates of weathering. Depending on the weathering products (coarse-grained arkose versus fine-grained laterite soils, for example), this exhumation layer, if reburied, has the potential to form a productive reservoir, as well as an effective seal. In the absence of other well-defined plays, the weathered tops of crystalline rocks have often been the main targets in basement exploration (Areshev *et al.* 1992, see also Schutter 2003). In principle, an exhumation layer can be expected in any crystalline basement topped by an erosional unconformity (as seen in offshore Vietnam), but in practice this possibility is controlled strongly by palaeotopography (Areshev *et al.* 1992).

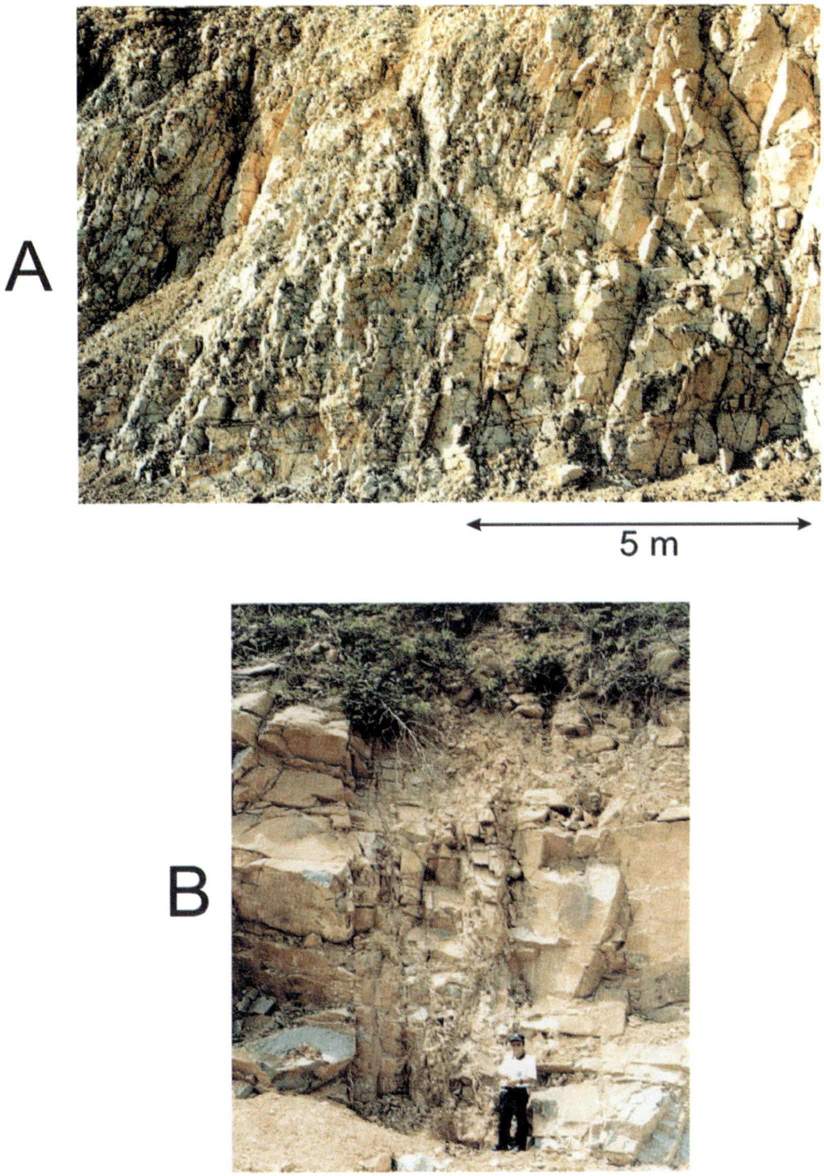

Fig. 5. Shear fractures (diffusely distributed) in rhyolite (**a**) and subseismic fault (**b**) in granite, both on Con Dao islands. Scale photograph in (**a**) is roughly 10 m.

Modelling of discrete fracture networks

General

Fractures were modelled using Fracture Generator, an interactive software tool that allows the simulation of various fracture sets (see Swaby & Rawnsley 1996). The length, growth, density and orientation of fractures and their interaction with other fractures or lithology can be controlled by numerical parameters chosen by the user, or by numerical maps of such attributes as strain or curvature. These can also be generated using structural modelling software. In this way, fracture growth is simulated in a geological environment and guided by structural concepts and data. The resulting discrete fracture networks can then be correlated and improved against well data or other observations.

Primary fracture modelling

Primary (cooling) fractures are difficult to simulate, especially for offshore basement, as there are few data to constrain the variations in fracture density and orientation. The nature of a long-lasting active magmatic province such as Vietnam (or the Andes), is one of multiple intrusions and magma pulsing. Individual magma pulses differ in both composition and intrusion temperature, and thus in material properties. This in turn will control the style of primary fracture formation, thus making prediction of primary fractures difficult. However, since they play a significant role in the overall fracture network, and are present in all intrusive rocks, they are best simulated in a purely stochastic way. Only a large amount of well data will give any insight into the spatial distribution of the magmatic rocks and the parameters that control them, underlining the need for more accurate descriptions of primary fracture formation.

Exhumation fracture modelling

Exhumation fractures can be expected in the top 50 m or so below an erosional unconformity, and can be modelled accordingly. However, reliable data to constrain the complex porosity and permeability variations, due to the combined effects of fracturing and weathering and the thickness of the affected rock layer, are not generally available. It is not unreasonable to expect large, lateral variation due to palaeotopography, vegetation and the original mineralogy of the rock. Given the uncertainties in these parameters, for the scope of this paper they are not considered further. This leaves the secondary (or tectonic) fractures as the main topic of this paper. From field observations they seem to dominate the porosity and connectivity throughout the basement (apart from the weathered cap). Sufficient control over crucial parameters exists to create a comprehensive model that can be tested against well data. The simulation of tectonic fracture networks thus helps to create new insights into fracture distribution in crystalline basement, and informs technical risk assessment of fractured basement plays.

Modelling tectonic fractures

In crystalline basement under brittle conditions (the upper 10 km of the crust, approximately) it is reasonable to assume that all deformation is accommodated by fracturing (e.g. Ranalli 1987). Therefore, control over the deformation processes forms the main key for tectonic fracture appraisal in crystalline basement. Control on the deformation processes in turn, can be well constrained with two-dimensional and three-dimensional structural modelling and restoration of a geological model built from seismic interpretation. During extensional basin formation, several deformation processes may take place at the same time. Faulting and associated movement of fault blocks forms the most obvious process (Fig. 2), but large-scale flexing of the lithosphere may be another process to consider (Beek 1995). Thus, firstly, the large faults along which the movement takes place can be identified on seismic imaging. These faults are usually not planar, two-dimensional structures (as they are often represented), but are made up of a fault damage zone of a certain width, a zone of high fracturing around the main plane of movement. Secondly, due to the movement of the hanging wall block along the fault, the hanging wall block usually needs to change shape and deform internally (Hauge & Gray 1996). Well-known examples are the roll-over anticlines above listric normal faults. Finally, there will be footwall deformation related to the extensional tectonics.

Each of these three processes is now looked at in more detail. We do not give a complete description of all the possible deformation processes in extensional settings (we ignore much of the syn rift magmatic processes, for example). Rather, we show that when individual deformation processes can be singled out, analysed and compared, it is possible to predict their associated fracture network and analyse their importance for hydrocarbon exploration and production strategies. The description below thus helps to give an insight into the apparently complex fracture networks often observed in extensional settings.

Seismic faulting

The most obvious structural features seen in extensional settings are large faults visible on seismic images. These faults, although usually represented as a line feature in two-dimensional sections, are in reality made up of a zone of high fracturing, displaying numerous individual displacement planes roughly parallel to a major plane of fault movement. The width and intensity of the fracture zone, or fault damage zone, is not necessarily symmetrical around the main fault plane. In general, the fracture density gradually decreases away from the fault, and several studies have reported that the width of the damage zone is related to the amount of slip as observed along the fault plane (e.g. Sleight *et al.* 2001).

Hanging wall deformation

During extensional basin formation, hanging wall blocks move over faults to accommodate the overall stretching of the lithosphere (Lister *et al.* 1986). Because the faults are usually not planar in three-dimensional space and because of the commonly observed variable slip along these faults, the hanging wall needs to deform internally. Under brittle conditions this deformation is accommodated by fracturing (or subseismic faulting). Well-known examples are roll-over anticlines which are commonly observed above listric faults (Bruce 1973; Gibbs 1983; Dula 1991). However, even if faults are planar at depth (a fair assumption for the upper 5 km or so in typical crystalline basement, as in the Cuu Long basin in Vietnam), they are commonly irregular in plan view and show significant laterally variable displacement. This deformation is simulated with three-dimensional structural modelling using kinematic restorations. In the geological model, the fault block is moved over the fault with the observed variable slip until the hanging wall top basement horizon matches the footwall top basement horizon. During this movement, the hanging wall block deforms internally (Dula 1991). The internal deformation in the fault block is simulated by a numerical algorithm that mimics the deformation of the rocks in a geometrical way with preservation of volume (Hauge & Gray 1996; Kane *et al.* 1997). When the restoration is successful, the hanging wall's movement is reversed to forward model the deformation history from the undeformed (restored) state to the deformed (present-day) situation. During this process, the amount of strain in the hanging wall block is calculated and colour mapped on the surface. Strain is calculated as the difference in geometry between an initial state and a deformed state at any time in the deformation history (see below). The resulting strain map forms a guide to the spatial variation of deformation in the hanging wall (if necessary for separate time phases), which forms a key to fracture density assessment. The advantage of this technique over, for example, curvature analysis is that strain calculations integrated with kinematic restorations take the deformation history (strain path) into account, whilst curvature analysis only represents a static situation. Furthermore, because of the irregular erosional top surface of the crystalline basement, static analysis of curvature is heavily biased by irregular topography caused by erosional processes. For the same reasons, non-kinematic restorations will greatly over-estimate the strain in the basement, since the topography was originally not a flat surface (a common assumption used for non-kinematic restorations of sedimentary beds).

Footwall deformation

The footwall block is the stable block during extensional deformation and therefore does not change shape (unless it forms a hanging wall block for its own underlying fault, as in en echelon trains of extensional fault blocks, see Fig. 2). Extensional fault block movement is therefore not expected to create any deformation fractures in extensional horst blocks.

The fracturing observed in these horst blocks has been ascribed to: exhumation processes and weathering (e.g. Areshev *et al.* 1992; Tandom *et al.* 1997), fault damage zones (Ngoc pers. comm. 2000), or due to the general term 'footwall uplift' without efforts to specify footwall uplift or the process that caused it. In general, footwall block uplift is hard to accomplish, since it has to work against the main regional stress vector s1, which is directed vertically downward in Andersonian theory for extensional settings (Anderson 1951). Therefore footwall uplift in an extensional setting takes place predominantly due to flexural isostatic uplift, controlled by buoyancy forces acting deep in the lithosphere and asthenosphere (e.g. Kusnir *et al.* 1991; Egan 1992; Beek 1995). Uplift due to flexural isostasy takes place due to unloading (thinning) of the lithosphere by extensional fault block movement. However, because of the strength of the lithosphere at depth, the unloading does not create local uplift and the effects are instead distributed over a regional scale. The strength of the lithosphere (commonly expressed as EET, Effective Elastic Thickness) (Egan 1992; Beek 1995), defines the wavelength of the rebounce. Typical EET values for extending lithosphere are between 5 km and 20 km (Beek 1995), where the lower values are typical for weak lithospheres with high heat flow. Note that if the lithosphere has no strength at all (EET = 0), we have effectively Airy isostasy. Unloading creates a lithospheric flexural rebound causing uplift in the flank of the extensional basin (therefore 'footwall uplift' or 'flank uplift' (see Fig. 6). The uplift pattern

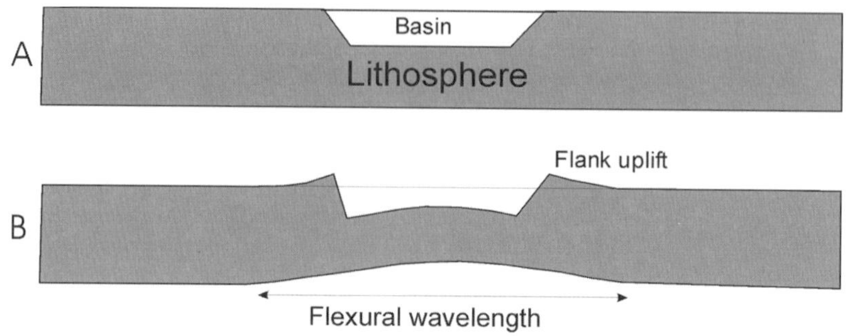

Fig. 6. Lithospheric scale diagram explaining the process of lithospheric flexure due to unloading. (**A**) Basin formation with thinning of lithosphere. (**B**) Flexural rebound due to unloading. The flexural wavelength is defined by the strength of the lithosphere, usually expressed in EET values.

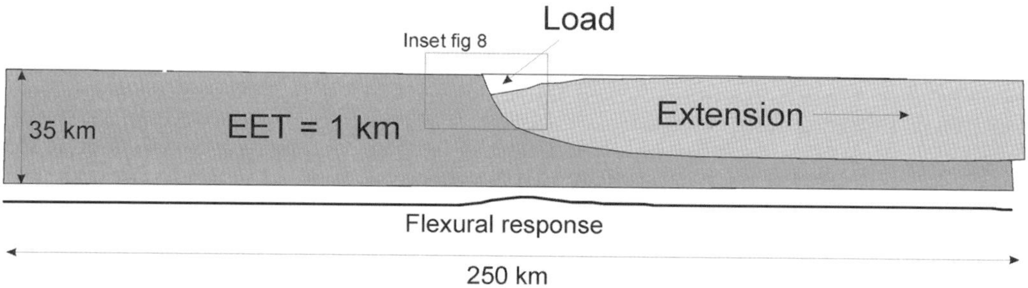

Fig. 7. Two-dimensional model showing crustal-scale extensional fault creating a basin. EET is set at 1 km. Flexural response has small wavelength. Inset shows enlargement for the analysis of the flank uplift (Figs 8 and 9).

depends on the amount of unloading and the strength of the lithosphere.

To analyse the effect of flexural isostatic uplift on extensional basin, a conceptual two-dimensional model has been build (Fig. 7). The lithosphere extension is modelled by a crustal scale extensional fault. Density of the load of the crust is set at 2.8 g/cm^3 and the density of the asthenosphere at 3.3 g/cm^3 with sub-marine conditions. Due to extension and basin formation the lithosphere is unloaded and the flexural isostatic rebound calculated similarly as in Egan (1992). Uplift itself (as epeirogenetic uplift) does not generate deformation unless laterally variable uplift takes place, which would be expressed by curvature in the upper crust. With a low EET, we generate a short wavelength for lithospheric flexure and thus the highest possible curvature (deformation) in the footwall.

To assess its potential as a fracture forming process, the EET has been kept low to generate as much deformation as possible in the footwall. With fault displacement of up to 5 km (which has been observed in the Cuu Long basin), as much as 620 m of uplift can be generated, but the curvature still takes place over a distance of approximately 12 km (Fig. 8). Further examples of the influence of unloading and EET are shown in Figure 9. It is noteworthy that for all scenarios, most of the uplift resulting from flexural response of the lithosphere takes place in the hanging wall where most of the unloading has taken place.

Strain analysis

Strain analysis was carried out to compare the amount of deformation caused by the hanging wall and footwall processes. Strain is calculated for the deformed state at any time in the deformation history. In order to compare the footwall uplift mechanism with the kinematic hanging wall deformation, a three-dimensional model has been build from six two-dimensional lines (Fig. 7), with a variety of fault offsets and with an irregular plan view to simulate a typical large-scale fault bounding a high horst block in the Cuu Long basin (Fig. 10). A large scale is

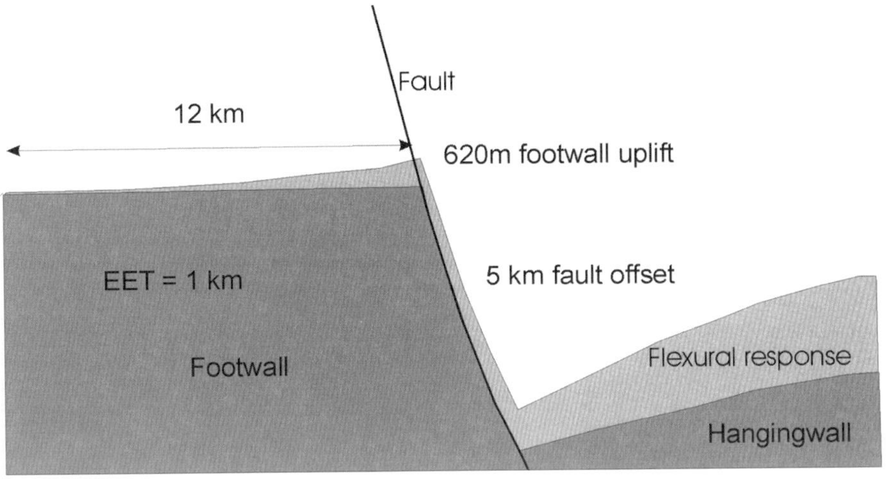

Fig. 8. Enlargement of inset, Figure 7. In dark grey is the original shape of the crust after fault movement. Light grey shows the flexural uplift. In this case a offset of 5 km was used, EET = 1 km. This setting results in a maximal flank uplift of 620 m while the curvature in the footwall is distributed over roughly 12 km. Note the strong flexural response in the hanging wall.

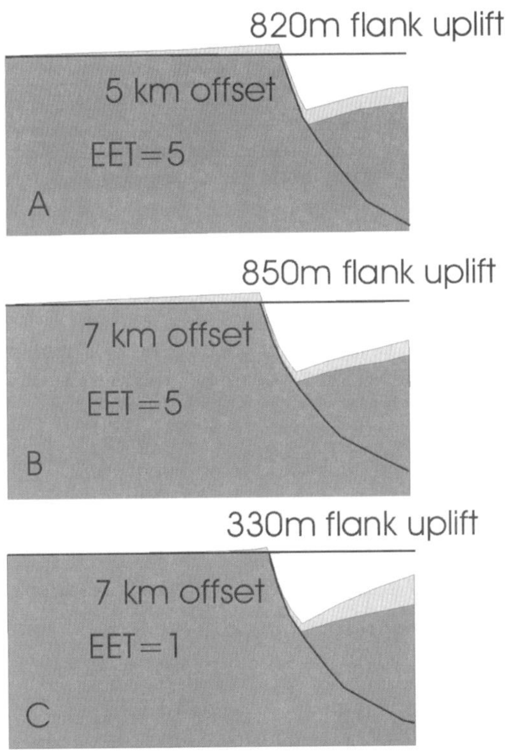

Fig. 9. Three scenarios with different EET and fault offset. (**A**) 5 km offset, EET = 5 km results in a flank uplift of 820 m. (**B**) EET = 5 km and fault offset of 7 km results in a maximum flank uplift of 850 m. (**C**) EET = 1 km and fault offset 7 km results in a flank uplift of 330 m. Note the different uplift patterns as a result of different unloading patterns and EET.

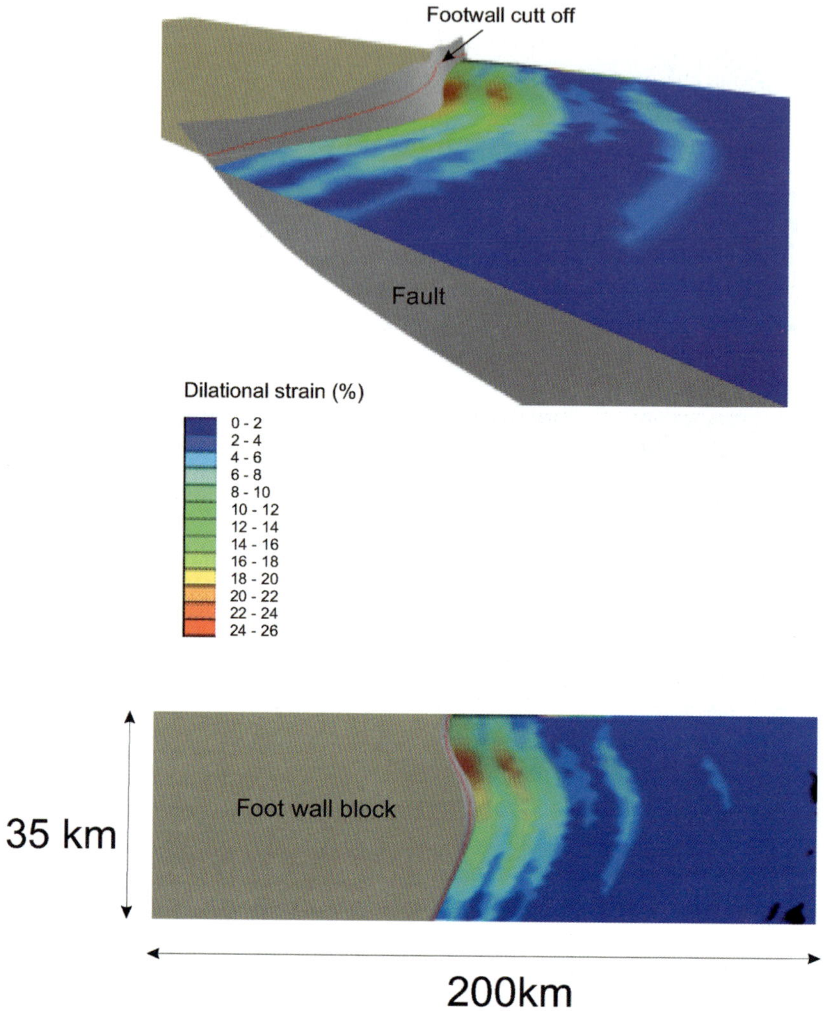

Fig. 10. (**A**) Three-dimensional geological model showing variable offset along a fault plane (note the footwall cut off). A colour strain map on the hanging wall surface highlights the distribution of dilational strain due to kinematic hanging wall deformation. (**B**) Map view of the same model.

necessary in order to analyse the regional character of flexural behaviour of the lithosphere. The main extensional fault is planar to a depth of 10 km, but becomes listric deeper in the lithosphere, as is likely to happen for a Cuu Long basin setting (Lister *et al.* 1986). The hanging wall block has been restored kinematically using an inclined shear algorithm (60° shear antithetic to the main fault, and movement orthogonal to the main trend of the fault). It is then forward modelled with the same shear vector while the dilational strain on the top surface is calculated. The dilational strain calculated in this way for the hanging wall deformation amounts to up to 25% locally (note that for this ideal flat-bedded model kinematic and non-kinematic strain calculation yields the same result).

Figure 10 shows the inhomogeneity of the strain throughout the hanging wall, caused mostly by the variable offset along the fault. Note the high strain values near the fault although in that locality the strain is controlled by the planar fault surface at depth. This is usually the scale at which the exploration geologist will be interested. The range of shear vectors that resulted in realistic restorations yielded very similar strain values but with slightly different spatial patterns. Control over the seismic reflector of the top basement, and the variable

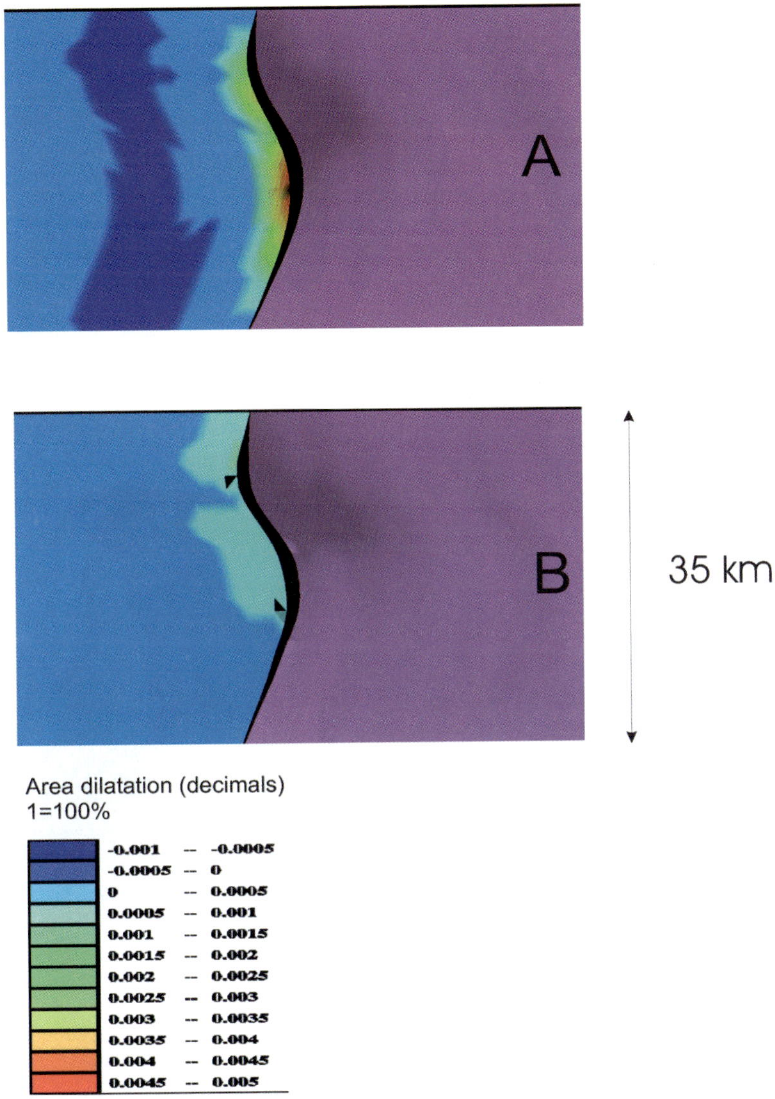

Fig. 11. Map view of three-dimensional model of Figure 10, showing strain values due to footwall uplift, (**A**) for EET = 1 km, (**B**) for EET = 5 km.

offset along the faults, form the two main parameters that control the deformation in the hanging wall for any realistic example of the Cuu Long basin. These two parameters can usually be well defined with seismic imaging.

Deformation in the footwall due to flexural uplift was calculated using current dilational strain, comparing the undeformed state (flat bed for the footwall) with the deformed state (flexured state), using a vertical shear non-kinematic restoration. Figure 11 shows that for a model of EET = 1 km, the footwall creates no more than 0.5% strain (while the hanging wall can suffer as much as 10% dilational strain due to flexural isostatic rebound). With a more realistic EET of 5 km, the strain in the footwall is no more than 0.1% (Fig. 11b), while the strain in the hanging wall is 3%.

Hanging wall versus footwall deformation

If we compare the deformation caused by the flexural response in the footwall with the deformation from kinematic movement of the hanging wall, we can compare the relative importance for fracture generation. The structural modelling shows that deformation in the hanging wall is an order of magnitude greater than deformation in the footwall. With smaller amounts of offset, the hanging wall still suffers significant strain of up to 10% due to kinematic movement, while footwall deformation due to isostatic flexural uplift is negligible. When only the flexural uplift mechanism is considered, the hanging wall is still significantly more deformed than the footwall. This indicates that the hanging wall is more prone to fracturing than the footwall for both kinematic and flexural isostatic processes. The strain values in the footwall are generally so low that it is unlikely that any fractures will be formed here due to flexural isostatic uplift.

Discrete fracture network results

The results from modelling and field observations can be integrated to simulate fracture growth in a conceptual model (but also for real case models). Discrete fracture networks are

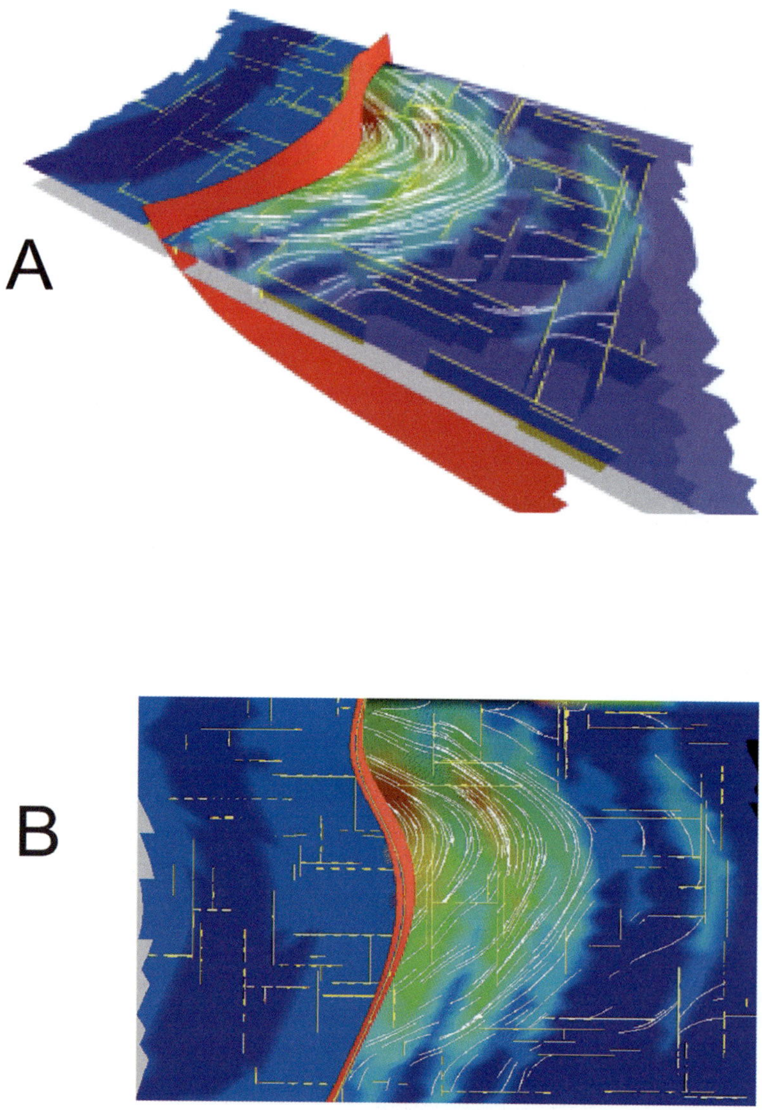

Fig. 12. Fracture patterns from the simulation (see text): (**A**) three-dimensional view; (**B**) plan view. Same model as in Figures 10 and 11. Primary fractures in yellow and shear fractures in white. The colour map shows areas of high strain which control the intensity of shear fracturing.

simulated interactively using the Fracture Generator. For the model described above, as an example, the fracture density is controlled by the strain value maps. The strain map for the kinematic hanging wall deformation is used for the hanging wall and the strain map for the flexural uplift for the footwall, to represent the dominant deformation processes in the two blocks. The orientation of the tectonic fractures is controlled by the local strike of the top surface of the hanging wall, mimicking the gravitational collapse of the hanging wall due to extension. The result is shown in Figure 12. In both hanging wall and footwall, two fracture sets of random density and orthogonal orientation are simulated to represent the primary cooling fractures. A proportion of the primary fractures is defined to stop growing upon reaching other primary fractures (typical for joint sets, see Price & Cosgrove 1990). The primary fractures are simulated before the extensional movement of the fault block (as defined by the geological history). During the subsequent extensional movement the primary fractures are moved (deformed) as material objects in the hanging wall. At this stage, the tectonic shear fractures are simulated. The fracture modelling shows that shear fractures do not form in the footwall because of the low strain values compared to the hanging wall. The footwall is therefore characterized by primary fractures only, while the hanging wall is characterized by both shear fractures (in areas of high deformation only) and primary fractures. Shear fractures are allowed to cross other pre-existing fractures (since they are in effect small faults). This simple fracture modelling exercise allows complex spatial fracture density and orientation to be controlled by a few and relatively simple geological parameters.

If the fit between well data or field analogues and the model is poor, we can reconfigure the simulation until a good fit is established between model and observation. Thus, fracture networks can be generated that match with well data and field observations, allowing the development of more realistic models of the geological processes that have controlled fracture development.

Shear fractures have been found in horst blocks in the Cuu Long basin (like the Bach Ho field) but these are believed to be related to the fault damage zones (Ngoc pers. comm. 2001). Wells are often targeted close to faults visible on seismic sections. In a similar way, the fracture patterns of fault damage zones can be modelled if we assume that the density of the fractures is controlled by the amount of slip along the fault, and the width of the fault damage zone is equally related to the local amount of slip along the fault. The fracture density gradually decreases away from the main fault plane, and the fracture orientation is usually parallel to the strike of the main fault plane. Because of the very high strain in fault damage zones (>100%) the fracture density is expected to be very high compared to fracturing from kinematic hanging wall deformation. A possible fracture model is shown in Figure 13. The damage zone can be located predominantly in the hanging wall (as

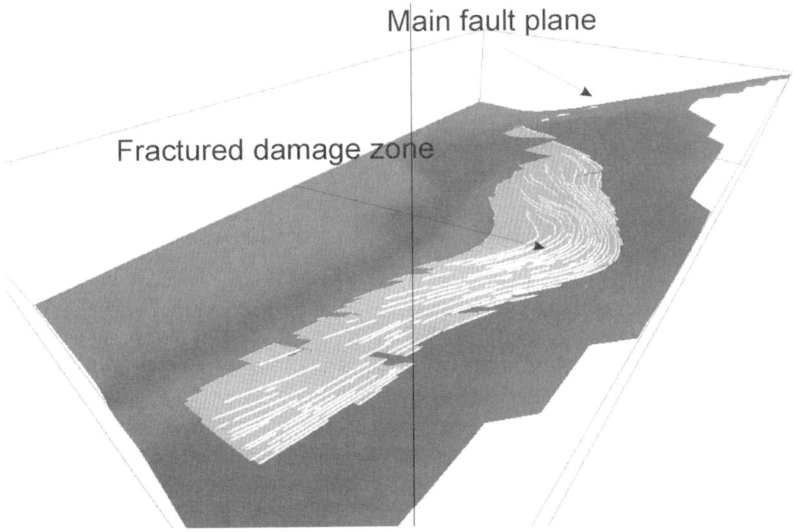

Fig. 13. Main fault plane with associated fracture damage zone as modelled in Fracture Generator. Note the gradual decrease of the fracture density away from the fault (see text for further explanation). This model is independent of scale.

shown here) or symmetrically distributed over the hanging wall and footwall. Note for scale that these fracture damage zones are typically 50 m wide, as for example in the Bach Ho field (Ngoc pers. comm. 2000) and it is thus not possible to view this damage zone at true scale in Figure 12.

When incorporated into our earlier model, we now have in place a model of fractured crystalline basement in an extensional setting, containing primary cooling fractures in both hanging wall and footwall, shear fractures predominantly in the hanging wall and a fault damage zone around the main extensional fault in both hanging wall and footwall. A zone of exhumation fractures can also be added to the top surface.

Discussion

We are not claiming that our fracture network model in this way gives a true picture of fracture systems in extensional basements. Rather, it provides a guide, based on structural analysis, that requires further correlation and validation against well data and seismic attribute maps. The fracture networks can then be exported to reservoir simulation codes for comparison with production data or well tests. Initially, a limited number of scenarios may be appraised simultaneously to define a range of possibilities and risk. In this way, understanding will grow over the field, based on the quantitative assessment and integration of all available data and geological concepts.

As such the proposed workflow, and the modelling specifically, can be thought of as a learning tool that will help to gain insight into complex fracture systems characteristic of extensional basement settings. It allows simulation of petroleum plays and reservoir behaviour and characteristics in a geologically realistic manner, and provides testable concepts for exploration strategies. This allows for better prediction of fracture patterns in areas of sparse data, while reducing the technical risk of exploration and production strategies in fractured reservoirs. The techniques described here have also been applied successfully to fractured sedimentary rocks with extensive well control.

To summarize, our results suggest that extensional hanging wall blocks form a lower risk prospect for petroleum exploration than the conventionally drilled horst blocks. They not only contain the intense shear fracturing related to kinematic hanging wall deformation, but also have fracture populations relating to flexural isostatic uplift and primary (cooling-related) fractures. In contrast, the footwall blocks will probably only host primary fractures. Finally, the fractures related to relatively narrow fault damage zones and the equally thin weathered cap of the basement may occur in both footwall and hanging wall blocks. Several new well results indicate the likelihood of this hypothesis (e.g. Tandom *et al.* 2001; Koning 2001) with wells drilled in the hanging wall showing higher fracture density and better oil production.

This paper does not address the behaviour of fractures once they are formed. For example, reactivation of primary fractures during the extensional basin formation phase will probably take place in areas of high strain. As such, they may influence the orientation of the shear fractures that are prone to form here (if they form as reactivated primary fractures). Also, the influence of the present-day stress regime on the opening and closing tendency of existing fractures forms an important parameter, especially for connectivity issues. Such primary fractures are more sensitive to closure because essentially they form hairline structures in contrast to the commonly brecciated shear fractures. A correct simulation of the three-dimensional fracture network is fundamental to further studies into the behaviour of fractures in relation to fluid flow, stress regimes, secondary mineralization or enhanced oil recovery techniques such as steam injection.

Conclusions

Data on fractures in offshore crystalline basement are sparse and often apparently chaotic. Geological concepts, integrated structural modelling and interactive fracture simulation using seismic interpretations, geological data and field observations were used to evaluate and understand complex fractured crystalline basement petroleum reservoirs. Apart from simulating realistic three-dimensional discrete fracture networks, the method described significantly aids the understanding of the geological processes responsible for the fracturing, and as such forms the best technical risk reduction for exploration and production strategies of these petroleum plays. Extensional hanging wall fault blocks are much more likely to be intensively fractured than footwall fault blocks. Therefore they are more likely to have higher porosity, connectivity and conductivity and form a lower risk prospect than the conventionally drilled footwall horst blocks. The techniques proposed here allow various deformation processes to be analysed and compared, which may result in several fracture simulation scenarios. Comparison with well and production data makes it possible to validate these scenarios and define likely concepts to represent the actual fracture network. The simulated fracture networks can be further modified to take into account the effects of secondary mineral fill, changing stress regimes and fluid pressures, hydrocarbon migration and production enhancement techniques.

We would like to thank the staff of Japan Vietnam Petroleum Company, Cuu Long JOC, Hoang Long JOC, Vietnam Petroleum Institute, BP and Schlumberger HRT for discussions during the development and application of these techniques. C. McKeown and N. Petford are thanked for thorough reviews of the paper and editorial assistance.

References

ANDERSON, E. M. 1951. *The dynamics of faulting*. Oliver and Boyd, Edinburgh.

ARESHEV, E. G., TRAN LE DONG, NGO THUONG SAN & SCHNIP, O. A. 1992. Reservoirs in fractured basement on the continental shelf of southern Vietnam. *Journal of Petroleum Geology*, **15**, 451–464.

BEEK, P. VAN DER 1995. Tectonic evolution of continental rifts, inferences from numerical modelling and fission track thermochronology. PhD thesis, Vrije Universiteit, Amsterdam.

BRUCE, C. H. 1973. Pressured shale and related sediment deformation: mechanism for development of regional contemporaneous faults. *AAPG Bulletin*, **57**, 878–886.

CLOOS, H. 1922. Tektonik und magma. Bd. 1 Abh. d. Preuss. Geol. Land, 89. *In*: Price, N. J. & Cosgrove, J. W. (eds) 1990. *Analysis of Geological Structures*. Cambridge University Press, Cambridge, UK

DULA, W. F. 1991. Geometric models of listric normal faults and roll over folds. *American Association for Petroleum Geometry Bulletin*, **75**, 10.

EGAN, S. S. 1992. The flexural isostatic response of the lithosphere to extensional tectonics. *Tectonophysics*, **202**, 291–308.

GIBBS, A. D. 1983. Balanced cross section construction from seismic sections in areas of extensional tectonics. *Journal of Structural Geology*, **5**, 153–160.

HAUGE, T. A. & GRAY, G. G. 1996. A critique of techniques for modelling normal-fault and rollover geometries. *In*: BUCHANAN, P. G. & NIEUWLAND, D. A. (eds) *Modern developments in structural interpretation, validation and modelling*. Geological Society, London, Special Publications, **99**, 89–97.

KANE, S. J., WILLIAMS, G. D., BUDDIN, T. S., EGAN, S. S. & HODGETTS, D. 1997. Flexural slip based restoration in 3D, a new approach. *AAPG Annual Convention Official program*, A58.

KOENDERS, M. A. & PETFORD, N. 2003. Thermally induced primary fracture development in tabular granitic plutons: a preliminary analysis *In*: PETFORD, N. & MCCAFFREY, K. (eds) *Hydrocarbons in Crystalline Rocks*. Geological Society, London, Special Publications, **214**, 143–149.

KONING, T. 2001. Oil and gas production from basement reservoirs in Indonesia. (abstract). Hydrocarbons in Crystalline Rocks (convenors K. McCaffrey, T. Murray & N. Petford), p. 19.

KUSNIR, N. J., MARSDEN, G. & EGAN, S. S. 1991. A flexural cantilever simple shear/pure shear model of continental lithosphere extension: applications to the Jeanne d'Arc basin, Grand Banks and Viking Graben, North Sea. *In*: ROBERTS, A. M., YIELDING, G. & FREEMAN, B. (eds) *The geometry of normal faults*. Geological Society, London, Special Publications, **56**, 41–60.

LISTER, G. A., ETHERIDGE, M. A. & SYMONDS, P. A. 1986. Detachment faulting and the evolution of passive continental margins. *Tectonics*, **10**, 1038–1064.

MATTHEWS, S. J., FRASER, A. J., LOWE, S., TODD, S. P. & PEEL, F. J. 1997. Structure, stratigraphy and petroleum geology of the SE Nam Con Son Basin, offshore Vietnam. *In*: FRASER, A. J., MATTHEWS, S. J. & MURPHY, R. W. (eds) *Petroleum Geology of Southeast Asia*. Geological Society, London, Special Publications, **126**, 89–106.

MCCAFFREY, K. J. W., BEACOM, L. E., HOLDSWORTH, R. E. & SLEIGHT, J. 2001. Characterisation of fracture systems in onshore analogues of crystalline reservoirs. Hydrocarbons in Crystalline rocks, Abstract book, London.

PRICE, N. J. & COSGROVE, J. W. 1990. *Analysis of Geological Structures*. Cambridge University Press, Cambridge, UK.

RANALLI G. 1987. *Rheology of the earth, deformation and flow processes in geophysics and geodynamics*. Allen and Unwin, London.

SANDERS, C., NGOC, N. H., STURT, D., HERRIES, R., KATROSH, M. & NAM, N. M. 2001. Regional basin modelling, south Cuu Long basin, offshore Vietnam, constraints on tectonic basement activity by sequential backstripping of basin sediments. *In*: Hydrocarbons in Crystalline rocks, Abstract book, London.

SCHUTTER, S. R. 2003. Hydrocarbon occurrence and exploration in and around igneous rocks. *In*: PETFORD, N. & MCCAFFREY, K. (eds) *Hydrocarbons in Crystalline Rocks*. Geological Society, London, Special Publications, **214**, 7–33.

SLADEN, C. 1997. Exploring the lake basins of east and south east Asia. *In*: FRASER, A. J., MATTHEWS, S. J. & MURPHY, R. W. (eds) *Petroleum Geology of Southeast Asia*. Geological Society, London, Special Publications, **126**, 49–76.

SLEIGHT, J. M., MCCAFFREY, K. J. W. & HOLDSWORTH, R. E. 2001. Micro to regional scale fracture characteristics from the More Trondelag Fault Complex, Central Norway. Hydrocarbons in Crystalline rocks, Abstract book, London.

SWABY, P. A. & RAWNSLEY, K. D. 1996. *An interactive 3D Fracture Modelling environment*. Society of Petroleum Engineers, SPE 36004.

TANDOM, P. M., TRAN XUAN NHUAN, TJIA, H. D. & SPAGNUOLO, S. A. 1997. Fractured basement reservoir characterisation, offshore Vietnam. *Third well logging symposium, Japan, 1997, Proceedings*.

TANDOM, P. M., NGOC, N. H., DIN, T. H. & LLOYD, P. M. 2001. Identifying and Evaluating Producing horizons in Fractured Basement, Hydrocarbons in Crystalline rocks, Abstract book, London.

THORPE, R. & BROWN, G. 1985. *The field description of igneous rocks*. Geological Society of London Handbook.

Index

Note: Page numbers in *italic* refer to figures, those in **bold** refer to tables.